'War is a topic perennial and urgently current. The insightful contributions to this discussion here published capture many of war's psychological complexities and will help the questioning reader to think more clearly about a topic both fascinating and horrifying.'

Murray Stein, Ph.D., *author of* Jung's Map of the Soul

'The important thing about this book is how real it is. Sure, it is full of Jungian, post-Jungian and spiritual reflections on war. And these include challenges to a great deal of orthodox psychosocial and psychoanalytic thinking. But I truly felt the smells, sounds, wounds and sheer mortality of war thrusting themselves at the reader. It is the kind of book that should have a "trigger warning" on it, that it might upset some readers. And a good thing too.'

Professor Andrew Samuels, *author of* A New Therapy for Politics?

'*War as Reset* is a big slow cooked stew with many ingredients including reflection on wars in Ukraine/Russia, Israel/Palestine, Argentina, Italy (Sicily) Ireland, China and issues of gender, identity, trauma, displacement, terror, and the presence and/or absence of the gods in the world in general and in wars in particular. *War as Reset* is most ambitious in scope and depth. It has in mind a specific focus – the intriguing notion of reset – of war as an attempt to "restore a deteriorating order and set of values, striving to revive the world of yesterday. against the fear of the "world of tomorrow".'

Thomas Singer, *Co-Creator/Editor* Mind of State

War as Reset

In an age continuously shaped and shocked by wars and societal crises, this book serves as an antidote to superficial media frenzy. Exploring the interplay between the insights from analytical psychology and global dynamics, it unravels the meanings behind our shared fears and invites readers to confront challenging truths shaping our present and future.

Part I of this book explores the multifaceted aspect of war, as Stefano Carpani interviews authoritative figures from the fields of Jungian psychoanalysis, sociology, history, and religion. Their insights shed light on the meaning of war, the concept of fatherland, the masculine nature of war, and the potential for total conflict. In Part II of the book, Jungian therapists reflect on their experiences, offering insights into the impact of war on the field of analysis, presenting a comprehensive exploration of war from interdisciplinary perspectives. The contributions touch upon themes like the Israeli–Palestinian conflict, healing through arts-based research, violence practiced by the state in Argentina, sexual violence, and the effect of the Irish Republican Army (IRA) on Irish society.

This book proposes that war can serve as a reset mechanism, and that our era can be termed the one of hypocrisy. It will be of interest to academics, scholars, and students within the fields of analytical psychology, psychosocial studies, psychoanalysis, and sociology.

Stefano Carpani, Ph.D., psychoanalyst and sociologist (graduate, member and lecturer of the C.G. Jung Institute Zürich, and post-graduate of the University of Cambridge), initiated the YouTube series *Breakfast at Küsnacht, Lockdown Therapy and War as Reset*, co-created *Psychosocial Wednesdays* and currently chairs it, curates *Jungianeum: Initiatives for Contemporary Analytical Psychology and neo-Jungian Studies*, the book series titled *Re-covered Classics in Analytical psychology* and *JUNGIANEUM/Yearbook*. Among his recent books: *Breakfast at Küsnacht* (Chiron, 2020 – IAJS Best Edited Book nominee); *Anthology of Contemporary Classics in Analytical Psychology: The New Ancestors* (Routledge, 2022 – GRADIVA Best Edited Book nominee), and *Absolute Freedom* (Routledge, 2024).

Ludmilla Ostermann, M.A., is a German journalist and editor based in Berlin and North Rhine-Westphalia. She studied German and English studies as well as history at the University of Bielefeld. Ostermann writes for various German media outlets, including dpa (German Press Agency) and Westfalen-Blatt, covering political, social, economic, and scientific topics. She contributed to an interview series by the University of Bielefeld on the war in Ukraine and currently writes for the university on a range of academic and societal issues. She is a member of *Jungianeum: Initiatives for Contemporary Analytical Psychology* and *Neo-Jungian Studies* and serves as a co-editor and regular contributor to the *JUNGIANEUM Yearbook*.

War as Reset

Insights from Contemporary Analytical
Psychology on the Age of Hypocrisy

Edited by Stefano Carpani and
Ludmilla Ostermann

LONDON AND NEW YORK

Designed cover image: Getty Images

First published 2025
by Routledge
4 Park Square, Milton Park, Abingdon, Oxon OX14 4RN

and by Routledge
605 Third Avenue, New York, NY 10158

Routledge is an imprint of the Taylor & Francis Group, an informa business

British Library Cataloguing-in-Publication Data
A catalogue record for this book is available from the British Library

Library of Congress Cataloging-in-Publication Data
Names: Carpani, Stefano, editor. | Ostermann, Ludmilla, editor.
Title: War as reset : insights from contemporary analytical psychology on the age
    of hypocrisy / edited by Stefano Carpani and Ludmilla Ostermann.
Description: Abingdon, Oxon ; New York, NY : Routledge, 2025. |
    Identifiers: LCCN 2024046428 (print) | LCCN 2024046429 (ebook) |
    ISBN 9781032486437 (hardback) | ISBN 9781032486420 (paperback) |
    ISBN 9781032486420 (ebook)
Subjects: LCSH: War—Psychological aspects. | War and society.
Classification: LCC U22.3 .W348 2025 (print) | LCC U22.3 (ebook) |
    DDC 355.0201/9—dc23/eng/20250115
LC record available at https://lccn.loc.gov/2024046428
LC ebook record available at https://lccn.loc.gov/2024046429

ISBN: 978-1-032-48643-7 (hbk)
ISBN: 978-1-032-48642-0 (pbk)
ISBN: 978-1-003-39003-9 (ebk)

DOI: 10.4324/9781003390039

Typeset in Times New Roman
by Apex CoVantage, LLC

To Sasha, Rita, and Kyril,
three children from Kharkiv/Ukraine,
who, like many others
in countless wars around the world,
were forced to leave their home,
this book is dedicated.
In unconditional support of all *Davids* around the globe,
and against all *Goliaths*' barbaric aggressions.

QUANDO A DITATURA É UM FACTO
A REVOLUÇÃO È UM DEREITO

WHEN DICTATORSHIP IS A FACT
REVOLUTION IS A RIGHT

Tombstone in memory of
the anniversary of the operation VAGÓ
(Cemitério de Prazeres, Lisbon/Portugal)

Contents

Disclaimer   xii
Acknowledgement   xiii
Preface   xiv
STEFANO CARPANI

Introduction   1
LUDMILLA OSTERMANN

PART I

1 **Day 22 of War**   11
   TINE PAPIČ

2 **Day 27 of War**   23
   MURRAY STEIN

3 **Day 33 of War**   32
   GEORGE B. HOGENSON

4 **Day 37 of War**   45
   CATERINA VEZZOLI

5 **Day 40 of War**   53
   GIUSEPPE BETTONI

6 **Day 42 of War**   61
   JOSEPH CAMBRAY

7 **Day 52 of War**   71
   DMYTRO ZALESKYI

8 **Day 55 of War**   79
   DMITRY KOTENKO

9   **Day 59 of War**   86
    IRYNA SEMKIV

10 **Day 69 of War**   92
    VICKIE SIMS

11 **Day 70 of War**   100
    VERENA KAST

12 **Day 84 of War**   108
    LUIGI ZOJA

13 **Day 92 of War**   116
    ELANA LAKH

PART II

1   **Analysis in the Shadow of Terror: Clinical Aspects**   127
    HENRY ABRAMOVITCH

2   **Donbas in the Battle for Cultural Identity, or Cultural Identity
    in the Battle for Donbas**   136
    NATALIA BOLYCHEVA

3   **Sicily's Infinite War: A Neo-Jungian Point of View**   154
    CHIARA CAPRI

4   **The Preserved Moment Through Art: Looking at Jungian Arts-Based
    Research and the Articulation of Inherited War Traumas**   164
    ROULA-MARIA DIB

5   **Embodied Analysis: The Recovery of Early Psychological Functions
    Interrupted by an Experience of Early Trauma Due to State Terrorism**   173
    KARIN FLEISCHER

6   **Dream With the Heart, and the Heart of Dream**   185
    HEYONG SHEN

7   **The Sacrificial Murder of Palestine: Grinding Bones to Dust**   201
    HEBA ZAPHIRIOU-ZARIFI

8 The Northern Ireland Conflict: From IRA, to Sinn Fein, to Peace
   Ireland's Cultural Complexes Transformed   212
   KATHLEEN KIRGIN

9 When Our Shadow Makes Us Blind and Deaf to Suffering   222
   ELANA LAKH

10 Insight Into an Analysis With a Patient Who Became Frozen in Fear
   Because of the War   226
   MARIANNE MEISTER-NOTTER

11 Destructiveness, Complexity, and Archetypal Epistemology: Critical Reflections   232
   RENOS K. PAPADOPOULOS

12 Tales of Trauma, Terror, and Awe: Counter-Trauma, Counter–Adversity
   Activated Development, and Mutual Transformations in the Clinical
   Setting With Survivors of Collective Violence   246
   ELIAS WINTERTON

   Outro: The Age of Hypocrisy – From the Suspension to the
   End of Certainties: War as Reset   260
   STEFANO CARPANI

   List of Contributors   275
   Index   282

Disclaimer

An attentive reader, especially while perusing Part I and Carpani's outro, may find echoes of Jung's *Collected Works*, Vol. 10, and in particular, the essays "Wotan" and "After the Catastrophe." Both of these essays are exquisite examples of applied Jungian and post-Jungian theoy, illuminating contemporary events from an archetypal and complex psychology. Simultaneously, as the editors are certain the reader would concur, both essays have been a subject of controversy and have suffered from various misinterpretations and misconstructions. Consequently, the editors are acutely aware of the potential for this work to be misconstrued.

We sincerely hope that the methodology outlined in the introduction (referenced in what follows) will assist in acknowledging the challenges faced by Jung and, by extension, the editors of this book.

We wish to explicitly disavow any one sided political intent in our writing, solely because the subject matter is highly contentious and prone to misinterpretation – possibly even deliberate misinterpretation by parties with their own agendas.

We firmly believe that the insights presented in this work are more crucial than ever. However, it is regrettable that a clear and unambiguous declaration of our methodological stance would go a long way in preventing misunderstandings and misrepresentations.

Acknowledgement

The editors would like to express their profound gratitude to all the contributors who generously agreed to participate in this book. Additionally, they extend their heartfelt thanks to Routledge for placing their trust in us.

This book is dedicated to Sasha, Rita, and Kyril, three children from Kharkiv, Ukraine. Like so many others affected by this conflict and countless others worldwide, they were tragically forced to leave their homes. Just hours after their departure, their houses were destroyed in a barbaric attack. We, the editors, had the opportunity to host them after they reached Berlin in March 2021. We didn't have much information, and we didn't even know if they spoke Russian or Ukrainian, or if their families were pro-Russia or not. They were children who had fled with their mothers, leaving their fathers and their entire lives behind. They were children in need of a place to stay.

Stefano Carpani and Ludmilla Ostermann
24 February 2025, in the anniversary of
the third year of the invasion of Ukraine

Preface

Stefano Carpani

On February 24, 2021, I found myself bedridden with COVID-19, and I couldn't help but find it somewhat ironic that I contracted the virus just as media outlets were on the brink of shifting their focus from the pandemic to the Russian invasion of Ukraine. However, at that moment, I was unaware of this development.

Battling boredom during my COVID-19–induced confinement, I turned on the TV and received a jolt – Russia had invaded Ukraine. This event quickly overshadowed the pandemic that had long held the headlines.

Following the invasion, I found myself drawn to engage senior Jungian thinkers, extending the themes explored in my YouTube series *Breakfast at Küsnacht* and *Lockdown Therapy*. My wish, once more, was to seek wisdom and psychagogia, and share it with a global audience.

I believe that these discussions offered an antidote to "media bulimia." Media outlets inundate us with superficial information through various channels, often falling short of providing a genuine understanding of complex issues. War, in particular, starkly contrasts with what could be described as a "vanity fair"; instead, it can be seen as an "atrocity fair." Sadly, when presented through the lens of media, it tends to regress into the former, becoming yet another form of shallow entertainment. We consume daily servings of war, yet substantial change eludes us.

As a few months pass in the conflict, war gradually morphs into just another news headline[1] – and when this immediate threat fails to materialize – we quickly revert to our routine, business-as-usual existence. The initial shock, fear, and paranoia that gripped us at the war's outset ceased. This pattern resembles our approach to life post–COVID-19 – a return to business as usual, as if nothing significant occurred. It is as if we gained no insights or allowed the reality of war to influence our lives, even in its absence. In my view, this represents a missed opportunity for profound transformation, a lamentable oversight that hinders both personal and collective growth.

Why write a book about war after creating a YouTube series on the same topic? My purpose aligns with that of my book *Lockdown Therapy* – to convert a missed opportunity into a path for depth, personal development, and profound change.

Furthermore, my aspiration is to continue the legacy of thinkers like Simone Weil, Paul Valéry, Joseph Roth, Stefan Zweig, James Hillman, and C.G. Jung, all of whom contemplated the World Wars that left an indelible mark on the past century. My wish is to affirm that, from a psychological point of view, this war could be regarded as another episode of the drama that shook Europe at the beginning of the 20th century with World War I. Metaphorically, it's like an aftershock following an earthquake that happened a long time ago but has not yet found stability – and, perhaps, the metaphoric seismic shifts are generated by the ongoing, opposing forces propelling the West toward the future and acceleration, initiated at the close of the 19th century with industrialization. These forces encounter resistance from traditional Russia as the West transitions through modernity, post-modernity (Lyotard, (1984 [1957]), late or second modernity (Beck, 2022), and second-late modernity (Carpani, 2023b and 2024). Three years after the invasion of Ukraine, these forces have become dominant in many Western countries, including Trump's America and several European nations that now embrace similar values and mindsets.

This book, therefore, can be viewed as a counterbalance to the prevailing media voyeurism, offering an opportunity to delve deep and counteract the deluge of superficial news regarding the Ukraine invasion (and the ongoing one-sided portrayal of war-related news around the world) disseminated by media outlets more concerned with their bottom line than with providing meaningful insights. This book presents an opportunity to challenge what can be termed "media bulimia," a reflection of our collective anxieties and paranoia. Hence our inability to differentiate, develop consciousness and learn to live more consciously.

Note

1 Since Russia's invasion of Ukraine, the geopolitical landscape has been marked by numerous significant events. These include Hamas' reprehensible attack on Israel and Israel's response, which in our opinion can be described as a genocide. Other major developments include the fall of the Assad regime in Syria and the re-election of Donald Trump – just to name a few topics that dominate Western media. The list of consequential events is far longer.

Introduction

Ludmilla Ostermann

This book explores war from a Jungian and psychosocial point of view. We examine the reasons for its outbreak and the complex realities of its aftermath. But what does "war as reset" really mean?

War by its very nature brings destruction, chaos, and the profound loss of human life. It displaces communities, destabilizes political and social systems, and devastates economies. The repercussions of war extend far beyond borders, disrupting the very foundations of societies and leaving a legacy of suffering and instability.

The concept of "reset" may metaphorically align with the idea of *tabula rasa.* But for us, it represents something different than a clean slate. It is a state of mind – a yearning to return to a bygone era, reclaim lost territories, or revive geopolitical systems and outdated agreements through war. Consider, for instance, the underlying motivations for World War I and World War II, or the 2014 annexation of Crimea by Russia. These acts illustrate a preference for regression rather than grappling with the complexities and opportunities inherent in the evolution of society. This perspective, rooted in resistance, frames war as the sole means to reclaim what is under threat (such as the status quo in Germanic countries facing the challenge of Bolshevik Russia) and/or what has been lost (e.g., Germany in 1933 and Crimea in 2014).

Part 1

During the first 100 days of the war in Ukraine, when the world was about to be freed from the restrictions and lockdowns imposed during the pandemic, Stefano Carpani recorded conversations entitled *War as Reset* and uploaded them to his YouTube channel. The interviews included here were conducted with leading authors in analytical psychology and other fields. They come from Ukraine, Russia, the USA, Switzerland, Slovenia, Italy, and Israel. Each provided a revised and updated version – which was then further adapted for this work. The hope was to engage with a more diverse and extensive group of interviewees. Unfortunately, many of those Carpani contacted were either unavailable or declined his invitation.

To get different insights of different people on the same topic, he built each interview on the same questions:

- **War:** What is war (and its ultimate meaning)? Is war masculine only? Is war "a love that no other love has been able to overcome," as underlined by Hillman (2005, p. 259)[1]? Can war be seen as a reset? If so, of what?

DOI: 10.4324/9781003390039-1

- **Compensation:** Should the Russian invasion of Ukraine be seen as compensation (in the same way that World War II might have been a compensation for the German defeat in World War I)?
- **Fatherland:** Each interview focuses on the concept of nation, father and fatherland, and past empires that aspired a glorious future but were defeated.
- **Total war:** Each interview discusses whether the Russian invasion of Ukraine is a prodrome for total war in Europe and/or the rest of the world, and what could be done to avoid it from an analytical psychological point of view.
- **Beasts:** The interviewees discuss why ordinary people can turn into beasts during wartime.
- **Putin:** When it comes to Russian President Vladimir Putin (or any leader, for that matter), interviewees examine whether it is inappropriate to diagnose him.
- **Reset:** In the light of former U.S. President Obama's desire to reset relations with Russia (see the Clinton-Lavrov attempt in 2008), the discussion shifts to U.S. foreign policy after World War II and reflects on the difference between this war and those waged by the United States (Korea, Vietnam, Iraq, Afghanistan, etc.).
- **Empires:** Nobel laureate Svetlana Aleksievich has drawn a very clear picture of the "imperialist" impulses not only of Putin but also of a large part of the Russian population. She calls it the long wake of Homo Sovieticus and, before that, of the tsarist tradition. Applying this theory, we look at the imperialist impulses of the United States and the British crown. The question then arises: Are their actions any different from those of Stalin and/or Putin?
- **Europe:** Each interview focuses on Jürgen Habermas and Jacques Derrida, who, in the wake of the massive street demonstrations against the Iraq war in 2003, asked themselves: "What binds Europeans together?" in relation to the international order. The values described by these philosophers are, according to Carpani, the reason why Russia is not interested in an independent and liberal Western European Ukraine. He suggests that reset means a move toward co-construction of meaning and a declaration of interdependence globally.

Part 2

The second half of the book consists of chapters that address the broader issue of war from a Jungian psychosocial perspective. We invited contributors to either write an original chapter or to re-publish a paper that has gained new relevance and now requires attention. These essays outline the subjective experiences of Jungian therapists in relation to war and conflict. They are a collection of reflections from a theoretical, technical, clinical, and practical point of view, including various techniques: dream analysis, sand play, drawing and painting, etc. The second part of the book deals with this in more detail. The topics are the following.

- The Israeli-Palestinian conflict in many ways: the *dream worlds and treatment of Palestinian patients*, the *compromise of the transference-countertransference matrix* between analyst and patient due to politics, the *collective shadow and cultural complex* in the Israeli context as well as by *challenging the ethical and psychological implications of equating the two parties in the conflict as mere opposites*.
- *Destructiveness as a complex, multi-dimensional phenomenon* that extends beyond a purely mental health perspective.
- How can war reset the analytical field and the *analyst's psychic life*?
- The fact that not all wars are experienced in a direct way, and on how *arts-based research* can be a method that heals as much as it teaches and creates.

- (Clinical) cases when *violence is practiced by the state*, as was the case with the dictatorship in Argentina between 1976 and 1983.
- The *dreams related to war today in China*, the paintings of Chinese artists related to war, and we study the images of war in Chinese culture.
- The *catastrophic effects of Russia's "special operation" on an analyst* and at a vulnerable point in her life, her descent into absolute darkness and the arduous path back to the light, incorporating dream material and images.
- The *Donbas region defending its cultural identity* and how this is reflected in contemporary war poetry.
- The silent war against *the Mafia in Sicily*, a war as dangerous and terrifying as any other.
- *Ireland's reclaiming of its culture and the transformation* of a once colonized nation into a sovereign country whose people, from both Northern Ireland and the Republic of Ireland, are committed to lasting peace and reconciliation for all who live on the island nation.

War as Reset: Perspectives on Restoration and Transformation

These examples of war and conflict underscore the point made at the beginning of this introduction. Let us take this opportunity to define our concept of *war as reset* further.

We propose that, from a psychosocial and psychoanalytical perspective, war can be understood as a reset. A reset typically involves a *tabula rasa*: a clearing of the slate to reaffirm the previous status quo in opposition to novelty and development, effectively returning to a known state or default condition.

In this context, we propose the desired reset is not necessarily tied to a specific historical moment. Instead, it may represent a state of mind – a restoration of values and ideals perceived as threatened, buried, overthrown, or lost.

We suggest that the concept of a reset is not universal; it initially resides in the realm of the collective unconscious before being perceived and brought into consciousness by an individual or group. The example of Vladimir Putin is particularly illustrative in this context. This concept is shaped by the goals, values, and historical contexts of those who experience it and ultimately invoke it.

To elaborate on an example introduced at the beginning of this introduction, consider Russia's annexation of Crimea in 2014 followed by the ensuing insurgency in eastern Ukraine. These events marked – to Western eyes – a turning point in regional and international dynamics, changing geopolitical alignments, straining relations between Russia and the West, and prompting Ukraine to seek closer ties with the West while prioritizing its sovereignty and military capabilities.

Behind these strategic goals, it is worth considering Putin's own perspective. In 2005, Putin famously described the collapse of the Soviet Union as "a major geopolitical disaster of the century."[2] While some Russians express nostalgia for the perceived stability and superpower status of the Soviet era – more than 65% in 2020 rated the dissolution of the Soviet Union negatively, according to the Levada Center[3] – it does not necessarily translate into a concrete policy goal of recreating the USSR. Putin's actions may be seen as an attempt to reset Russia's trajectory by selectively revisiting elements of the Cold War era – elements that align with his vision of a strong, heroic fatherland. This, in turn, can be interpreted as a rejection of the transformative changes attempted by Mikhail Gorbachev.

Meanwhile, from a Western perspective, the world has moved far beyond the Cold War and World War II and few of us in Europe have experienced an attack like the one Russia is waging against Ukraine. Former German Chancellor Olaf Scholz has aptly described the current era as a "Zeitenwende"[4] or a "turning point." Since the end of the Cold War, there has been a dominant narrative in the West, one that envisioned a world becoming progressively more democratic, with Russia as a partner. The American political scientist Francis Fukuyama theorized that the entire world would eventually evolve into an interconnected liberal democracy (1992). He was proven wrong – and even admitted it in his latest book in 2018.[5,6] It is clear now that the invasion of Ukraine and the ongoing war as well as other conflicts since 2022 are shattering this vision and signaling a new, harsher reality.

However, Russia and China as well as other countries perceive the past three decades quite differently. From their perspective, this period has been dominated by Western pressure and force, with the West imposing its worldview on the rest of the world. They argue that the world is moving toward multipolarity, with several states possessing significant power potential.[7] The possible reset brought about by the current war and conflicts is therefore a matter of perspective. Each side's understanding of what should be restored – or reimagined – is shaped by its own historical experiences and aspirations as well as cultural complexes.

War and Gender

A central theme of this book is the role of men and women, the male and female aspects of war and conflict from a social and psychoanalytical view. These can be diverse and multifaceted, influenced by historical, cultural, and societal contexts. Many women take on humanitarian and caring roles during conflict, such as nursing, providing medical aid, and supporting families. Women have historically played an important role in resistance movements and activism, both during and after wars, advocating for peace and social justice.

Recognizing and addressing the complex realities of women's experiences of conflict is essential to promoting gender equality, human rights, and peacebuilding. Inevitably, the issue of the nature of war becomes a crucial question in the dialogues. Statements touch on the idea of masculine aggressiveness and how it can be an instinctive and cultural factor that contributes to conflict and war. This perspective is in line with traditional gender roles that associate masculinity with traits such as dominance, competitiveness, and sometimes aggression.

Beyond any gender debate, war is seen as a failure of conflict resolution methods. Diplomacy and peaceful negotiation are the preferred alternatives in our world, but when these do fail, violence can erupt, bringing suffering, loss of life, and destruction. We know that war is a familiar and persistent aspect of human history, with conflict and violence recurring over centuries, woven into the fabric of mankind.

War and Trauma

In exploring the many-sided nature of war and its consequences, we journey into the depths of the human experience, unearthing stories of conflict, resilience, and transformation. While the interviews in the first part of this book give us valuable insights into the origins of war, we focus in the second part on the profound effects of war on societies – effects that extend far beyond the battlefield.

"There are some loud wars, with explosions, bombs, and migration of civilians, and others that are silent and subterranean," writes Chiara Capri in the introduction to her Chapter 3 on the Italian Mafia, revealing critical psychological factors that contribute to the perpetuation of Mafia involvement. One very poignant example of the lasting effects of conflict is the Israeli-Palestinian conflict, where the story of Jamilla, a Palestinian patient, told by contributor Henry Abramovitch in Chapter 1, emerges as a powerful illustration of collective trauma. Here we see how an entire community or society can bear the weight of trauma passed down through generations like an unrelenting legacy. The patient's experience sheds light on the intergenerational transmission of trauma and the daunting challenges faced by individuals in conflict-torn societies.

Heba Zaphiriou-Zarifi vividly depicts the trauma endured by Palestinians now and then, likening their suffering to the biblical figure of Job. This analogy underscores the profound psychological and physical devastation inflicted by war, displacement, and oppression. The invocation of Job symbolizes endurance and spiritual resilience amidst immense suffering, emphasizing the psychological scars of violence and dispossession.

On the other side of this complex narrative, we encounter a "collective traumatic condition" within Israeli society – a condition that has persisted because of past conflicts and wars. This collective trauma, deeply rooted in historical wounds, serves according to the author as fertile ground for the perpetuation of cycles of conflict. Each new conflict seems to trigger and reinforce this trauma, making the path to genuine reconciliation or meaningful new beginnings more difficult.

The concept of liminality offers us valuable tools for understanding the psychological landscape of Donbas, a region that became the epicenter of a protracted conflict in 2014. Here, Ukrainian government forces clashed with Russian-backed separatist groups, leading to the establishment of the self-declared Donetsk People's Republic (DPR) and Luhansk People's Republic (LPR) in parts of the Donetsk and Luhansk oblasts. This conflict has left significant political, social, and economic instability in its wake, deepening the sense of disorientation and insecurity.

While the international community recognizes the Donbas as part of Ukraine, the self-proclaimed republics persistently seek autonomy or full independence – a contentious issue with far-reaching consequences, as Natalia Bolycheva portrays in Chapter 2. In this tumultuous environment, we observe the potential for psychological reset – a phenomenon that often occurs as individuals and communities grapple with the disorienting effects of war and seek to make sense of their experiences in an effort to adapt and heal.

The universal nature of the psychological consequences of collective trauma becomes evident when we shift our focus to a different corner of the world – Argentina, the home of analyst Karin Fleischer, in Chapter 5. Here, the term "state terrorism" evokes a dark period in the country's history, often referred to as the "Dirty War" (Spanish: "Guerra Sucia"). During the late 1970s and early 1980s, Argentina witnessed systemic and organized state-sponsored violence, including force, torture, disappearances, and extrajudicial killings, targeting perceived political opponents, left-wing activists, and individuals labeled as subversive.

In a context that differs from Ukraine's war, Argentina's state terrorism inflicted its own brand of collective trauma. As we consider the experiences of those who lost loved ones and saw their homes and communities ravaged, we recognize the echoes of unresolved grief and trauma. Despite the distinctions in context, both Argentina and Ukraine share a common thread – the pressing need to restore psychological functions.

Connecting these insights to the motif of war as a reset, we find that trauma – despite its overwhelming pain – can also carry the potential to be a catalyst for change. In the case of state terrorism in Argentina, the trauma experienced by the population eventually played a pivotal role in driving political changes and facilitating the transition to democracy. In the context of war, traumatic experiences can have the power to motivate societies to embark on introspection and re-evaluation. This process may extend to their values, policies, and leadership, potentially leading to shifts and reforms that can help break the cycle of violence and pave the way for a more peaceful future.

War and Identity

The intricate relationship between individual identity and collective identity forms the foundation of our understanding of the profound impact of war and conflict. Individuals draw their sense of self from the groups to which they belong, a dynamic process that shapes their values, beliefs, and behaviors while simultaneously contributing to the construction and evolution of collective identities. This interplay between individual and collective identity holds immense significance in shaping societal norms, values, and group dynamics.

In the context of war and its far-reaching consequences, this intricate dance between personal and collective identities becomes particularly relevant. War disrupts and challenges individual and collective identities, casting individuals into a turbulent journey of self-discovery and redefinition. The essays presented in this collection illuminate how the experience of war can even blur the boundaries, presenting unique challenges for individuals and communities alike.

We witness how collective trauma can profoundly affect a society's identity. Long-standing conflicts can fracture the sense of self, especially when cultural, political, and psychological identities collide. These narratives shed light on the ethical dilemmas faced by therapists dealing with patients involved in violent acts during wartime. Additionally, we explore how war can redefine the cultural and regional identity of a place, as seen in the shifting symbolism of the Donbas in the context of the Ukrainian conflict, evolving from sacrifice to heroism.

The "ecology of destructiveness" discussed by Renos K. Papadopoulos in Chapter 11 reveals how war's destructive nature can affect various facets of an individual's identity, including language, faith, and a sense of belonging. Preservation and defense of cultural identity become paramount during wartime as a response to perceived threats and challenges.

War disrupts, reshapes, and challenges – and these texts provide profound insights into how individuals and societies navigate the complexities of identity during and after periods of conflict and trauma.

War and Healing

Amidst the chaos and upheaval of war, there is also a pursuit of reconciliation and healing. Trauma, while a consequence of war, can serve as a catalyst for re-evaluating and strengthening. Artistic expression turns out to be a powerful tool to explore, express, and preserve their identity during times of crisis and conflict, contributing to the healing process. War and resilience are intrinsically linked, as individuals and communities facing the adversity of conflict often demonstrate remarkable adaptability and strength in coping with trauma and upheaval. This resilience manifests as psychological recovery, post-traumatic growth, community support networks, and the preservation of cultural identity amidst war's challenges. Ultimately, war underscores the extraordinary capacity of people to endure, recover, and rebuild – even in the most trying circumstances.

The concept of adversity-activated development (AAD) suggests that individuals can experience positive changes and personal growth as a result of trauma and adversity, going beyond traditional notions of resilience. And even in the face of fear and trauma, individuals can tap into their inner resources to regain a sense of control and purpose, as exemplified by Lisa's dream in Marianne Meister-Notter's Chapter 10.

Various coping mechanisms are explored, such as seeking therapy, engaging in activism, and storytelling, serving as tools for individuals to navigate the psychological and emotional toll of war. Symbols and cultural identity also act as coping mechanisms for societies amidst conflict, reflecting how cultural elements help people endure and find meaning in challenging times.

As many contributors in this collection work in the field of dream analysis, dreams serve as a powerful window into the psyche's response to trauma and adversity during war. The psychological consequences of living in a terror-stricken environment, where the boundaries between victim and perpetrator blur, and maintaining one's sense are challenging. The dreams and experiences of individuals affected by conflict are shared, revealing the profound psychic turmoil resulting from cycles of violence and revenge.

A study on war-related dreams is discussed in Chapter 6 by Heyong Shen, emphasizing that such dreams predominantly evoke negative or mixed emotions, reflecting the emotional toll of war-related experiences. These dream analyses shed light on deep-seated concerns about humanity's fate, the cultural symbolism of war, and modern attitudes toward conflict, underlining the importance of addressing war's psychological and cultural impact on individuals and societies. Dream analysis illustrates how dreams serve as a powerful tool for understanding the unconscious mind during times of conflict. They symbolize various aspects of the psyche, including confinement, depression, and encounters with archetypal figures. This process of introspection and personal transformation highlights the complexity of an individual's psychological journey during war and conflict, ultimately leading to a form of renewed strength.

After war and conflict, healing becomes a crucial process for individuals and societies alike. Contemporary artists utilize creative expression to delve into traumatic histories, offering a means of understanding and preserving moments overlooked by conventional historical narratives. Art serves as a bridge to convey profound emotions and the essence of historical events, fostering personal and collective healing.

Psychotherapy and therapeutic approaches, such as embodied active imagination, play a crucial role in fostering psychological recovery and self-restoration. This process mirrors the broader need for societal healing and transformation in the aftermath of war. Confronting and addressing the truth about the war's impact are an essential step in this journey, serving as a form of reset that allows individuals and societies to face and integrate historical realities. As contributor Kathleen Kirgin observes in Chapter 8 a move toward reconciliation and healing in Ireland after a complex and troubled history with the paramilitary Irish Republican Army (IRA), the essays highlight the intertwined journey of healing, self-discovery, and resilience after war and conflict.

The work of Carl Gustav Jung emerges as a guiding light in understanding the intricate web of war and conflict. Jung's contributions to psychology, particularly his concepts and methods, provide valuable insights into the psychological and cultural aspects of these complex phenomena. Jungian psychology – with its exploration of the collective unconscious, archetypes, and symbolism – offers a framework for comprehending how individuals and societies grapple with the psychological toll of war. It emphasizes the importance of addressing the collective psyche, dreams, and the interplay between the personal and collective unconscious in the context of conflict.

Jung's exploration of the human heart and soul in times of global crises contributes to a deeper understanding of the emotional and cultural dimensions. Concepts such as the shadow, trauma, and the wounded healer from Jung's body of work offer valuable insights into the intricate process of psychological healing and transformation after war and conflict. In this collection, Jung's work serves as a rich psychological perspective for exploring the multifaceted nature of war and its impact on individuals and societies.

As we reflect on the stories and insights shared within these pages, we are reminded that the consequences of war are not confined to the battlefield but extend deep into the collective psyche of nations and communities – and that the complexity and ambiguity of war and conflict are characteristic of the post-modern era, when traditional narratives and certainties are often challenged or deconstructed.

It is our hope to shed light on the enduring legacy of trauma, to acknowledge its profound and often painful effects, and to explore the possibilities for healing and transformation. In doing so, we hope to contribute to the essential and difficult dialogue about the nature of war, its human cost, and the ongoing quest for a more peaceful and compassionate world.

Notes

1 Hillman, J. (2005) *A Terrible Love of War*. Penguin Publishing, London.
2 President of Russia: Annual Address to the Federal Assembly of the Russian Federation (25.4.2005), URL: http://en.kremlin.ru/events/president/transcripts/22931 (accessed 6 October 2023)
3 Bundeszentrale für politische Bildung, Umfrage: Sehnsucht nach der UDSSR (20.09.2021), URL: https://www.bpb.de/themen/europa/russland-analysen/nr-406/340834/umfrage-sehnsucht-nach-der-udssr/ (accessed 6 October 2023)
4 Foreign Affairs: The Global Zeitenwende (5.12.2022), URL: https://www.foreignaffairs.com/germany/olaf-scholz-global-zeitenwende-how-avoid-new-cold-war (accessed 6 October 2023)
5 Fukuyama, F. (1992) *The End of History and the Last Man*. Free Press, New York.
6 Fukuyama, F. (2018) Identity: The Demand for Dignity and the Politics of Resentment. Farrar, Straus and Giroux, New York.
7 Modern Diplomacy: China and Russia Renew Anti-Western Confrontation and Continue Building Multipolar World (accessed 20 December 2024), URL: https://moderndiplomacy.eu/2024/04/11/china-and-russia-renew-anti-western-confrontation-and-continue-building-multipolar-world/?utm_source=chatgpt.com

Part I

Chapter 1

Day 22 of War

Tine Papič

Abstract:

Tine Papič, a Slovenian psychoanalyst, observes the proximity of war in today's world, drawing parallels between Ukraine today and Slovenia in the early 1990s. Back then, the West represented a dream for Yugoslavians, akin to how Ukrainians desire individuation now. Papič applies Anthony Stevens' theory to war, highlighting the transformation of values – murderers become heroes in wartime, when the unimaginable suddenly becomes acceptable. This shift in archetypes occurs effortlessly during war, fostering polarization and rapid judgment of good and evil. However, Papič doesn't perceive the same level of polarization between Russians and Ukrainians as was seen in the Balkan wars. Unlike the intense hatred among Serbians, Croatians, and Bosnians, most Russians don't inherently despise Ukrainians; instead, they seem compelled into the conflict. Tine Papič and Stefano Carpani explore potential explanations for this war, delving into concepts like the father complex and compensation. They consider whether Putin, motivated by shame toward political figures like Gorbachev and Yeltsin, feels compelled to become a hero. Papič doesn't pathologize Putin alone; instead, he views figures like Hitler, Trump, and Putin as manifestations of the collective psyche. He suggests that Russia, as a whole, might be collectively traumatized.

Carpani: *Tine, many hope the pandemic and the Russo-Ukrainian war are just bad dreams, yet they persist as harsh realities for much of the world. According to psychoanalysis, nightmares only fade when their meaning is understood. This series, a continuation of* Breakfast of Küsnacht *(2020) and* Lockdown Therapy *(2022), delves into the symbolism of war. It's an attempt to infuse depth into the superficial war narratives prevalent in today's media. I hope this series can offer depth and counteract the media voyeurism, akin to media bulimia, whereby war is sensationalized for commercial gain, often at the cost of truthful information. In the early stages of the pandemic, we saw a similar lack of depth leading to what I term "media bulimia," characterized by fear, distress, and selfishness. What are your thoughts?*

Papič: I truly value your profound insights. My understanding of war goes far beyond media voyeurism. It's not something distant or abstract to me – it's a stark reality. With friends in Ukraine seeking refuge with us, the war is literally at our doorstep. You can sense it in the collective consciousness, even in the faces of refugees on our streets.

Carpani: *You are 500 kilometers away?*

Papič: Roughly, yes. There's some physical separation due to Hungary being between us and Ukraine, but the psychological distance is similar whether you're in Berlin or East Central Europe. What troubles me about mainstream media is its superficiality. Just like during the [COVID-19] pandemic, you need to delve deeper to find nuanced

DOI: 10.4324/9781003390039-3

insights and substantial content. A quick glance won't suffice to grasp the complexities of the situation.

Carpani: *An example: The latest* New Yorker *issue showcases a tank in the snow. While the entire issue contains almost 80 pages, only 20 focus on war (probably due to its pre-production timing). The next issue, though, will likely delve entirely into war. If we observe major news outlets like the BBC and CNN, they repetitively present the same information every half an hour, lacking in-depth coverage. Let's aim for a more profound understanding. Now, Tine, given your Slovenian background and childhood war experiences: What does war mean to you?*

Papič: It's a thought-provoking question. I observe several parallels between the current situation in Ukraine and my experiences in Slovenia, though there are important distinctions. When the Berlin Wall and the Soviet Union fell, Slovenians exhibited a rare unity, setting aside their typical political divisions. This unity was evident during the Spring Movement, which aimed to secure Slovenia's independence. Notably, Yugoslavia, to which Slovenia belonged, was part of the Non-Aligned Movement, a group of countries that included India, initiated by former Yugoslavian President [Josip Broz] Tito, which did not align with either the Western or Eastern bloc.

We used to travel west, to Trieste or Graz. It was frustrating to see the West thriving while our shops lacked goods. We felt poorer, and life seemed duller. In our minds, the West appeared like a heavenly ideal. We believed that breaking away from our system and embracing the Western world would bring us this idealized life. It seemed like the obvious choice back then. I still think it was the right decision, despite the disillusionment that many experienced. I believe things are much better now than they were.

When we aimed to join Western organizations like NATO [the North Atlantic Treaty Organization] and the European Union [EU], the United States was initially hesitant. Ambassador Lawrence Zimmerman even cautioned that leaving communist Yugoslavia might lead to the U.S. not recognizing our independence. The European response was mixed, with support from countries like Germany, the Vatican, and the Baltic states, who were in a comparable situation.

We were somewhat naïve, believing that Western powers would intervene militarily if the Yugoslav army started a conflict. It was a common assumption that someone would come to our rescue. However, it didn't materialize. Some countries, particularly the USA, opposed our independence from the outset. While some nations supported us, there was an arms embargo aimed at ending the war, which ironically disadvantaged us, as we were unarmed while facing one of Europe's largest armies. It seemed unfair, but in the end, we prevailed.

Unlike the Slovenian conflict, Ukraine has received more substantial international support in terms of military aid and resources. Slovenia was fortunate to come out of the Yugoslav conflict with relatively minimal damage, partly because Yugoslav forces were preoccupied with other conflicts in Croatia and Bosnia.

While there are notable parallels between the Ukrainian situation and Slovenia's history, the most significant distinction lies in the varying degrees of international involvement and support. In our case, we often felt isolated, whereas Ukraine benefits from a more unified European and American stance.

Carpani: *In the 1990s, Ukraine leaned towards the West, embracing Western culture and values. However, today, the focus has shifted towards building a strong nation with resilient institutions. Despite political changes and even attempted revolutions, these institutions remain stable. The United States, while dealing with internal tensions, provides an example of institutional resilience, especially during the transition from [President Donald] Trump to [President Joe] Biden. Ukraine now seeks closer alignment with Europe over the United States, moving in a western direction.*

Papič: I see similarities between what you're describing and Slovenia's historical aspirations. We shared a desire for an independent nation free from the corruption that had tainted the late years of the communist system. A ruling elite held all the power, and Milovan Đilas, a former prominent communist, criticized this system in his book *The New Class: An Analysis of the Communist System*. He revealed how a bureaucratic class had effectively become the new elite, controlling everything while the majority had very little.

     The situation also hampered personal growth, particularly in terms of entrepreneurship, which was largely prohibited. The extent of these restrictions varied among different Yugoslav republics, with Slovenia being comparatively more open and Bosnia more restrictive. This lack of individual autonomy ties into a broader psychological perspective I've explored, addressing whether one can fully realize their potential and individuate in Jungian terms, or if they're constrained by a repressive system. It delves into the plurality of the psyche: Can I become who I want to be, embracing whatever emerges from my psyche, or is there someone or something suppressing me?

     There's also what I refer to as a collective "mother complex" to consider. Under communism, people experienced a sense of security – unemployment was extremely rare, even though the available jobs were often far from perfect. It was a system that pledged to "take care" of its citizens. Nevertheless, this came with numerous limitations on individual freedom and personal growth, which I like to term "individuation."

     In contrast, capitalism requires perpetual competition and lacks a safety net. Under communism, it felt like the system said, "I'll provide for you, but you must conform to my rules." The concept of the "Homo Sovieticus" embodies this mindset – an individual molded by a system that provides security but restricts personal freedom and growth. I contend that a nation matures, or "individuates," through its institutions and its people.

Carpani: *Let's discuss "Homo Sovieticus," a concept by Svetlana Alexievich, highlighting the changing role of women in conflict. Returning to Habermas and Derrida's views on Europe, an article by Italian journalist Ferrara in* Corriere della Sera *notes the Iraq war's impact on Europe. Habermas and Derrida identified a unique European political mentality, distinct from the American one, characterized by aversion to force, commitment to law, support for multilateral institutions, and human rights. This mentality drove European foreign and security policy within the EU. Ferrara questions if this explains Europe's past indifference to Ukraine. I believe Ukraine has aligned with these values since the early 2000s, although often seen through the lens of "Homo Sovieticus." As a Slovenian, you understand the historical European ties shared by Slovenians and Galicians, formerly part of a multicultural empire. This is why your Prime Minister and counterparts from the Czech Republic and Poland visited Kyiv. These regions were once united within a multicultural-ethnic-lingual-religious empire.*

Papič: Absolutely, the grassroots pursuit of democracy in Ukraine is captivating and educational. As Dmytro Zalesky highlighted recently, the Ukrainian people have displayed unwavering determination in their resistance against corrupt regimes, a stark
contrast to the situation in Moscow. Their aim has been to align closely with European norms and institutions.

Conversely, the attitudes in Europe and the U.S. seem to reflect a degree of ignorance. It's not limited to the general public; politicians have frequently been misin-
formed, viewing Ukraine as an extension of Russia or the Eastern Bloc. I believe
this ignorance arises from a deficiency in education and a limited perspective on the
geopolitical landscape.

Similarly, there's a reluctance among Western leaders to provoke Russia. We
encountered a comparable quandary in Slovenia, wondering why no one was coming to our aid. The explanation is multifaceted: There may be verbal support, but
deploying troops and risking lives is a different matter altogether. This mirrors the
current challenge facing the West, particularly in Europe. The political considerations of engaging in such a conflict are immense and almost inconceivable in today's
European context.

This becomes even more pronounced when dealing with formidable adversaries,
alongside widespread ignorance or overly simplistic viewpoints that might suggest,
"Maybe Ukraine should be part of Russia." What's becoming increasingly apparent is the intricate web of interconnections that people are just beginning to grasp.
Decision-makers are gradually acknowledging the complexities, but there's still a
considerable distance to cover.

Carpani: *Recently, I had guests – a family from Lviv and, until yesterday, a family from
Kharkiv. Interestingly, their native language is Russian, and I'm uncertain about
their political leanings regarding Zelenskyy or Putin. However, what I do know is
that their home was destroyed by the Russians, forcing them to flee and endure a
five-day journey. It was a group of two mothers with three children, one of whom has
autism and requires assistance. There are more than three million refugees already.*

Papič: Europe, for the first time since World War II, is experiencing the reality of war. It's
not just in the news; we can feel its proximity. Perhaps we should revisit what war
truly means.

Carpani: *Tine, what is war?*

Papič: From a psychoanalytic perspective, Anthony Stevens provides intriguing insights.
He discusses two societal modes – peace and war – and how distinct archetypes
influence our behavior in each. In a peaceful society, killing is seen as murder, but in
a wartime context, killing young boys can elevate someone to hero status. Actions
that would typically be criminal can become somewhat 'acceptable' during war,
though not universally, but within a specific societal framework. Values are quickly
inverted.

I believe this shift between these archetypes is occurring within us psychologically, and we're experiencing it in Europe, as well. There's a psychological tension, where Melanie Klein's theories can offer insights into the current European
psyche. Typically, we reside in a depressive position, which is generally neurotic
but stable. However, due to current tensions, we are frequently transitioning into a
paranoid-schizoid position. Essentially, our collective psyche is fluctuating between

neurotic and borderline states, a change that is keenly felt across Europe these days, much like during the [COVID-19] crisis.

Renos K. Papadopoulos, with his extensive work with refugees worldwide, notes how archetypal energies often polarize our perceptions, categorizing people as either "good" or "bad," with minimal room for gray areas. This polarization seems less severe in the Ukrainian conflict compared to the Balkan wars, where *deep-seated* hatred defined relationships. In Ukraine, it appears that societal pressures are driving people into conflict rather than innate hostilities, which provides me with a glimmer of hope.

The Russian–Ukrainian relationship differs fundamentally from what we observed in the Balkans, where deeply entrenched archetypal hatreds had accumulated over centuries. Historical memories were like tinder waiting for a spark. Tito, with his utopian socialist vision, tried to extinguish these enduring feuds and unite these distinct nations as brothers. However, he couldn't erase what was deeply rooted in the collective unconscious.

Consider the ethnic cleansing by the Croat fascist state against the Serbs during World War II. That collective trauma left an indelible mark on the Serbian psyche and resurfaced during the Balkan conflict. These are not isolated incidents; they are echoes of a protracted historical narrative. Even when concerted efforts are made to foster unity and erase the past, these *deep-seated* grievances persist.

Ukraine presents a somewhat different scenario. While war tends to breed hatred as it unfolds, it didn't commence with the kind of profound animosity we often see in other conflicts. Most Russians do not harbor hatred towards Ukrainians, and vice versa; they've been pushed into this situation more by systemic forces than by intrinsic hostilities.

In war, we often witness a game of projections. The enemy is dehumanized, no longer perceived as people but as embodiments of evil. This psychological mechanism allows otherwise "normal" individuals to commit acts of violence. It taps into fundamental human instincts that have existed since the dawn of time, extending beyond this conflict.

We can delve into the anthropological aspect here, acknowledging the fundamental "us versus them" dichotomy that arises in conflict situations. Anthony Stevens presents a compelling comparison of human behavior in war to that of animals. Animals in conflict with their own kind engage in restrained combat, stopping short of causing lethal harm. However, when faced with another species, this restraint vanishes, and they kill. This primal mechanism may still be deeply embedded in our psychology. Dehumanizing the "enemy" allows us to project our dark aspects onto them, making it psychologically easier to harm them.

It's worth noting that many *rank-and-file* Russian military personnel do not appear fully committed to this conflict, similar to the experience in Slovenia. Yugoslavian soldiers often wondered why they were on the brink of battle in tanks when they could be at home enjoying a meal cooked by their mothers. Offering them a peaceful way out was an effective strategy, with many choosing surrender and returning home over needless confrontation.

This reluctance seems to apply to many ordinary Russian soldiers today. They appear to participate reluctantly, in contrast to career military individuals who might possess psychopathic tendencies and view war as a familiar and even desirable environment.

While the engine of war often runs on hatred and dehumanization, there is reason to hope that this conflict won't escalate into an *all-consuming* blaze. The absence of intrinsic, *deep-seated* hatred among the majority of those involved could serve as a mitigating factor.

Carpani: *Now, let's explore the negative father complex. Some historians see World War II as an attempt at reparation rather than a continuation of World War I, with Hitler's promises resembling Trump's "Make America Great Again." Prussia was more inclined toward conflict than Austria, which had avoided warfare since regaining territory from Italy in the mid-19th century. Is there a parallel today?*

  Let's consider Freud and Jung, who initially idolized their fathers but later experienced disillusionment. This may have driven them to achieve greatness, such as founding psychoanalysis. Could Putin be similarly motivated? He might be ashamed of [Boris] Yeltsin, the alcoholic political father figure, and [Mikhail] Gorbachev, who dismantled the Soviet Union with Perestroika. Is a negative father complex influencing Putin's current actions?

Papič: It's an intriguing thought. You mention that Hitler aimed to "Make Germany great again," and I believe the same applies to Putin, who seeks to "Make Russia great again." This is quite frustrating.

From my perspective, we should consider the dynamics of groups. Wilfred Bion's idea was that the group behaves like a single entity. Jung also touched on this, noting that the more people you bring together, the less intelligent the conversation becomes. He remarked, "One hundred intelligent people in a room make one big idiot." This occurs because in larger gatherings, we can only converge on common themes and shared traits. As more people come together, the commonalities diminish, making it challenging to engage in profound conversations, leading to a collective reduction in intellectual depth.

Jung suggested that larger nations can regress to more primitive states of consciousness, similar to that of a lizard. This, he believed, made smaller nations more appealing due to their seemingly more humane nature. One can perceive a nation as an individual entity. Consider Germany after its humiliations or Russia in recent times. Both could be associated with what might be called a "negative father complex." Russia, as a nation, harbors aspirations of being a significant power – a sentiment rooted in historical truth – but this position was compromised after the Soviet Union's dissolution. Understanding the nation's leadership is crucial in this context. Leaders often reflect the consciousness of their nations, an idea aligned with the works of Bion and Jung. In a mature nation, certain types of leaders wouldn't thrive. For instance, a leader like Putin might not achieve the same success in a country like Sweden and could possibly face expulsion from his political faction.

Carpani: *Let's talk about psychopathology. Is it useful to psychopathologize Putin? Is it inappropriate? What is your take?*

Papič: As I mentioned earlier, figures like Hitler, Putin, Boris Johnson, and Trump don't seize power; they reflect the collective psyche of their societies. It's the collective mindset that allows or even chooses them for positions of authority. As I've noted, this dynamic likely wouldn't occur in a country like Sweden, for instance.

I do believe that the Russian population is collectively traumatized. The nation's history is replete with tragedies – wars, revolutions, the era of the Soviet Union, all periods of profound collective trauma. Consider Putin's own background; he was born in St. Petersburg, his parents were survivors of the [World War II] siege, a time marked by unimaginable horrors and the loss of *millions* of lives. His parents lost two other sons, making Putin the sole survivor. What does this signify for a society that has endured such turmoil?

In the 1950s, the only thing that seemed to matter was power, largely due to the circumstances of that era. This is how trauma operates at a collective level. It was a society akin to a prison. Putin was born into this environment, a society [in which] power was paramount.

Communism, despite its flaws, held some humanistic ideals. However, it appears that even these ideals have vanished, leaving an obsession with power. Collective trauma shapes the collective mindset, and that mindset, in turn, empowers and shapes its leaders.

Carpani: *I view this war as a potential reset, but what would it reset, exactly? In 2008, [U.S.] President [Barack] Obama symbolically proposed a reset to Putin, and Secretary of State Hillary Clinton famously handed Lavrov a reset button on TV. Yet, that attempt failed. Could this war potentially serve as a reset, and if so, what would it reset?*

Papič: The concern is undeniably real. You mentioned Jung's "After the Catastrophe," which explores the concept of a societal "reset" following a disaster. World War II served as such a reset for Germany and Europe, paving the way for reconstruction and a new direction.

However, when we zoom out and examine history as a whole, spanning the last 4,000 *years*, statistics reveal that for every *year* of peace, there are approximately 13 *years* of war. This ratio is a stark reminder of a certain inherent aggression in our species.

If we narrow our focus to the latter half of the 19th century and the entire 20th century, we observe that virtually all major nations have been involved in wars about every 20 *years* – roughly once a generation. This recurring pattern suggests that aggression and conflict are deeply ingrained in our collective DNA, almost archetypal in nature.

If extraterrestrial beings were to observe humanity from space, they might conclude that we are a species characterized by endemic violence and war. This raises the crucial question: How do we break this cycle, particularly when the global stakes are higher than ever? This is what concerns me deeply.

This inherent inclination toward conflict indeed seems to be deeply ingrained in our nature. When you posed the question, "Is war masculine?" it initiates a profound exploration. While historically, men have primarily been at the forefront of physical warfare, on a symbolic or archetypal level, the essence of war and violence might be rooted in nature itself. Drawing from Neumann's theories, war could be seen as resonating with the 'mother' archetype – nature in its rawest form – while the aspiration for peace aligns with the 'father' realm, representing Logos. It's a challenging perspective to grasp, as it appears to invert common associations.

Certainly, men are often the ones directly involved in combat. Yet, if we delve into the symbolism, it's the Logos – the embodiment of order and reason represented by the "father" archetype – that strives to bring order to the chaotic and often ruthless realm of nature, where only the fittest thrive.

In the current Russian context, Putin embodies this *survival-of-the-fittest* doctrine. It's nearly inconceivable to imagine a leader with a more humanistic approach under the present circumstances. If Putin were to be replaced, it's probable that another "strongman" figure would take his place. Thus, the real transformation required is a shift in values.

One perspective posits that the East missed transformative eras like the Renaissance and the Enlightenment, which originated in Europe. Interestingly, Ukraine appears to be actively pursuing this *Enlightenment-inspired* change, a transition toward democracy, plurality, and a richer tapestry of life. In contrast to the monotonous landscape of socialism, Ukraine's journey serves as a beacon of possibility. However, it's clear that this progress unsettles certain entrenched structures in Russia, represented by figures like Putin.

The main issue seems to revolve around Ukraine's shift towards Europe. Within Russia's political circles, there's a fear that Ukraine's success might lead to demands for more democratic institutions, potentially endangering those in power.

Carpani: *Putin embodies a narrow-minded, patriarchal figure suppressing individual growth and transformation. Let me share a passage from Hillman's "A Terrible Love for War,"[1] because I want to hear your take on it:*

There is no practical solution to war, because war is not a problem solvable by the practical mind, which is better equipped for its conduct than for its avoidance or conclusion. War belongs to our soul as an archetypal truth of the cosmos. It is a human work and an inhuman horror and a love that no other love has been able to overcome. We can open our eyes to this terrible truth and becoming aware of it, devote all our passionate intensity to undermining the enactment of war strengthened by the courage the culture possesses, even in the dark ages to continue to sing as it resists war. We can understand it better postpone it longer, work to gradually remove it from the support of a hypocritical religion, but the war as such will remain until the gods themselves leave.

Papič: Stevens' idea that gods are inherent archetypes in our biology raises a deep question. How can we transcend these archetypes to attain true peace in the realm of Logos? Mere intellectual understanding won't cut it; a broader evolution of consciousness is essential.

You rightly noted the need for a conscious shift towards plurality in Ukraine, and the same is crucial in Russia. It's not about universal harmony or love, but about attaining mutual respect and coexistence. Žižek succinctly states that it's not about loving each other, but about reaching a level of mutual respect that embraces differences in religion, nationality, or culture.

The European example is apt. After enduring devastating wars, Europe underwent a transformative phase whe[n] the idea of conflict between nations like Germany and France now seems inconceivable. The challenge is to foster a similar evolution in other parts of the world, making respect and coexistence the foundational principles of society.

Carpani: *These values align with discussions by Habermas and Derrida. Mauro Magatti, in an article in the Italian Catholic newspaper* Avvenire *on March 1st, emphasizes that Putin's actions have disrupted the global liberal order established after the fall of the Berlin Wall. He suggests that Putin is testing the West's resilience. German sociologist Ulrich Beck calls for a declaration of interdependence, highlighting our shared vulnerability and collective responsibility, promoting a cosmopolitan perspective.*

However, I question my own initial optimism. At the beginning of the [COVID-19] pandemic, I expressed hope for positive change, encouraging introspection, spiritual reconnection, and transformative creativity.[2] But now, I'm beginning to think that the world may have become worse post-pandemic, in line with Michel Houellebecq's views. What are your thoughts on this?

Papič: Ending on these profound reflections, I hold mixed feelings about a world without nations. While the idea of universal harmony is enticing, preserving one's identity also holds value. Should we instead advocate for plurality? Just as an individual psyche contains various facets and feelings, acknowledging and accepting them can lead to psychological balance. Conversely, denying or suppressing aspects of oneself can result in neurosis. Similarly, tensions within or among nations might be seen as collective neuroses.

The escalation from tension to war appears to occur when societal structures break down, akin to an individual's descent into psychosis. Preserving distinct national identities adds vibrancy to the global tapestry with diverse cultures, traditions, and histories. The real challenge lies in fostering a world where these nationalities coexist harmoniously, mirroring post-[World War II] Europe's achievements.

So, what is the essence of the conflict we're discussing? Ukraine, in my view, isn't driven by nationalistic fervor but is a *multi-ethnic* nation aspiring to establish a democratic setup where[by] citizens live in plurality. This war may be less about territorial claims and more about differing states of consciousness. One side aspires to a society built on mutual respect and plurality, while the other prioritizes power above all else. The future, I believe, hinges on which of these consciousnesses will prevail.

Carpani: *Let's delve deeper into the issue of power, as it's easy to point fingers at Putin right now. Europeans are mostly unanimous on this, though there are some who argue for considering historical context. However, it's a fact that Putin's army invaded Ukraine. Now, the challenging question is: How does this war differ from the actions of the United States since World War II: Vietnam, Iraq, and Afghanistan. Consider the U.S. interventions, occupations, or political destabilizations dating back to 1946, supposedly in the name of liberation and, since 1991, under the pretext of promoting liberal democracy.*

Maurizio Ferrara, as I mentioned earlier, noted that some commentators saw the U.S. invasion of Iraq as a premeditated bloodbath. Similarly, the recent events in Ukraine share a troubling resemblance. Svetlana Alexievich's Nobel Prize–winning literature vividly portrays the imperial tendencies not only of Putin but also a significant portion of the Russian population, referring to it as the "Homo Sovieticus." But this raises the question of the imperialistic inclinations of the United States and Great Britain.

Whenever the British invaded and occupied nations, they left behind turmoil, as seen in Kenya, Cyprus, and Palestine. The British occupation of the northern part of Ireland persists. This raises the question: How do the actions and resulting casualties of the British crown and the USA compare to those of Stalin, Putin, Hitler, Mussolini, or Franco, when colonial Britain occupied countries and took everything, or [when] the capitalist U.S. invaded countries for geopolitical reasons and for their goods?

Maurizio Ferrara quotes your compatriot Žižek, who said: Putin feels himself to be the patriarch of an organic community in which being free means staying in one own place. Žižek added that Aleksandr Dugin, Putin's court philosopher, convinced Putin that truth consists [of] what one believes. So, if the United States [doesn't] believe the Russian truth, the world must be made to decide. But what to tell the Ukrainian people? Žižek wonders: Are they not entitled to their own truth? Or is Ukraine [merely] a battlefield for the use of those who want to rule the world? So, the division in two spheres of influence is still a fact of the contemporary international system, with which we must reckon. Europe should follow its vocation and act through diplomacy and mediation; however, when war is an established fact on our borders, it's very difficult. What is the difference between this invasion [and] what the U.S., the British and many other countries did in the past few decades centuries to perpetrate their imperialistic impulses?

Papič: Recognizing parallels and distinctions when analyzing the actions of major nations is crucial. Looking back at the United Kingdom's history, a more aggressive – and, to some extent, racist – culture was evident. However, we should consider these behaviors in their historical context, as many expansive empires have shown imperialistic tendencies at various points in their histories.

Drawing a parallel with the current Ukraine situation, there appears to be a psychological disconnect in the Russian perspective. Putin may have assumed that Ukrainians would welcome rapid integration into the Russian federation, similar to the American misjudgment in invasions like Iraq. Both cases reflect an egocentric or even narcissistic attitude common among major powers, often operating under the presumption of superiority and the expectation that others should eagerly align with them.

On the other hand, there's a prevailing narrative suggesting that NATO is encroaching on Russia gradually. Personally, I find this perspective somewhat *one-sided*. The complexities of these situations and their underlying motives are more nuanced.

Reflecting on Slovenia's journey, it appeared that the international community was initially hesitant to embrace or support us, with a preference for Slovenia to remain within Yugoslavia. Similarly, in the current dynamics involving Ukraine, while NATO may feel a duty to assist, many of its leaders might have preferred a less confrontational approach. If Ukrainians had uniformly expressed a desire to align with Russia, it could have been more acceptable. The notion that NATO was actively seeking to expand its influence into Ukraine may be somewhat overstated. While Ukraine may have looked to NATO for potential security, NATO's hesitation likely stemmed from the complexities of such an association with Russia.

The USA has a history of exerting influence through *non-military* means, exemplified by brands like McDonald's in Moscow. These commercial ventures can be as influential, if not more so, than military alliances in promoting influence.

Speaking of individuation, whether for individuals or nations, the process is complex. Slovenia's ongoing transition illustrates our aspiration to become a fully realized nation, albeit with obstacles. Some have misconstrued the concept of individuation, linking it closely with the idea of breaking free from the "mother complex." While there's debate about which nation plays the "mother" role in the Russia–Ukraine dynamic, the essence of individuation is realizing one's full potential. It involves embracing plurality, fostering freedom in all forms, including artistic, literary, entrepreneurial, and more, without being bound by restrictive complexes.

The American perspective is intriguing. Their interventions often stem from a genuine belief in spreading freedom. The documentary *The Unknown Known*, centered on Donald Rumsfeld, offers insight into this mindset. Rumsfeld's character, with psychopathic traits, is captivating not just for clinical aspects but also for his unwavering belief in his philosophical constructs, however questionable they may be. It provides a glimpse into the psyche that often influences international dynamics.

Carpani: *Rumsfeld and Cheney, already prominent, were present during George Bush's father's presidency on September 11, 1989. His declaration, "I bring you a new order," aimed at global peace and the spread of liberal democracy. The truth is that what came after is the invasion of Iraq and more. While I believe the world follows its course, I genuinely hope this war can serve as a reset, fostering interdependence in shaping meaning. Despite the horrors of war, it may offer an opportunity for positive change, new beginnings, and inner reflection. The ideas I held for the post–[COVID-19] era remain relevant.*

Papič: I've been contemplating the concept of co-constructing meaning, frequently encountered in therapeutic settings. When individuals experience trauma, therapy helps them find meaning in those experiences. It's a process that integrates trauma, aiding in coping, reconciliation, and potential healing. Most people wouldn't choose to relive their trauma, yet deriving meaning from adverse experiences can be transformative. Whether these events inherently hold meaning or if it's a product of our human need for coherence is a profound question.

Drawing parallels with the current geopolitical landscape, nations like Russia, Belarus, and others are at a critical juncture. They require an evolution or revolution of consciousness that embraces plurality. Europe underwent this transformative phase during the World Wars, reshaping its collective consciousness. This makes me apprehensive about our present times – it seems that profound changes in collective psyche often emerge only after enduring catastrophic events. The challenge is finding ways for societies to foster such transformations without going through such devastating upheavals.

Carpani: *Unfortunately, yes, but that's why I call myself naïve because the catastrophe that we went through with [COVID-19] didn't seem to help.*

Papič: The post–[COVID-19] world has indeed ushered in significant changes. While we shouldn't expect a radical overhaul in every aspect of life, certain areas have been undeniably transformed. One notable shift is digitization, allowing us to connect without physical proximity, such as our conversation between Berlin and Ljubljana. This not only offers convenience but also contributes to a more sustainable ecological footprint. There's a growing awareness of its environmental costs. Remote work, reduced commuting, and fewer physical interactions may be lasting legacies of the pandemic, but they also pose challenges to our social nature.

Turning our attention back to Russia's stance on Ukraine, it differs from historical instances like Nazi Germany's treatment of the Jewish population. The Nazis held an archetypal narrative rooted in hate and racial purification. In contrast, the Russians don't harbor the same level of animosity towards Ukrainians; they see them as brethren. Their intent isn't to eliminate, but to unite. However, this desire for union isn't universally reciprocated by all Ukrainians, creating a disconnect that Russians struggle to comprehend. The nuances of these relationships are deeply rooted in complex historical, cultural, and psychological factors.

Carpani: *Like: I am the man in the house, I run. So, you have to be under me, otherwise I kill you, otherwise I exterminate you. My feeling of what is happening in Mariupol now is exactly this: You don't want to stay under my roof – I kill you all.*

Papič: Exactly. The linguistic and cultural interweaving in Ukraine, where nearly half the population speaks Russian, underscores the profound intricacies of their relationship. It's reminiscent of the Yugoslavian dynamic, where[by] Serbs couldn't fathom why other constituent nations desired independence. The common theme, whether in Serbia's case or Russia and Ukraine, is this genuine perplexity: "Why wouldn't you want to remain united with us?"

The American intervention in Iraq offers a parallel, albeit with a different context. Americans expected to be embraced as liberators, but they were met with resistance. Similarly, Russia's struggle to understand Ukrainian sentiments reveals a significant comprehension gap.

However, there's a glimmer of hope in the current situation. The fact that a substantial portion of Russians, approximately 30%, opposes the war and protests in the streets is significant. In contrast to Nazi Germany's *all-consuming* obsession, which left little room for dissent, the presence of vocal opposition in Russia, even amidst crackdowns, suggests that the situation may not devolve into the catastrophe Europe witnessed in the 1940s. These pockets of resistance and reflection may serve as buffers preventing a *full-blown* disaster.

Carpani: *Let's hope not. Thank you very much, Tine. Good night.*

Papič: Thank you, Stefano, for this opportunity.

Notes

1 Hillman, J. (2005) *A Terrible Love of War*. Penguin Publishing, p. 214.
2 Carpani, S. and Luci, M. (2022) *Lockdown Therapy*. Routledge, London.

Chapter 2

Day 27 of War

Murray Stein

Abstract

Murray Stein, an American psychoanalyst, identifies both masculine and feminine aspects within war, attributing strategic dimensions to the animus and emotional facets to the anima. He notes the historical male dominance in warfare due to physical prowess, acknowledging the contemporary shift whereby women can perform effectively. Stein suggests that Vladimir Putin's actions stem from a theme of compensation, describing him as a "mana personality" feeling consistently undermined by other leaders, leading to a distorted perception of reality. Stein explores war as a reset, citing shifts in military thinking, increased military investments, a united European response to Russian aggression, assistance to refugees, and Ukraine's pursuit of a unique identity. Discussing Russia's view of the West, Stein and Stefano Carpani reflect on the turbulent U.S.–Russian history despite diplomatic efforts, emphasizing the lingering Cold War mindset in the Russian psyche.

Carpani: *Murray, last time we spoke was two years ago, at the beginning of the pandemic. Now, war is back in Europe. What is war?*

Stein: War, as General Marshall put it, is hell. I've been living with it since college. The United States has seen continuous conflict, from the Korean War to Vietnam, Iraq, Afghanistan, and now Ukraine. War becomes a familiar part of life, a dark backdrop to our existence. The core of war lies in the failure to resolve conflicts through discussion, diplomacy, and policy – the very principles the United Nations was created to uphold. While we strive to manage conflicts without resorting to war, events like the Ukraine crisis remind us of the inherent challenge. War has been a recurring theme throughout history, from the Trojan War to the medieval conflicts in Italy. It's an undeniable part of the human experience, shaping our past and, in some ways, determining our future. But, as we witness the horrifying images from Ukraine, we are reminded that war is indeed hell on earth.

Carpani: *Let me let me read you and to those listening to us what James Hillman had to say about war, this comes from* A Terrible Love for War:[1] *"There is no practical solution to war, because war is not a problem solvable by the practical mind, which is better equipped for its conduct than for its avoidance or conclusion." War belongs – as you said, Murray – to our soul as an archetypal truth of the cosmos. It is a human work and an inhuman horror and a love that no other love has been able to overcome. We can open our eyes to this terrible truth and becoming aware of it, devote all our passionate intensity to undermining the enactment of war, strengthened by the courage that culture possesses, even in the dark ages to continue, to sing as it resists war. We can understand it better, postpone it longer, work to gradually remove it from*

DOI: 10.4324/9781003390039-4

the support of a hypocritical religion, but the war as such will remain until the gods themselves leave. Haven't the gods left already, if we look at Nietzsche, if we look at Freud?

Stein: The gods remain intertwined with war, like the famous Greek god Ares, whose origin lies in Hera's anger and resentment towards Zeus, birthing him parthenogenically due to her spite. This reveals a common thread in the origins of war, often rooted in resentment and revenge. The gods and goddesses of war reflect aspects of the collective human psyche and are deeply ingrained in our archetypal consciousness. James Hillman's insights draw from his profound understanding of the archetypal psyche, suggesting that war has been an integral part of human nature since time immemorial.

War comes in various forms, from wars of liberation to religious, defensive, ideological, and cultural conflicts. Regardless of the type, war galvanizes energies, focuses intense forces, and sparks human creativity. Some of the most significant technological advancements, such as the internet, missiles, and space exploration, have originated from military research efforts driven by the anticipation of war. The atomic bomb and atomic energy are prime examples, with the need to stay ahead of the other side serving as a catalyst for scientific innovation.

Throughout history, wars and conflicts have propelled cultural and societal changes. Japan's Meiji period was influenced by Western technological superiority, leading to its rapid modernization. Human beings, as animals deeply ingrained with a territorial defense instinct and expansionist tendencies, are predisposed to engage in wars and territorial disputes.

The ongoing war in Ukraine reflects a larger ideological battle between Russian and Western worldviews. It revolves around the question of whether the world should be unified under one brand, leader, or religion, or if diversity and the presence of multiple gods should be allowed. This ideological conflict, which touches on monotheism and polytheism, encapsulates the fundamental differences between Russia and the West. The West advocates diversity and democratic systems, while Russia favors a single leader and a more unified vision of the world.

Carpani: *This also raises a crucial aspect: the duality of masculinity and femininity, a significant consideration for Jungian analysts like us. Svetlana Alexievich, a Nobel laureate, characterizes war as masculine. But is war exclusively masculine? We can explore this by examining figures like Hera and Ares from mythology. What truly defines war? We can draw examples from the Vikings or the female Kurdish warriors and other strong women engaged in combat, not just wielding rifles. For me, the enduring image is not solely that of men in military attire but rather women in civilian clothing, accompanying their children to school while carrying rifles provided by the Ukrainian government. Does this challenge the notion of war being exclusively masculine? Is it perhaps too simplistic to categorize it in such terms?*

Stein: In Jungian psychology, we encompass both masculine and feminine aspects, represented by animus and anima. Every individual carries both elements, and their dominance can shift over time. This means that a person's disposition may manifest in various ways. Men can exhibit feminine qualities, while women can display strong animus characteristics, such as aggression and authority. The division between masculine and feminine isn't clear-cut. The concept of war involves strategic elements (animus) like weapon development, ideology, and planning, while also encompassing emotional and relational aspects (anima), such as a sense of belonging, attachment, and territorial defense. Throughout history, warriors were predominantly men due to physical strength playing a significant role, but in today's world, technology has largely neutralized physical differences.

Women can be as capable in operating the tools of war as men, but the psychological attitude towards war is what truly matters. Some women are exceptionally aggressive, seeking dominance and control, just as many men can be. The gender divide in the context of war is not straightforward.

Ultimately, the animus and anima work together when war arises. The animus provides strategy, while the anima contributes emotional energy. Men and women can collaborate in warfare, and it's not confined to one gender or attitude. Andrew Samuels and Jung have both recognized the need to separate animus and anima from gender, aligning with our more flexible and inclusive contemporary perspectives on masculinity and femininity in relation to both genders.

Carpani: *Let me shift our focus from masculinity–femininity and anima–animus to profound archetypes: the mother and the father. Historians often argue that World War II continues World War I, and from a Jungian viewpoint, I see it as compensation. Some view war as an ongoing cycle. In reference to ancient Greece, Hera's role makes me ponder if current wars are echoes of the past. Could this war compensate for the Soviet Empire's fall and reset it? It's interesting that Putin blames Lenin, but historical accuracy points to his own political father, Yeltsin, who oversaw the Soviet Union's dissolution, granting Ukraine autonomy and independence through treaties.*

One can also consider compensation within Putin's personal history, perhaps stemming from his parents' traumas. This evokes Jung and Freud, who initially regarded their fathers as heroes. Recall Freud's father's hat incident, where he fled rather than confronting a bigot, or Jung's priestly father losing faith. Both shifted from perceiving their fathers as heroes to recognizing their fallibility, motivating them to seek heroism themselves. Could this war be a generational compensation, a psychological endeavor to attain what one's fathers – biological or otherwise – could not achieve?

Stein: A friend and colleague suggested analyzing Putin from an Adlerian perspective, highlighting his potential inferiority complex and its compensation with a superiority complex. This dynamic involves a desire to inflate one's self-worth to compensate for perceived inadequacies. Putin might have felt humiliated by past interactions with Biden and Obama, especially when they belittled Russia's global status, economy, and industry. In response, Putin worked on building up Russia's military and enhancing his own image. He has become a billionaire with an estimated net worth of $100 billion and developed a larger-than-life personality.

Putin appears to have fallen into a psychological trap described by Jung as the "mana personality." In this state, the ego identifies strongly with an archetype, losing touch with reality and believing that its constructed version of reality is the absolute truth. Putin's worldview includes mystical thinking, emphasizing the "Russian World" and his role as a heroic figure destined to restore Russia's dignity and power in the world.

Leaders who surround themselves with sycophants, identify with inflated archetypal figures, and gain total control over a formidable military, like Putin, can

become dangerous, not just to the world but to their own people. They may prioritize their grand vision over economic realities and the *well-being* of their citizens.
This mindset can make them nearly impenetrable and resistant to dialogue.

Putin's alliance with the Russian Orthodox Church and their shared vision of
Russia's dominance further reinforces his impenetrable aura. It's akin to trying to
engage in dialogue with dogmatic individuals, be they fundamentalist Christians
or Putin himself. At this point, Putin may see himself as a vessel for divine will
and may believe he is helping the Ukrainian people return to a state of grace by
de-Nazifying Ukraine and removing leaders who advocate *Western-style* democracy. It's unlikely that anything will dissuade him from pursuing this path.

Carpani: *He's been also working against gays, for Kirill.*

Stein: Yes, that's part of the West; LGBTQ [lesbian, gay, bisexual, transsexual, and queer
             identity] is part of Western sinfulness.

Carpani: *Two weeks ago, I organized this talk with our colleague Dmytro Zaleskyi that
             received great success. Yesterday, I noted there was a comment under the YouTube
             video saying: "Jung supported National Socialism 100 years ago, now post Jungi-
             ans support Ukrainian neo-Nazis." I responded that is historically wrong. This is
             part of not being able to see what is really happening.*

Stein: Yes.

Carpani: *What comes to mind is Jung's exploration in "After the Catastrophe," where he
             delved into the psychopathological aspects of individuals like Hitler. This proves
             to be a fascinating area of study. In the 21st century, leaders like Putin and vari-
             ous populist figures engage in a curious phenomenon – they propagate falsehoods,
             genuinely believing in their own deception, while their followers blindly trust and
             adhere to these lies. This often leads to a state of collective hysteria. It's intriguing
             to note that some individuals, including certain YouTube commentators, draw par-
             allels between Jung and his association with National Socialism in the 20th century,
             and contemporary Jungians who focus on neo-Nazism in Ukraine. However, what
             seems overlooked is the examination of their own shadows in this context. This
             raises important questions about the dynamics at play in such discussions.*

Stein: Historical situations are inherently complex, and oversimplifying them into terms
             of good versus evil can obscure important nuances needed for a balanced and real-
             istic understanding. It's a fact that Ukraine has neo-Nazi groups. It's also true that
             many Ukrainians did not oppose Hitler when he came into the region. In fact, some
             of them actively supported Hitler's cause and were known for their anti-Semitism.
             Ukraine, therefore, cannot be seen as entirely free from this historical stain.

Carpani: *We have to be also historically clear because anti-Semitism was a big problem at
             the beginning of the century and especially at the end of the Austria-Hungarian
             Empire. I always say I have a picture of Emperor Franz-Josef upside down. He was
             the first fascist, an emperor of anti-Semitism. Perhaps this is not perfectly histori-
             cally correct, but the region of Galicia, where everything started in [World War I],
             was very anti-Semite. Europe was very anti-Semite because there was the biggest
             population of Germanic-speaking Jewish in Europe. Look at the Austrians now, so
             right-wing. Look at the Hungarians, so right-wing. So is historically in their psy-
             chological DNA. Sorry to interrupt. I just wanted to add this.*

             *Historical accuracy is vital here because anti-Semitism was a significant issue
             at the turn of the century, particularly from the end of [the] 19th century. Galicia,*

where World War I began, was profoundly anti-Semitic, and Europe exhibited strong anti-Semitic sentiments due to the large population of Germanic-speaking Jews. This history may be linked to the current right-wing inclinations in Austria and Hungary, reflecting deep-seated psychological influences. I wanted to interject this point.

Stein: When addressing the topic of Nazis in Ukraine, we must exercise caution and avoid dismissing it entirely. However, it's crucial to recognize that this isn't the central narrative. Notably, the President of Ukraine is Jewish, which adds an ironic dimension. Putin is exploiting these claims as part of his propaganda to pursue his ultimate goal of reintegrating Ukraine into Greater Russia. His invasion is not about liberating Ukraine from neo-Nazis; it's about fulfilling his vision and his role as a heroic figure executing what he believes to be divine will.

Carpani: *Is it useful or inappropriate to psychopathologize Putin – or any other leader?*

Stein: It's challenging to resist the urge to analyze public figures from a psychopathological perspective. When Trump was elected, there was a notable focus on his potential psychopathology, with some experts pointing to a severe case of narcissistic personality disorder. It's essential to recognize that psychopathology alone doesn't encompass everything about these individuals. In clinical practice, we don't merely label patients with a psychological disorder and leave it at that. There's a broader aspect to the human psyche, including cognitive elements, beliefs, philosophy, and even mystical accounts, as seen in Putin's vision of Russia.

   While these beliefs may seem unconventional or even irrational, many religious belief systems have led to religious wars throughout history, often involving figures who could be seen as more or less psychopathological. Therefore, psychopathology is not the sole explanatory framework. Those in our field may naturally gravitate towards this analysis, but we must also explore the archetypal and cultural layers that contribute to the overall understanding of such figures.

Carpani: *We deliberated on Putin's ambition to unite people of shared origins and secure control over strategically vital regions. He pointed the finger at Lenin for Ukraine's independent trajectory, yet historical precision dictates that it was Yeltsin who, following Gorbachev's removal, inked agreements with Lithuania and Ukraine in 1991, endorsing their independence from both these regions and the Soviet Union. This prompts me to contemplate what the current landscape of Russia and Europe might resemble today if Gorbachev had been permitted to persist in his efforts back in 1991.*

Stein: History is full of "what ifs." What if Hitler had won World War II? What if Trump had been re-elected instead of Biden [being elected]? Trump has suggested that if he were in power, there might be no war in Ukraine right now. However, it's possible that his approach could have led to a different outcome, such as yielding Ukraine to Putin's control, effectively saying, "That's yours, and we won't support NATO or impose sanctions." This might have prevented the current destructive and stalled conflict in Ukraine, but it's uncertain. Putin seems determined to continue and doesn't appear concerned about flattening cities.

Carpani: *. . . like Grozny and Aleppo.*

Stein: Putin's actions in Ukraine are undeniably brutal. Some argue that if Trump had been re-elected, such extreme measures might not have been necessary, as he might have taken a more accommodating stance towards Putin. However, there's also the argument that NATO's eastward expansion, including the inclusion of Eastern European countries that were once part of the Soviet system, played a role in escalating tensions. Russia has historically been sensitive about its borders and quick to feel threatened. The West's perceived aggression, along with Ukraine's desire to join NATO, may have been a tipping point for Putin.

But it's essential to consider whether this argument holds much weight. It's possible that Putin's ambitions to restore Greater Russia would have persisted, with or without NATO's expansion. I've visited Lithuania and the Baltic countries and witnessed the *deep-seated* fear of Russia among their populations. They vividly remember the trauma of Russian occupation. These countries sought NATO membership as a means of *self-defense* and protection, rather than it being solely imposed by NATO. Their desire to join NATO was driven by the urgent need to prevent another Russian takeover, making their perspective entirely understandable.

Carpani: *Hence, I've titled this series* War as Reset. *But what exactly does this reset entail? In 2008, with Obama's election, there was an attempt to reset the U.S.-Russia relationship. You may recall the YouTube video [in which] Secretary Clinton handed a reset button to Lavrov, symbolizing a fresh start. Regrettably, it didn't achieve its intended outcome. I propose that many wars, including this one, serve as resets. Consider the World Wars in the context of what Jung wrote about Wotan. Now, what does a reset mean? Building upon your insights, I believe that Ukrainians are not drawn to the West solely for its capitalist or commercial aspects, such as McDonald's, Hollywood, or BMWs. Rather, their aspiration aligns with what you've eloquently stated: a desire to thwart Russian aggression and authoritarianism. They yearn for self-determination, for a nation where robust institutions can weather tensions, rebellions, and uprisings. Take the United States, for example. The past four or five years were marked by considerable turmoil during the Trump era. However, the essential point is that the U.S. remains institutionally healthy, as evidenced by the peaceful transition of power through democratic elections. Institutions endure even in times of strain. Ukraine sends a powerful message: [Ukrainians] aspire to belong to a world where institutions thrive, exemplified by countries like the United States. I'm eager to hear your thoughts on this perspective.*

Stein: The Clinton/Lavrov reset attempt did not effectively end the Cold War. While many in the West believed the Cold War was over, it has continued from the Russian perspective. This ongoing Cold War mindset in Europe is prompting countries like Germany, Switzerland, and others to increase their military investments for defense. They are becoming more militarized due to their perception of Russia as a threat.

This is not what Putin aimed for, as it has resulted in a stronger and more unified NATO, along with greater cohesion within Europe and a broader alliance, including countries like Australia and Japan, against authoritarian regimes, mainly China and Russia. This signifies a global reset in our collective thinking, marking the end of Western complacency following the Cold War victory. The current crisis in Ukraine has shattered that complacency, highlighting the need for vigilance and resilience.

Ukraine's aspiration is for a *Western-style* democracy, characterized by a rule of law and strong institutions capable of managing conflicts without resorting to violence, even though violence exists in democratic societies. The key lies in strong institutions and legal systems that can contain potential violence and facilitate constructive dialogue, debate, and gradual progress. Ukrainian leaders like Dmytro Zaleskyi and President Zelenskyy seek to safeguard their country from external domination and avoid a one-party system.

Carpani: *Maurizio Ferrara, an Italian journalist, wrote on "Corriere della Sera" on March 7th:*

The war in Iraq gave an important shock to Europe. In the wake of the massive street demonstrations, two great European intellectuals, Jürgen Habermas and Jacques Derrida, asked themselves the question: What binds Europeans together with respect to the international order? Their answer was: a political mentality, different from the American one, and based on some common traits. An aversion to the use of force, first of all, and therefore an insistence on law and respect for international legality; support for a global system based on "liberal" multilateral institutions and on human rights. Fruit of a past inspired by opposite principles and characterized by centuries of bloody carnage, precisely, this mentality had to push Europe to build a foreign and security policy anchored to the EU, overcoming the "stupid" and simplistic opposition between war and peace.

I ponder if Ukrainians are, in fact, conveying a crucial message: "Listen, we are also part of Europe, and we reject Soviet imperialistic influence." However, I question whether we Europeans truly paid attention to them during key moments, like [Viktor] Yanukovych's departure and the Maidan protests in 2014. Therefore, [the] "War as Reset" concept also encompass[es] this aspect: asserting their European identity despite the perception of lingering Soviet influence.

Stein: It's true that, historically, Ukraine was often considered a part of Russia, with numerous ties, connections, and shared elements. Dmytro Zaleskyi, who received his education in Moscow under the Soviet system, highlighted in his interview with you how there was a sense of unity, despite many not genuinely believing in Marxism and communism. I remember going to conferences in Moscow and Kyiv, and there were Russians and Ukrainians and they were singing the same songs.

However, the recent war has illuminated Ukraine's unique identity and long history. Prior to their connections with Russia, Ukrainians had 1,000 *years* of history, interacting with various cultures, including Lithuanians, Poles, and others. In 988, Vladimir the Great's conversion to Christianity was a significant milestone. This growing recognition of Ukraine as a distinct people – with their values, history, language, culture, and identity, separate from Russia – is a new perspective for many of us. It represents a reset in the European mindset.

Carpani: *Absolutely. Certainly. I'm deeply interested in Ukraine's historical context, especially within the former Austro-Hungarian Empire. World War I's roots are tied to this region, with Cossacks, the Russian army, and the Austria-Hungarian army on the verge of conflict. With your Canadian birthright and U.S. citizenship, how do you distinguish this ongoing war from the actions taken by the United States post-World War II?*

You brought up Korea, and we can include Vietnam, Iraq, and Afghanistan. The U.S. has a history of invading, occupying, and politically destabilizing countries since 1946, initially claiming liberation and later promoting liberal democracy, as Fukuyama noted. Ferrara, as I mentioned earlier, examines the U.S. invasion of Iraq, highlighting that some saw it as a premeditated conflict, much like the recent 20 day events in Ukraine, following months of preparation.

Ferrara, referencing Svetlana Alexievich, highlights the "imperial impulses" in both Putin and a significant portion of the Russian population, rooted in the legacy of Homo Sovieticus and the tsarist tradition. However, painting Russia solely as the villains would be overly simplistic.

We should also weigh the U.S.'s imperialistic tendencies. The British, historically, have led to issues and civil conflicts through their invasions and occupations, as seen in Kenya, Cyprus, and Palestine. Furthermore, their occupation of Ireland persists. Are the British crown's actions and the lives affected any less egregious than those of Stalin, President Putin, or the United States, when considering British colonization and exploitation, or the U.S.'s capitalist-driven invasions for geopolitical gains and resource acquisition?

Stein: I'm an American citizen, and I deeply believe in the American experiment that began with the noble American Revolution, where the colonies threw off colonial rule and established themselves as an independent nation. I have studied American history extensively. The U.S. Constitution, with its ingenious balance of powers that influenced the Swiss Constitution, is a work of genius. I hold great respect for the American experiment, but I strongly detest imperialism.

Imperialism can be understood to some extent as a response to the fear of being surrounded and dominated by other ideologies or powers. I recall when George Bush Jr. [George W. Bush] met Putin in 2001 and spoke of looking deeply into Putin's eyes and seeing a good man with a genuine concern for his people. Bush, an Evangelical Christian, saw a fellow believer in Putin, who also had a Christian ideology with a vision of Christianizing the world. It seemed like they shared a similar worldview.

Putin gave Bush the space he needed to invade Iraq and Afghanistan, perhaps because he saw Bush as a *like-minded* believer in Christianity's destiny to rule the world and spread its mission. American imperialism often carries a missionary aspect – the idea of making the world safe for democracy, which is rooted in a Christian belief system.

In comparing America's imperial wars, if you will, and what Putin is currently doing, the differences are not always stark. I don't see a substantial distinction between what the United States did in Iraq and what Putin is doing in Ukraine, although the contexts and leadership figures like Saddam Hussein and Zelenskyy are different. Saddam Hussein was a brutal dictator who mistreated his people, using horrific weapons against them. Removing Saddam Hussein can be framed as a pursuit of righteousness and justice, but it also led to many innocent casualties and generated significant chaos.

Carpani: *Geopolitical imperialism – I differ on Bush and Putin. Bush saw Putin as a supportive older brother, and his statement, "I saw his soul, and he would do everything to protect his country," is crucial. Putin indeed prioritizes Russia's protection, garnering support, rightly so, from an imperialistic standpoint.*

Allow me to conclude with a mea culpa: Two years ago, during our discussion on [COVID-19], I expressed hope that the pandemic could usher in fresh beginnings, balance, spirituality, and a deeper exploration of the soul. We also discussed how [COVID-19] thrust death into our conversations, a grim daily presence on our screens. My mea culpa arises from my initial optimism, hoping for profound change through newfound inner–outer harmony. I knew I risked sounding naïve, recalling Michel Houellebecq's view that "the world will be the same, only a little worse." It seems he was right, and I was naïve. Naïveté is tough to shed. So, here's something even more naïve, hoping it contributes to our search for meaning. I genuinely believe this war, this reset, can foster a shared sense of purpose. Following the German sociologist Ulrich Beck, I advocate moving from declarations of independence to declarations of interdependence – cooperate or perish on a global scale, acknowledging the vulnerability we all share and our collective responsibility for survival. Beck contends that nationalism is harmful, promoting a cosmopolitan perspective rooted in co-creating meaning. You may label me as naïve, but I hold onto hope that this war could signify just that.

Stein: Globalization raised hopes of increased interdependence, but it primarily led to significant economic gains for the very wealthy while impoverishing the poor. To make the concept of interdependence a reality, we need substantial economic adjustments and a reduction in the extreme wealth disparities caused by free-market capitalism and neoliberalism. These disparities foster conflict, envy, and hatred rather than true interdependence.

For instance, Putin is worth $100 million, but he's not the richest person globally. The situation in the United States is concerning, with around 30% of American workers earning less than $15 an hour, which is insufficient for a decent standard of living. Economic inequality at this level can be likened to a form of slavery. Achieving the dream of interdependence is a commendable goal, but it requires substantial changes in the economic system.

European countries seem to address these issues better than the United States and some other nations. Despite disparities in wealth, ordinary workers in Switzerland enjoy a good quality of life, with the ability to take vacations, live in comfortable homes, send their children to school, maintain good health, and access health insurance. Having lived in both the United States and Europe, I find the European experience more appealing.

Carpani: Thank you Murray, for your psychological vision on what is going on, on war and beyond.

Stein: Thank you, Stefano.

Note

1 Hillman, J. (2005) *A Terrible Love of War*. Penguin Publishing, p. 214.

Chapter 3

Day 33 of War

George B. Hogenson

Abstract

Psychoanalyst Georg B. Hogenson, a military veteran who once worked for the Pentagon in 1978, aligns with Carpani's observation that there is a lack of in-depth discussion in both media and politics regarding Russia's profound attachment to land. Hogenson asserts that comprehending the significance of land in Russian culture is essential to grasping the complexity of this conflict. In his analysis, Hogenson delves into the concepts of democracy and authoritarianism, pondering which offers a superior approach to societal organization. Drawing from his firsthand experiences in Russia before and after 1990, he questions whether Western decadence has permeated Russian leadership to an extent that hinders cooperation. Hogenson also emphasizes the role of history in understanding Vladimir Putin's actions today, particularly citing Putin's narrative leverage of the Nazi siege of Leningrad, where he himself hails from, to mobilize support against a perceived Nazi-infested Ukraine. Addressing the transformation of individuals during wartime, Hogenson offers an intriguing perspective. He suggests that the strict regulations, discipline, and numerous rules within the military can often serve to temper and channel violence. He draws parallels to General William T. Sherman's approach during the American Civil War, where the objective was total destruction, but executed in a controlled manner. Hogenson criticizes Putin for aiding in the development of Russia's kleptocratic society, which has eroded the country's military capabilities due to rampant corruption. Furthermore, Hogenson discusses the concept of war as a total reset, noting the present convergence of multiple crises, including climate change and the COVID-19 pandemic. He posits that we may be witnessing the decline of the Christian Era in this context.

Carpani: *George, it's been over a month since the Russians started the war, or "special operation," as they call it. What does war mean to you, given your military and analyst background?*

Hogenson: There are two that I think capture different aspects of what we're talking about with war. One comes from Carl von Clausewitz, a great theoretician of war around the time of the Napoleonic wars. His classic definition is that "war is a continuation of politics by other means" and it's important to keep that one in mind – that it's not something that simply springs spontaneously out of the world. The other definition by General Sherman [Union General William Tecumseh Sherman] in the American Civil War is: "War is at best barbarism. . . . Its glory is our moonshine. . . . War is hell." And that's a very different definition. Both of them capture critical aspects of what it means. It's both a formal undertaking that nations can decide to do in pursuit of interests and political objectives that are not otherwise resolved. And once entered into, it becomes something that most people don't want to really pay attention to. You talk about the television. It does take on some of the same qualities as what people do when they pass a serious car accident on

DOI: 10.4324/9781003390039-5

the highway, to pause and look at it. It's a fascination with situations like that. There are lots and lots of pictures that people will see of destroyed tanks. It's worth keeping in mind that the basic tank crew consists of three people and if those tanks are hit with missiles, whoever's inside that tank is incinerated along with it. Which is the way in which wars take place, and that's where Sherman's remark becomes relevant and that part even in the graphic depictions that you'll see these days, which is much more so than in previous wars, frankly. You rarely get to see the full dimensions of the human destruction that's taking place.

Carpani: *We moved from talking about [COVID-19] to being obsessed with war. I believe that war contradicts what Truman Capote referred to as the "Vanity Fair." War is the Fair of Atrocities, which is brought to us by the media, such as television, that only need to fill their commercial breaks. I am greatly troubled having to view increasingly distressing images on a continuous basis. One week ago, 30 soldiers died; two days ago, 80 soldiers died; yesterday there were ten bombings; today there are 50 bombings. The news feels like an escalation that is truly repulsive.*

Hogenson: Are you asking if I think there would be any change in the dimensions of the war if media were not involved in it?

Carpani: *No, I'm trying to ask for a consideration or the reflection of the power of the media. There is no allowance for depth. That's why our conversation will be long.*

Hogenson: In the current situation, the circumstances tend to be very narrowly constructed in order to try to simplify things for people. In the case of war and international conflict, there's a major school of thought in international relations going back certainly to the 19th century that views all of these conflicts is simply contests between great powers. During the Cold War, we are trying to avoid a direct conflict between the United States and the Soviet Union. That could lead to an incomprehensible catastrophe, but in order to manage the great power conflict we have proxy wars like Vietnam and Afghanistan. But on that way of thinking about it, you leave out chunks of what's actually taking place – what actually the motivations are. There are people commenting on it, but they're not getting into the larger media networks about the conflict within the Orthodox Church, the global structure of the Orthodox Church between the Patriarch of Moscow [and] the Patriarch of Constantinople, the attempts on the part of the Orthodox Church in Ukraine to separate itself from the Patriarchate of Moscow and come under the Patriarchy of Constantinople, and that the Patriarch of Moscow has got interests in Africa. That's completely lost in discussions, from a geopolitical standpoint. And it's the dimensions of its full extent are not very well understood – even, I suspect, by some of the people who do more classical international relations thinking. They may be aware of it, but it doesn't factor into their thinking. People talk about Vladimir Putin thinking that the fall of the Soviet Union was one of the worst things that ever happened in the 20th century. Mind you, in a century that had quite a number of worse things happening. There are other ideological things going on; there's an intellectual underpinning to a lot of this that doesn't find its way into the spectacle, and frankly doesn't seem to find its way into a lot of the strategic thinking [and m]ay indeed have far more relevance to a particularly a Jungian interpretation of some of this. I can say something about that from some of the work that I did do when I was in the military at the Pentagon. In 1978, I was back at graduate school from my time in the Air Force and doing reserve duty down at the Pentagon, working in an area of an office in the Pentagon that most people have never heard of called the Directorate of Net Assessment – a sort of internal think tank. Two questions came up that I was particularly working on. One was an overall issue of how to enhance deterrence, nuclear deterrence; another one was "why were the Russians intent on building such large dirty bombs?" In the United States, we were building very small, very clean bombs. I was running this is in parallel with my work on Jung and writing my dissertation on Jung and Freud at Yale, and took a step back and said, "well what are the governing myths here?" I wrote two papers. On the dirty bomb question, I said that if you look at the history of Russian warfare that the Russians relied so heavily on the land [and] being able to retreat from Moscow. In fact, in World War II, there are stories about the Germans advancing rapidly, so rapidly into Russia as the Russian army pulled back that they ran off the edges of their maps and they didn't know where they were any longer. The Russians now had an advantage by simply making use of the space. My speculation in the paper is if your presumption is that the land is what is most important in war, you would try to develop ways in which to prohibit the use of land. You would sow the soil with salt. On the other hand, in the West, what is the central strategic element of conflict? In the West, it's the city busting that's what we did in World War II, because cities were much more strategically important. Well, how would you do that with a nuclear weapon? You have this dichotomy between the two strategic arrangements in terms of what the sort of fundamental idea in your strategic equation is, whether it's the land or it's the city. It was the first time we had some kind of a plausible explanation from a strategic standpoint as to why they were doing this. The other paper that I wrote was a deterrence paper, where I said the last thing anybody wanted to do was to start a war. You're always looking for ways to convince the other side that not only would they be destroyed but that you would not be destroyed. There was a sort of constant playing with quite sophisticated and intricate ways of signaling one another about ways to do this. Our concern at that time was that the Russians had been significantly improving their civil defense capabilities to the extent that we no longer could pose the kind of threats that we needed to do in order to keep this balance going. I wrote that if you have this retreat into this vast hinterland as a way of conducting war, what would happen if you signaled that you had the capability to cut European Russia off from that vast space in the back by the use of nuclear weapons, to attack that land concept that they rely on and put that concept under threat. When we look at what's going on right now, we talk about the integrity of the land that extends from Russia into Ukraine and even beyond.

Carpani: *What is – according to your experience as a former military – the ultimate meaning of this war? Let me share my psychological perspective on the matter, which concerns land, fatherland, and fathers. I believe this war is not about Ukraine or Ukrainians desiring to join West Europe or the U.S. for their luxuries like BMWs, McDonald's or Hollywood. It's about joining a philosophy of life [of] democracies where[by] the institutions and the nation are sturdy and stable, as evident in almost all European countries. And although attacks were made on U.S. foundations and institutions during the Trump years, and there was a great deal of tension, the democracy has survived and moved forward. I believe this is what Ukrainians have been trying to convey to us since the early 2000s, particularly in 2014 during the Maidan revolution when peaceful demonstrations were suppressed. What is your opinion on this matter?*

Hogenson: I completely agree with you how Ukraine is becoming a reasonably well-functioning democratic state. What we do know about reasonably well-functioning democratic states, Western Europe, the United States – is that they are in a precarious position right now. We need to ask the counter question about authoritarianism, not simply as the impulse of one particular dictator or a strong man, but as a genuine theory of the appropriate way to organize a society. Going back to Russian intellectuals who influence Putin – Ivan Ilyin or Mikhail Yuriev. His novel called *The Third Empire: Russia as It Ought to Be* seems to be almost a textbook for what Putin is doing right now in terms of these moves. Amongst this group, Dugin is a very important figure in this area. It seems to me that the authoritarian structure is in fact a superior or the more appropriate or the correct way to organize a society and democracy is fundamentally corrupt and degenerate. Why is it corrupt and degenerate? Well, if you listen to a lot of what's being talked about in Russia, all of these social movement homosexual rights, transgender rights, one thing and another in the USA – these are the defining features of a democracy that allow these kinds of things to happen. I visited the Soviet Union twice in the late 1980s before the fall and I was back again about six years ago to lecture to the analyst group over there. It was unbelievably different. In the late 80s, I was a faculty advisor to a group of students from Yale and we were going to spend about a week with students from Moscow State University. They were very cordial; we had some wonderful time there. We were staying sort of in central Moscow just off of Red Square, which was all very nicely put together, but I wanted to go visit a couple other places. You got off the main streets in Moscow and the roads weren't paved, the streets weren't paved and I actually went back, made the comment that what in heaven's name are we so worried about [with] this country. When I went back six years ago, you could not find bread in the supermarket. When I went back six years ago, it was like a Disneyland of products, an unbelievable change. It seems that for some elements in the leadership in Russia, moving down that road towards Western material abundance was somehow not properly Russian or not the way that the world was supposed to be organized and you were getting way too much Western decadence.

Carpani: *Let me tell you about my experience. In 2000, I went to Budapest as an Erasmus student. I was one of the first Western European students to go there and I felt like I was part of something developing, with a vast cultural divide. I wonder what the current conflict is over? You said that the roads were unpaved and there was no bread, but now there's extravagance with Prada, Gucci, Armani, Ferrari, and Lamborghini. Is it a matter of compensation? When we delve into the notion of compensation, our thoughts naturally turn to World War II – a response to defeat and humiliation. But what's unfolding today? Is it akin to Hitler's Germany? Could it be more about seizing a moment (Kairos) rather than simply marking time (Kronos)? Perhaps we should also delve into the concept of the father, or fatherland. And about Patriarch Kirill – is this war a manifestation of opposition to LGBTQ rights and a struggle over purity? You have written an incredible book delving into Jung and Freud, both of whom grappled with complex relationships with their fathers. Initially, they regarded their fathers as heroes, but the moment they recognized that their fathers were not heroes, an unconscious urge, in my view, drove them to compensate for this perceived lack by becoming heroes themselves. They ventured into groundbreaking territory, emerging as monumental figures. Freud's involvement in an anti-Semitic incident in Vienna paralleled Jung's father, a pastor who had lost his faith. Their stories converge when they realize their parents aren't heroes, compelling them to seek heroism to fill that void. Could this same narrative apply to Putin? Does it resonate with Russia today? Reflecting on Jung's insight in "After the Catastrophe" regarding leaders as mirrors of the collective unconscious, I wonder: Is Putin compensating? It's not Lenin, whom he blames for Ukraine's autonomy, but rather his political father Yeltsin, who was ineffective due to alcoholism. How significant is the concept of the father and fatherland alongside the notion of land in this context?*

Hogenson: I think one has to be a little careful about trying to get inside his head. I suspect that given what we've seen of Putin's behavior over the last two years, he seems to have a truly paranoid fear of infection. Putin has these bizarre pictures of him sitting 60 feet away from his advisors, it's bizarre. My guess in the short term is that he really isolated himself for the last two years with a very small group of protectors and a very small group of influencers. We can talk about the Russian army and how it operates; they're not likely to try and cross the leader very much. Trump had generals who would tell him that he couldn't do certain things and George W. Bush's defense people told him before we had the Iraq war. A couple of them got fired but they didn't get shot, which can happen in Russia from what I gather.

Carpani: *Or public humiliation like two weeks ago with this chief security advisor.*

Hogenson: About the father figure: I'm not sure about Yeltsin. It's a little bit unclear to me whether Yeltsin would be the father figure that he would be reacting against. There's been speculation about Hitler that his father was Jewish and he was reacting. I would not want to try [to] validate that; it's not my field. What I've seen about Putin's background seems to be fairly salient and it surprises me a little that he is doing what he's doing in Ukraine. When I hear him talking about de-Nazifying Ukraine, I think that to lot of people, it doesn't make any sense. But if you want to motivate particularly the older people who were alive during the siege of Leningrad, it makes sense. One of the funny things in Russia that I could see was that there remains to this day a very pronounced, very strong, almost cult of Stalin. He was one of the most vicious destructive human beings of the 20th century and along with Hitler and Mao Tse-Tung, but he is revered by a substantial part of the population in Russia to this day.

Carpani: *Until now. You mentioned this statue: a huge woman with a sword. Svetlana Alexievich and others underline that war is masculine. It's worth contemplating whether that still holds true. Contemporary armies consist of both male and female soldiers, including transgender individuals, as seen in the U.S. military. In this ongoing conflict, the sudden influx of four million refugees encompasses grandparents, mothers, and children. Remarkably, many women have also chosen to remain and actively participate in the fight. Considering your earlier point about the significance of the statue, it prompts the question: Is war inherently masculine? Would warfare take on a different character if women were in charge?*

Hogenson: When I was in middle school and in high school, there were quite a few teachers who had been in World War II and they rarely talked about the war. I had one history teacher who had been in the Marines, in the Pacific, and one teacher in eighth grade who had been with the OSS [Office of Strategic Services] in Yugoslavia. The OSS was the predecessor to the CIA [Central Intelligence Agency]. He was in Yugoslavia with the resistance forces in Yugoslavia and he did tell us a story that every once in a while, these little cells of resistance fighters would get together out in the woods to coordinate. He says they were almost all men but the most terrifying group was a small group of women resistance fighters and the men were scared of them because they were known for the viciousness with which they would attack the Germans. I think one can ask the question whether war is just a masculine enterprise. In this instance, one almost gets the feeling that the men would rather the women weren't there because [when] they're even given the opportunity or the circumstances, they can be as aggressive and as dangerous – if not more so – than the men are in terms of regulating aggression.

Carpani: *Men are instinctively more aggressive and women – it is a generalization – are instinctively pondering and reflecting. I strive to engage with both men and women in my discussions. In the case of* Breakfast at Küsnacht, *I've noticed an interesting distinction in responses. Men often express eagerness to participate, similar to your willingness. However, women tend to approach it differently, often responding with a reflective tone, such as "Perhaps later" or "Not right now." This makes me wonder whether this hesitation might manifest differently in the context of warfare. Consider instances like the Kurdish freedom fighters, where women ardently fight for their nation, or historical examples like the Vikings and the Valkyries. My last question about war before we switch to something else: Why does war have the capacity to transform ordinary individuals into something akin to beasts? Lastly, regarding non-Ukrainians who volunteer and travel to Ukraine to join the fight, do you view them as heroes, individuals with a penchant for self-destruction, or perhaps modern-day Davids facing Goliath?*

Hogenson: What's important to keep in mind is that military operations can get out of control very quickly and become nothing but destruction. One of the reasons for the way in which militaries organize themselves and the discipline that is imposed on people and the regulation that's imposed on people wearing uniforms is to restrain violence and direct it. That's lost on people who've never been around war or around the military. I mentioned General Sherman who says "war is barbarity; it cannot be refined." In Sherman's march to the sea, he cut through Georgia and destroyed the city of Atlanta, but first told the people in the city: You have to evacuate the city. I will take care of you once you've evacuated the city, but I'm going to destroy Atlanta because it's a railroad center, it's a manufacturing center supporting the Confederacy. But he executed any soldiers that pillaged. He had a policy of destruction several miles wide and anyone who randomly started to pillage or rape or anything else was held to account. The policy was in one way "total destruction," but that it had to be managed in a particular way. It couldn't be random destruction; it couldn't be individuals just acting outside of that space. It's a paradoxical situation that genuine military operations are undergoing. One of the problems with the Russian military has always been is that Putin thought that things would operate the way they had in the past where[by] sheer mass overwhelms the enemy. Putin relied upon his kleptocratic society, that is so corrupt, that tires are rotting on their vehicles. There are many articles now showing the way the corruption worked. If an order went in for 100 bombs, the Air Force would get five and the whoever the oligarch was that had gotten the money for [them would get] the other 95. It seems to me that Putin has created a system within the oligarchy that has probably done more than anything else to undermine his ability to execute the war successfully.

Carpani: *And what do you think about non-Ukrainians who have enlisted?*

Hogenson: There's a movie called *The Hurt Locker*. It's about an explosive ordnance disposal team in Iraq. My first two years in the Air Force, I worked very closely with the explosive ordnance disposal people because I was responsible for trying to deal with serious weapons accidents. They are constantly on an adrenaline rush. One of the problems I think we have in the United States with a lot of our veterans right now is that we have conducted a war that went on for 20 years and a lot of these men – since it's a voluntary force – were back in going through cycle after cycle after cycle of this very high-tension environment. If you're in the cause, particularly if you're in combat, you're on an adrenaline high all the time, and what you'll see on *The Hurt Locker* is the main character [who] comes back from a period in Iraq from disarming bombs. He's with his wife and his kid in the grocery store and he simply can't take it; he goes back. He's going back into that environment. Probably some people have a genuine commitment to the idea of the Ukrainians, but I would suspect that some of them are looking for a return to that adrenaline high.

Carpani: *Junkies for adrenaline. It's an interesting view; thank you. Let's revisit Putin: At the outset of this conflict, the media attempted to psychopathologize him, often portraying him as obsessive or paranoid. However, is this approach truly useful when, in essence, Putin appears to be a calculated and cold-blooded strategist? His goals include uniting people of shared origins and potentially even resurrecting elements of the Soviet Union. He accuses Ukraine of genocide while simultaneously engaging in aggressive actions. So, the question arises: Is it appropriate to psychopathologize figures like Putin or Trump, and if so, what purpose does it serve?*

Hogenson: I do think that we can say quite a bit about Trump. There's a group of Jungians [who] wrote a collection of essays together on Trump as a malignant narcissistic personality disorder. I think there's every reason to believe that that's an accurate diagnosis for him. And the fact of the matter is, he is a coward and a bully and if you pardon me for saying so, he's a "draft dodger" because I'm the same generation. And Putin doesn't strike me as a coward or a bully in that sense of the word, as somebody who tries to compensate for some inadequacy by pushing people around. And we must use your expression – far more "cold-blooded" about it. There is some speculation right now that there's something wrong with him now, physically. I do happen to think that he may have been counting more on the possibility of Trump's re-election, in which case he would have had almost an open door to Europe, because it's very likely that Trump would have pulled the United States out of NATO. You can say what you want about NATO, but it worked for a long time and President Biden should stick it back together again, at least for the time being. There may be a calculus there. One of the things we're dealing with is that a couple of the personalities on Fox News are basically just repeating Putin's line endlessly and in fact are being broadcast in Russia to demonstrate that there's this American faction within the United States that is sympathetic to Putin and to the invasion. I have this suspicion that Putin has been watching too much Fox

News. The line of Fox News on Joe Biden is that he's senile, he barely makes it to the desk in the morning, doesn't know what he's doing. Fox News has been pumping that all the way through the election and through his presidency. But Biden's getting old, but he sure isn't losing his touch, it seems to me. But if you were listening to certain elements of the American press, you're going to come away with the idea that this will be a pushover. I would say, in terms of pathologizing Putin, I think it's a little bit tricky. He wasn't a real high-level KGB operative; most of his career he was in a sort of backwater in East Germany for quite a while. His background is a little bit elevated in some of the commentaries as if he was this sort of master spy – and he really wasn't.

Carpani: *When Obama became President in 2008, he suggested a symbolic reset to Putin. There was a moment when the former Secretary of State, Clinton, met with Exterior Minister Lavrov and presented him with a red plastic button that said "reset."*

Hogenson: It didn't work.

Carpani: *Is war a reset? Is it the great reset? What do you think? You have experience, you are one of the few Jungians with military background. I remember my mother or grandma saying: "Somehow, war is needed to reset."*

Hogenson: I don't want to exactly sound like there's an automatic cycle in foreign policy, but if you do look over a period, you'll see a major conflict, a war of some sort. And then you will have roughly a 75-year period [when] you'll have the Treaty of Westphalia, or you'll have the Treaty of Versailles. I'm of that school that sees [World War I] and [World War II] is a massive, prolonged conflict over a long period of time. Barbara Tuchman, in her book on *The Guns of August*, says that William of Germany and the Czar and a whole bunch of other minor ones – the crowns of all of Europe – get together and it's the last time they'll ever be together. And this entire regime that's existed from the Middle Ages collapses. Then you have this prolonged conflict and then you get the end of World War II and you get the establishment of a new regime, the new international regime United Nations and so on. We're now 75 years on roughly and no one from that period is alive anymore. When you talk about a reset, it is a way of thinking about it that a reset then will be generated in order to establish some new regime, going forward for a period. The problem right now is that I think we have an unusual confluence of crises coming at the world. I talked earlier about how we are ignoring the religious aspect of this war. We have a serious parallel process in the United States with right-wing evangelical Christians. One of the things that I think in Jung is very interesting that we really need to pay attention to is when he talks about symbols. When they first emerge historically, they're connecting the conscious with the unconscious; they carry this great numinous quality with to them – but over time, they become exhausted. The work on the symbol that takes place over time eventually reduces them to a sign to use his distinction between Freud and ideas and his own ideas. You can have different time scales here. When I talk about the 75-year time scale in international relations, you can sort of see that working itself out historically. You have a reset in that degree. Jung talked about and people get all very excited, all the way back to the 60s about the Age of Aquarius and this change. Astrology folks in the Jungian community are out there talking about the new gods. The Christian Era, the symbolism, is exhausting itself – and in consequence, you're getting these increasingly radical attempts to hold on to the symbolism. To my mind, it's a little bit like what probably happened in late Hellenistic Antiquity when the ancient pagan system was collapsing, because people just couldn't take it seriously anymore. You get a new set of symbolism coming in with Christianity. It does seem to me [that] where reset is concerned, it's a little bit of question of what time scale we are talking about resetting. We may be at an odd place where the reset is taking place at several different time scales and the Ukrainian War is part of that because it does have this apocalyptic view almost, coming out of some of these intellectuals in Russia that we need to expand in this way. The [COVID-19] thing – we may get this particular one under control, but I'm fairly certain that we're beginning to see what will probably be a cyclical process.

Carpani: *From pandemic to epidemic.*

Hogenson: One pandemic to another. Not least because some of that is tied up with climate change, which we're absolutely simply not getting any control over at all, as far as I can see.

Carpani: *Indeed, it's somewhat sidelined from the agendas and media attention. Your observation about the shift from a symbol losing its meaning and becoming a mere sign, particularly in the context of Kirill's assertion that this is a religious war, is quite intriguing. We've witnessed numerous religious conflicts over the past two decades. I wonder whether this transition from symbol to sign and back, as well as the broader shift from religious dogma to the development of spirituality, as we Jungians understand it, holds a significant role. This spirituality, while distinct from religion, could be a transformative force.*

Hogenson: I would point out that there are many religious wars, and they increasingly have this fundamentalist quality to them. Look at India and the fundamentalist Hinduism. There is a very powerful right wing in Japan that's very tied to the old emperor ideology. It's approaching a global phenomenon, and that raises some very curious questions about what we're dealing with globally, spirituality. My daughter Katherine now is a freshman in college and doing a double major in religious studies and environmental studies. Her project is to try to see if there's some way to connect the environmental issue up to indigenous spiritual traditions, and whether that has some traction here. I think things like that are a possibility. My original field was Buddhist Studies a very long time ago. I studied in Japan with Masao Abe as an undergraduate. I think that one must be a little bit thoughtful about undifferentiated spirituality without any distinctions. Spirituality too easily becomes something that is a little bit too easily just an emotive state that doesn't give you any traction. You have to have some way of distinguishing things. I gave a keynote to the folks that work with Wolfgang Giegerich. I confess I wasn't entirely happy with it. I'm very interested in him but it's hard to wrap your head around him in some ways, but I made the remark there that I had difficulty with James Hillman's idea of this sort of daemon,[1] this sort of guiding principle. It's because I got in an argument with him once about how this even goes back to Putin. How do we know that Putin isn't just following his daemon? How do we make ethical judgments without some sense of more than just the sort of romanticized spirituality? I do think that's an area where Jungians need to be a little careful, a little thoughtful when invoking things like the Self or soul, or the personal daemon. Theologically, it gets you into the question of theodicy: how do you justify horrible things happening if there's this superordinate guiding force in the world?

Carpani: *I want to examine the United States' foreign policy over the last 50–60 years. But before we do, I want to go to Hillman and Giegerich: Is war a love that no other love has been able to overcome, as underlined by Hillman?*

Hogenson: In terms of U.S. policy for the last 50 years – let's be more specific and say since World War II – that is a subject for another discussion. I don't really think I can construct a meaningful answer in a paragraph or two. I have not been in combat myself, although I know many men who have been, and I consider myself at least as well informed on the nature and horrors of combat as Hillman. I was moved to the plans and policy group at Headquarter 5th Air Force in Japan with responsibility for planning the defense of Korea and Taiwan. War planning is a very complex process that moves through many layers in the U.S. defense system and involves every element within the organization. My role was general oversight of the planning process at 5th Air Force, which controlled all U.S. Air Force assets in Northeast Asia, and primarily the Air Force position on command relations between all the services in Korea – an extremely complex network of responsibilities in the event of war. What was more notable, and this is the too complex part, was that this position was designated for a full Colonel and usually held by a senior lieutenant colonel. When I took the position, I was 25 years old and a 1st lieutenant. This position gave me an unusual perspective on war and paved the way for the work I did at the Pentagon once I was back in graduate school. Since Hillman quotes General [Dwight D.] Eisenhower at one point, "the force of war has no limits," I will give you another quotation from Eisenhower that reflects the role of planning in war: "In preparing for battle I have always found that plans are useless, but planning is indispensable."[2] What bothered me immediately when I read Hillman's book was that he started by quoting from the film *Patton*, where George C. Scott as General Patton almost mystically reflects on his love of war. First off, Patton was an unusual – some would say almost psychopathic – person. Along with MacArthur, he went to great lengths to create a myth around himself. But if you are going to start an analysis of war with a quote from Patton – and lines like Hillman quotes were in fact the kind of things Patton would say – you should perhaps spend some time with other commanders, as well. I did occasionally encounter Army generals who had served with Patton in the war, and some of them emulated him and bordered on crazy, from my point of view, as did some Air Force generals who were influenced by Curtis LeMay. But if you add in people like Eisenhower or [Omar] Bradley, you get a very different view of the whole situation. Here is something I think Hillman misses in his analysis – and many of the other commentators he cites. At one level, General Sherman's "war is barbarity, it cannot be refined. War is hell," is appropriate. But you can't lose sight of Clausewitz's war as "the continuation of politics by other means." Taken together, war is not simply violence, which would be the case if only Sherman's comment was relevant. And it is frequently all that the soldier on the ground experiences. War, as a complex combination of violence taken to the most extreme level and the pursuit of policy, requires a very different lens to understand. Patton may love war because he has these delusional ideas about being the reincarnation of great generals all the way back to Hannibal, but General [George C.] Marshall and President [Franklin D.] Roosevelt did not make Patton the Supreme Commander in the War. They chose Eisenhower, who was Marshall's chief of planning at the War Department.

Eisenhower had never been in combat – he spent World War I training men to go to Europe – but he understood the true complexity of managing a war at the highest level, which extended to keeping the Pattons and Montgomerys in line – no easy task – making the most critical decisions – D-Day – and managing the politicians – [Winston] Churchill and [Charles] De Gaulle. Eisenhower had a profound understanding of what his decisions meant for the common soldier in terms of the horrors of combat. It is not clear that Patton could appreciate the fact that most soldiers would rather be back home. I think that to understand war, you need to look at the quotation from Eisenhower that Hillman cites from a different angle – one that takes his god of war idea more seriously. If war is a phenomenon that "has no limits," how are mere mortals able to stand in its presence? The soldier on the ground experiences it one way; there are some commanders who lose the ability to control their force – see Hillman's example from the Civil War – some commanders go mad – see Patton – and some like [Ulysses S.] Grant and Eisenhower find a way to stand in the midst of the chaos and bring some modicum of order – to stand against the god, or perhaps with a deeper understanding of the ways of the god. To be a bit poetic about it, these are the great commanders. I do not think Hillman's analysis is as insightful as his exceptional rhetorical skills may lead one to believe. I will leave this answer with another quote from Eisenhower regarding the notion that there is no love that can overcome the love of war – somewhat reminiscent of Jung's hanging by a thread" remark:

Every gun that is made, every warship launched, every rocket fired signifies, in the final sense, a theft from those who hunger and are not fed, those who are cold and are not clothed. This world in arms is not spending money alone. It is spending the sweat of its laborers, the genius of its scientists, the hopes of its children. . . . This is not a way of life at all, in any true sense. Under the cloud of threatening war, it is humanity hanging from a cross of iron.[3]

(From the Chance for Peace address delivered before the American Society of Newspaper Editors, *April* 16, *1953*)

Carpani: *You studied Giegerich's work in depth. What is your take on his essays on nuclear weapons?*

Hogenson: My view is essentially the opposite of my opinion on Hillman's book. I first read Giegerich when I received the copy of *Spring Journal* containing his paper "The Invention of Explosive Power and the Blueprint of the Bomb." At the time, I was the associate director of the MacArthur Foundation's program on peace and international security, and I felt that Giegerich was the first person I had encountered outside a small group of intellectuals in the national security world who was really thinking deeply about the essentially metaphysical – Giegerich would want to say psychological, but only in the special way he uses that word – significance of nuclear weapons.

Carpani: *Maurizio Ferrara (in* Corriere della Sera, *7.3.22) reminded us – looking at the U.S. invasion of Iraq – that some commentators observed that it was a premeditated carnage; just like that of Ukraine, which took place after months of preparation. He reminds us that Nobel Prize winner Svetlana Alexievich has described the imperialistic tendencies of not only Putin, but also a significant portion of the Russian population: She calls it the long wake of the Homo Sovieticus, and before that of the tsarist tradition. Using this as a reference, what can we conclude about the imperialist tendencies of the British crown and the United States?*

Apart from the United States, every time the British took over a country, they caused trouble (including civil wars) when they departed – as in Kenya, Cyprus, Palestine. They continue to occupy a part of Ireland. Are their deeds any different to those of Stalin and/or Putin?

Hogenson: I need to respond to this question in parts. Beginning with Maurizio Ferrara's comment that the invasion of Iraq was "premeditated carnage" that required months of preparation my first reaction was "of course it was premeditated carnage; it was a war." At that level of abstraction, all military operations are premeditated carnage and the U.S. and the Russians and the Germans in both World Wars and the Japanese and everyone else have the same thing for centuries. To refine this, we could ask whether there are any distinctions that one can make and whether, in the end, they matter. There have been efforts – at least through the modern period – to come up with some codification of rules of war. One could start with a comparison of how different nations and different forces do or do not adhere to what international rules ostensibly exist. In Ukraine, the Russian army has forcibly deported thousands of people to Russia, purposely targeted civilian targets with no military significance, and allowed the widespread practice of rape and pillage. All of these are actions prohibited by international agreements on the conduct of war. The U.S. Army does make a concerted effort to avoid these practices and prosecute whe[n] possible those who break at least some of the rules. On the other hand, do refugee displacements, which are common, equate with forced deportations? Interesting question. Regarding imperialism, there is a very real – and curiously, 18th or 19th century – form of imperial impulse in Russia at this point. We discussed this along with the business about Putin and Kirill. One of the curiosities of imperial impulses, going back at least to the Romans, is the degree to which the imperial power doesn't entirely allow itself to think it is just being an imperial power. There seems to almost always be an element of bringing some form of cultural enlightenment to the benighted people who need to be incorporated into the empire for their own good. Of course, that is not what is really going on, but it makes for an interesting psychological adjunct to the imperial impulse. American imperialism, in terms of the international side of things only begins in the late 19th and early 20th centuries, particularly with the Spanish–American War of 1898. You could arguably go back to the Mexican–American War of 1846, but that was more a part of the Western expansion of the United States, which could also be viewed as imperialism directed against the indigenous population. With the Spanish–American War, the U.S. gained its first real colonial position outside the confines of North America. To some extent, this was viewed as simply joining the club of imperial expansion that most of the European countries has been engaged in for year – centuries, in some cases. In terms of the effects of imperialism, one can certainly argue that on balance, it is never really positive for the populations that are absorbed into the empire. If you have a relatively well-established national identity before the imperial takeover, for example [in] India, it is possible to recover fairly quickly and successfully after the imperial power leaves. In the Middle East, the aftereffects of French and British imperialism continue to this day. Most people, at least in the U.S., have no idea what the Sykes-Picot agreement of 1916 was or what its impact has been on the region.

Carpani: *According to Ferrara, the war in Iraq gave an important shock to Europe. He underlines that in the wake of the massive street demonstrations, two great*

European intellectuals, Jürgen Habermas and Jacques Derrida, asked themselves the question of what binds Europeans together with respect to the international order. Their answer was: a political mentality, different from the American one, and based on some common traits. An aversion to the use of force, first of all, and therefore an insistence on law and respect for international legality; support for a global system based on "liberal" multilateral institutions and on human rights.

Fruit of a past inspired by opposite principles and characterized by centuries of bloody carnage, precisely, this mentality had to push Europe to build a foreign and security policy anchored to the EU, overcoming the "stupid" and simplistic opposition between war and peace.

As an American citizen, how do you feel about this?

Hogenson: There were, of course, two wars in Iraq lead by the U.S. The first, under George H.W. Bush, was a model of military and political planning. The objectives were well defined and limited, the planning was comprehensive and meticulous, and a genuine international alliance was formed to execute the war. The second Iraq war, under George W. Bush, was a model of how not to plan or execute a war. The objectives were expansive and illdefined, planning was totally inadequate, and any senior officer who dissented from the plan, such as the Chief of Staff of the Army, was fired. The execution was haphazard and massively destructive to no particular end, and the political alliance – such as it was – was an illusion.

Carpani: *Could it be that Europe has averted its gaze until now precisely because Ukraine aligns more closely with the values expounded by Habermas and Derrida? I suggest that Russia's war against Ukraine stems from the fear that a Ukraine firmly rooted in liberal Western European values would profoundly destabilize Russia.*

Hogenson: In terms of how the Europeans reacted, at least to the second war, it seems to me that has been lost to more recent events. I have to say that I think too many people thought that Europe had passed beyond the old conflicts. Hillman makes lot out of failures of imagination. The attack on the World Trade Center was not imaginable – even though an attempt had been made just a few years prior. While I think Hillman wants to make more out of this to connect with his psychological views, I would have to say that the Europeans, including luminaries like Habermas and Derrida, seem to have suffered a failure of imagination. As an American, I suppose I would have to say "Welcome to the club." One thing I might add is that I think it is likely there will always be some country out there that aspires to take over as the hegemonic power in the world. This impulse goes back to ancient Egypt, Cyrus of Persia, the Romans, and up to the present. I do think that as hegemonic powers go, the U.S. hasn't [been] the worst, even in recent history. That might be debated – but in the end, one might want to consider who the next hegemon might be before getting too enthusiastic about displacing the current occupant of that position.

Notes

1 Hillman, J. (2005) *A Terrible Love of War*. Penguin Publishing, p. 214.
2 https://www.oreilly.com/library/view/the-little-book/9781292148458/html/chapter-079.html
3 https://www.eisenhowerlibrary.gov/eisenhowers/quotes

Chapter 4

Day 37 of War

Caterina Vezzoli

Abstract

Caterina Vezzoli views the war as a failure on multiple fronts: Putin's inability to handle opponents differently and the West's failure to understand his intentions. In her discussion with Stefano Carpani, they explore the gender dynamics of war, drawing inspiration from Greek mythology and Agamemnon's story. Vezzoli labels the masculine approach to war as "Wotanistic," echoing C.G. Jung's ideas. She examines the role of women in Troy, highlighting their suffering and dignity, drawing parallels to women's historical struggles in the West and making them more open to dialogue. For Vezzoli, war symbolizes the vital importance of democracy, emphasizing the danger posed by forces promoting illiberal values, with fascism as a lasting threat. She stresses the necessity to safeguard democracy and its principles at all costs, citing novelists like Tolstoy and Dostoyevsky to emphasize the soul-centric aspect of Russian culture. She admires their ability to delve into the complexities of the human soul, especially in the context of conflict, whereby relationships with others and oneself are crucial, believing this cultural facet still exists in Russian society today.

Carpani: *Caterina, what is, in your perspective, war?*

Vezzoli: To me, war represents a failure for Putin, as he resorts to it because he lacks the ability to handle his opponents in any other way. He wants to demonstrate strength and revive the image of the Soviet Union as a superpower capable of challenging the West. However, war is not just a failure for Putin; it's also a failure for the West. The West failed to understand Putin's intentions and didn't take steps to prevent the situation, despite being aware that the money paid for resources from Russia goes to the Russian mafia and Putin's corrupt oligarch associates.

   Psychologically, war represents a failure because instead of managing the tension between conflicting situations and seeking alternative solutions, it leads to a polarized outcome. This polarization revolves around possession, and becoming possessed ultimately results in widespread destruction. There's another personal reason I view war as a failure. My father was a soldier during [World War II] and spent two *years* as a prisoner in Germany because he was in Greece in 1943. He survived the atrocities committed by the Nazis in Greece and endured captivity in a German camp. When my father returned, he was a committed pacifist, believing that war is always wrong. This perspective profoundly influenced me.

Carpani: *Let me share with you what James Hillman wrote about war in* A Terrible Love for War*:*

   There is no practical solution to war because war is not a problem solvable by the practical mind, which is better kept for its conduct than for its avoidance or

DOI: 10.4324/9781003390039-6

conclusion. War belongs to our soul as an archetypal truth of the cosmos. It is a human work and inhuman horror and a love that no other love has been able to overcome. We can open our eyes to this terrible truth and becoming aware of it, devote all passionate intensity to undermining the enactment of war, strengthened by the courage that culture processes even in the dark ages to continue to sing as it resists war. We can understand it better, postpone it longer work to gradually remove it from the support of a hypocritical religion. But the war as such will remain until the gods themselves leave . . . .[1]

Vezzoli: I appreciate this perspective. Changing the god is a fundamental aspect. The gods of war are always active and present, but it's essential to deconstruct them. Agamemnon comes to mind when discussing the gods. He was willing to sacrifice his daughter to the goddess of the wind. Despite being a leader of warriors, he wasn't a true warrior, and he easily succumbed to his vices. He was the son of Atreus, who had killed his nephew. Agamemnon was generally despised, but the loyalty of the kings kept him in power. We need to deconstruct the notion of power that Agamemnon represents. Greek tragedy exposed this, but at the time, there wasn't a culture to deconstruct and differentiate one's feelings about the world. It's a task we must undertake. Greek tragedy provides valuable insights into this deconstruction process.

Carpani: *War, a lamentable failure, prompts questions about our instincts, akin to Freud's "Trieb" in German. It seems we've been drawn to war since its inception. Think back to the Gulf War in 1991 when we fixated on the TV, consuming and regurgitating information as if it were entertainment. Is war an inherent facet of our nature, echoing Freud's instinct theory? War sharply contrasts Truman Capote's "Vanity Fair" and resembles more of an "Atrocity Fair." Paradoxically, this atrocity spectacle has transformed into a "Vanity Fair" for profit-driven television and media outlets, prioritizing gains over genuine information dissemination.*

Vezzoli: From a Jungian perspective, the instinct and the spiritual aspects are interconnected. It's crucial to distinguish between the wish and the instinct, particularly the aggressive instinct related to violence. Instinct, in itself, can be a positive force, while violence represents its opposite. What's essential is maintaining the tension between the instinct to attack and be aggressive, and the instinct that isn't about aggression but acceptance. Both aspects are constantly present within us, and it's a matter of differentiation between the wish to harm and the wish to save or play the role of the savior.

Carpani: *Is war symbolically or mythologically masculine? Modern armies now feature both male and female soldiers, yet during the 2014 Maidan protests, men naturally shielded women from the police.*

Vezzoli: Indeed, when we consider mythology, we encounter figures like the Amazonians and the Valkyrie. However, when we revisit the story of Agamemnon, it appears that war is often associated with masculinity.

Carpani: *What about Helen of Troy? Some view her as the war's catalyst, while others attribute the conflict to a contagious influence that swept through men, prompting them to wage war.*

Vezzoli: The women of Troy are fascinating. Consider Hecuba, who displayed remarkable strength, even when she became a captive after the war's conclusion. Women in the context of the Trojan War demonstrated strength by maintaining their dignity, even in captivity.

Carpani: *This brings to mind the Kurdish women who fought against the Islamic State. Female warriors seem oriented toward progress, while the male approach leans more towards destruction. My apologies if this distinction appears minor.*

Vezzoli: Reflecting on the women of Troy, they possess a strong sense of self. Despite their impending fate as slaves, they maintain their dignity. Through a modern lens, women – who have historically experienced gender-based discrimination and the traumatic use of power on their bodies – are aware of these challenges. Even if they hold positions as princesses or queens, women are familiar with these experiences. Women might make excellent warriors, yet they are also more open to dialogue.

Carpani: *As if the purpose of warfare greatly differs. A thought occurred to me, one I haven't shared in previous interviews, perhaps due to our dual perspectives. I thought of Penelope. She isn't technically engaged in war, but she awaits her husband's return, sewing and undoing her work at night. It's an act of patience, understanding that Chronos and time are distinct. It's the art of waiting. I have a patient who escaped Ukraine with her daughters, while her mother opted to stay in Lviv, considering it her homeland and wanting to assist. She's engaged in crafting camouflage blankets for the military. This woman embodies a connection between Penelope and herself. Though involved in the war, her contribution to the country differs from the traditional masculine role. Her embodiment of Penelope's waiting, perseverance, and endurance is a vital aspect of her service.*

Vezzoli: My comments about women are not meant to diminish them; in fact, I consider the quality of being true to oneself and fighting for one's destiny, even in situations [when] women are considered inferior, to be highly significant. It takes immense strength to navigate such circumstances, and it represents a unique way of thinking.

Carpani: *Perhaps this signifies the distinction between male and female perspectives on war. Women seem to have a more direct connection with or a deeper understanding of the concept of providence, allowing them to navigate it effectively.*

Vezzoli: Women may be more differentiated in this regard, likely due to their lived experience of inferiority. They are constantly aware of their status and the challenges they face, which makes them more differentiated in their perspectives.

Carpani: *What's the ultimate meaning of this war? Allow me to look into the duality of Chronos and Kairos. Since the early 21st century, Ukrainians have been conveying a message to Europeans and Americans, one we've been hesitant to embrace. They aspire to self-determination and robust, less corrupt institutions. The storming of the [U.S.] Capitol in January 2021 showcased the resilience of U.S. institutions in enabling a peaceful transfer of power through an election, a challenge in Ukraine due to Russian influences and internal conflicts like Yanukovych and Maidan. I also ponder whether there's an element of compensation at play for the events of the late '80s, akin to the aftermath of [World War I] when the former Austrian-Hungarian Empire crumbled.*

Vezzoli: You're correct regarding Ukraine. I believe they are reminding us Europeans of the vital importance of democracy because it's something we often take for granted. Their ability to emphasize the significance of democracy and the fight for the ability to shape our lives has deeply resonated with me. In terms of the ultimate meaning of this war,

I think it's precisely about humiliating democracy and the associated values. These forces aim to showcase the superiority of illiberal values. Unfortunately, fascism never truly disappears, and we have witnessed its resurgence even in our own countries.

Carpani: *I listened to a radio program on RAI1 last night discussing a show called "M – the Son of the Century," focusing on Mussolini. During the discussion, a theater actor made an interesting comparison, likening fascism to a virus people contract similar to [COVID-19]. This idea aligns with Jung's perspective in "After the Catastrophe," [when] he suggests that Hitler reflected his nation. It's akin to saying that, as I criticized earlier, the media also serves as a mirror for the country. However, it's crucial to clarify that fascism isn't the virus itself; it's a separate entity.*

Vezzoli: Just as Hillman observed about war, fascism always remains latent, and the tension is ever-present. Italy has never entirely eradicated it, and populism has emerged as a modern iteration of fascism. Figures like Salvini embody this sentiment, but there are numerous others across various countries. The ongoing war highlights that this issue isn't confined to Russia or China; it also lurks within our own democratic nations and we must remain vigilant.

Carpani: *Jung's insights on fascism and Hitler apply to contemporary populist leaders who share traits of being emotive, deceitful, and adept at manipulating their nations, seen in figures like Hitler, Mussolini, and modern populists. This brings us to the themes of fathers, the father archetype, and the concept of the fatherland. In the context of the father–son relationship, sons often idealize their fathers, like Freud or Jung, as heroes. Eventually, they may realize their fathers are not heroes, potentially experiencing a loss of honor they themselves never possessed. To compensate, some strive to become heroes, too. Notably, both Freud and Jung achieved hero status in their own right. I'm intrigued by the spread of illness, current events involving Putin, historical leaders like Hitler, and how societies have often grappled with similar occurrences in the past.*

Vezzoli: When it comes to dealing with the father figure, deconstructing rather than destroying the parental imago is a crucial part of adolescence. Failing to do so can lead to the destructive possession of one's parents. An Italian colleague has authored an interesting book on Jung and his father, titled *Pro Bono Patris*. I believe it's important to recognize that, while I don't know much about Freud's father, Paul Achilles Jung had his values and capacities. Perhaps he wasn't an ideal father, but he had a sense of respect. Pier Claudio Devescovi discusses this in his book. Jung was right in emphasizing the need to deconstruct the father, not destroy them, during our individuation process. I must admit that I don't have an in-depth understanding of Putin or the Russian people. I know a few Russian colleagues, but my knowledge is limited, and I've never been to Russia. My knowledge is primarily through the works of authors like [Fyodor] Dostoevsky, [Leo] Tolstoy, and [Anton] Chekhov, especially during my time as a communist. The soulful exploration in their novels was profound, and I was particularly drawn to [Dostoevsky's] *The Brothers Karamazov* during my teenage years. It provided an incredible insight into the conflicts within relationships and the individuals themselves. I recall an experience at a European conference in St. Petersburg when I was in a group where everyone else spoke Russian. While I couldn't understand their language, I hesitated to leave the group because I didn't want to offend them. In that moment, I had a unique experience [when] I felt a deep unconscious participation mystique, sensing that my beloved novelists had spoken that very language. It allowed me to connect with something beyond the surface, a hidden aspect of their culture.

I'm not suggesting that we should establish *cross-cultural* literature groups, but understanding these cultural undercurrents can be a profound way to connect with the essence of a culture that isn't immediately apparent.

Carpani: *Wars can drive ordinary people to animal-like behavior. How do you view non-Ukrainians joining the fight in Ukraine? Do you find them heroic, risking their lives needlessly, or perhaps acting foolish[ly], even suicidal or as if they are making an unwise choice?*

Vezzoli: I believe they are fighting against oppression. I do think Ukrainians are justified in defending themselves. On February 22nd, when Putin initiated the war, I was working with a colleague and friend on translating a book by John Beebe. The last section we were translating dealt with the inferior function. Marie-Louise von Franz noted that all projections originate from the inferior function, the unknown aspect of ourselves. This is the crux of the matter. It's commendable that people want to help, but since the projection of the inferior function, what's unknown to us, operates universally, individuals should focus on being aware of this projection and practice differentiation. Becoming overly identified as the savior can lead to problems.

Carpani: *There are four functions. For one person, the superior function could be thinking. Someone experiences the world through thinking while the inferior function is feeling. When we get hit by the feeling something happened, that overwhelms us. An example is the events around Will Smith and Chris Rock. Smith's inferior function was hit by a joke and he responded in a violent way.*

Vezzoli: The inferior function is often unknown to the individual because, by its nature, it's considered inferior. This can be a challenge for people who go to fight in Ukraine. I'm not suggesting they are projecting, but it's a potential concern. Furthermore, even if they are here, there's still the issue of inflation. What's perilous is that those individuals – even if they embark on this journey for the right reasons – might confront an apocalyptic scenario and lose sight of their own identity. Without awareness, they risk becoming possessed. In a state of possession, they might act improperly. It's crucial that those on the frontlines receive the support they need. While we often talk about supervision, there's more to it than that – they require assistance that helps them maintain their humanity. Even if you identify as a savior, it can be problematic because you might believe you're the sole bearer of knowledge.

Carpani: *Many individuals attempt to diagnose Putin . . .*

Vezzoli: . . . I think the problem there is not psychopathology; it's authoritarianism.

Carpani: *Many view Putin as a dangerous narcissist, akin to Trump. In truth, he's a ruthless killer, reflecting an authoritarian mindset. Such personalities have persisted throughout history, from the Roman Empire onward. He's a calculated murderer with a strategic agenda. You mentioned forecasts. Putin accuses Ukraine of genocide, but isn't this a case of projection?*

Vezzoli: Of course.

Carpani: *Isn't he essentially repeating the same pattern, albeit on a grander scale? He employed a similar approach in Grozny, Chechnya, and Georgia might have faced a similar fate. Aleppo, Syria, bore the brunt of Russian intervention, with a seemingly relentless bombing campaign. While we should also examine American actions in Japan, it seems like Putin is projecting and perpetrating a genuine holocaust.*

Vezzoli: Putin is undoubtedly pursuing this course of action, but I believe his larger goal is to assert his authority by convincing the Russian people that he's in the right, defending their interests, and promising peace. However, it's crucial to understand that his version of peace might only exist in the grave. This aspect of authoritarianism isn't limited to Putin alone; it's a recurring theme. I've observed on Italian television the crowds that gather around him when he speaks and have seen how particularly young people react to his words. His strategy involves manipulation, implying to his audience that they will enjoy all these benefits, possibly as a means of persuading them that they'll have a better life in Russia than what the West offers. His aim is to showcase Russia as superior to the West.

Carpani: *Indeed, we should consider the impact of propaganda messages. Erich Fromm's 1941 statement reminds us that when people struggle with the responsibilities of freedom, they often turn to authoritarian figures. This pattern has recurred throughout history, even after we believed it ended following World War II. Major conflicts have emerged roughly every 75 years in Europe, with the Austrian-Hungarian Empire maintaining peace from 1848 until [World War I]. Obama sought to reset U.S.–Russia relations when he assumed office, symbolized by the reset button Clinton jokingly gave to Lavrov, the Russian foreign minister. However, the reset didn't succeed, and now there's a war – prompting a reset of what, exactly?*

Vezzoli: We have something to reset. There's a critical aspect we need to address, and that's the realization that while the Cold War may have ended for us in the West, it never truly concluded for others who ascended to power. You frequently mention Gorbachev, and that highlights the core issue.

Carpani: *Putin accused his political predecessors, Lenin and Yeltsin, the latter possibly seen as his political mentor, of wrongdoing and criticized him for corruption and alleged alcoholism, claiming he sold the country to the West. Had Gorbachev retained power in the summer of 1991, preventing the spread of violence, how might Europe and Russia have fared? What might have transpired in Europe and Russia if Gorbachev had successfully completed his initiatives, including Glasnost and Perestroika?*

Vezzoli: It's clear that our world would be in a much better state. I recently read an interview in the *Financial Times* with Andrei Kozyrev, a former foreign minister under Yeltsin. He was a moderate, aligned with Gorbachev's approach, advocating for change. Having lived in the United States for over a decade, Kozyrev made some remarkable points. He described Yeltsin as a weak man who – not to undermine Ukraine's significance – was known to say yes to everything when he was drunk. Some of his best decisions might have been influenced by alcohol. Kozyrev highlighted that while Putin can still be thwarted due to the presence of strong nationalists, the West must act to diminish his power. The crucial aspect is that Gorbachev recognized the presence of nationalists and the potential for Russian bloodshed. It's our responsibility in the West to do something about it, like refraining from providing financial support to Putin and his oligarchs. If we consider the increasing level of terror in Russia since Putin's rise to power, the urgency becomes even clearer.

Carpani: *This perspective extends to the CIA, often overlooked due to Western bias. Mauro Magatti, in his March 1st piece in* Avvenire, *highlights that Putin's return of war to Europe has shattered the global liberal order established after the Berlin Wall's fall. Magatti suggests that Putin is testing Western resilience, not just through war but also via gas supply. How much cold can we endure next winter? However, I'd like to reference the ideas of Ulrich Beck, a German sociologist, who advocates a shift from the U.S.'s "Declaration of Independence" to a "Declaration of Interdepend-ence" – a call for international cooperation or face dire consequences. Beck deems nationalism particularly toxic and promotes a cosmopolitan perspective. Could this be the true reset, or am I being overly optimistic? Just as I viewed the [COVID-19] pandemic as an opportunity to slow down, seek inner and outer balance, and explore the soul's creativity, I find myself navigating the post-pandemic world with a sense of naivety, further compounded by ongoing wars.*

Vezzoli: I completely agree with Magatti and your perspective. The end of the Cold War marked a significant moment, not just for Russia but for the countries that joined Europe. The fall of the Iron Curtain opened up the possibility of introducing democ-racy in these nations. Despite the destabilizing effects of the war in Ukraine, I believe we are learning the vital importance of the world we believe in. Take Poland; the internal conflicts there have momentarily subsided as everyone refocuses on what truly matters. Perhaps it sounds idealistic, but I'm hopeful that we're recognizing the value of democracy, its role in safeguarding people, and the willingness to offer support at any cost – akin to what we did during the [COVID-19] pandemic. While we still tend to think in terms of nations, environmental concerns are global. I see us growing more aware that, as Europeans, we need to foster greater independence and exercise our sense of responsibility as democratic countries to protect our people. One of the significant developments over the past 30 years has been our realization of the importance of universally accessible services like healthcare.

Carpani: *This brings to mind Pier Paolo Pasolini, a prominent 20th-century Italian intel-lectual, and his view of capitalism as a corrosive force. I believe our moral respon-sibility extends not only to examining events in Russia but also to scrutinizing the actions of the West. How does this war differ from the actions of the United States in Vietnam, Iraq, and Afghanistan? How to consider the U.S.'s invasion, occupation, and political interventions since 1946, often presented as liberation? Let's look at events since 1991, like the invasion of Iraq, ostensibly to promote liberal democ-racy. Italian journalist Maurizio Ferrara analyzes the Iraq invasion, suggesting it was a premeditated carnage, much like the recent events in Ukraine, occurring after extensive preparation. Ferrara also reminds us of Nobel Prize winner Svet-lana Alexievich's vivid portrayal of imperialism, not only in Putin's Russia but also among a substantial portion of the Russian population, stemming from the long legacy of Homo Sovieticus and the tsarist tradition, as we discussed earlier.*

Nevertheless, we must also scrutinize the imperialistic actions of the U.S. and Britain. Every time the British left a country they previously invaded and occupied, it led to strife and civil wars. Are the consequences of the British crown's actions and the lives it cost any less severe than those attributed to Stalin, the U.S., and now Putin? Whether it was colo-nial Britain exploiting resources or the capitalist U.S. invading nations for geopoliti-cal reasons . . . , however, a crucial question arises concerning the U.S. and Britain. While the U.S. played a vital role in liberating Europe, for which we are grateful, we must distinguish between current events and those since 1946. Addressing the nation on September 11, 1989, George Bush senior [George H.W. Bush] promised a new era of peace and bilateralism without top-down relationships. However, this "new order" ultimately led to conflicts in Iraq, Afghanistan, Syria, and more.

Vezzoli: The burden of responsibility from colonialism in Europe is substantial. Even though I won't live to see the end of it, your generation, too, is grappling with the lasting effects. We must recognize that what happened should never be repeated.

When discussing Russia or analyzing Putin's actions, we should always remember the West's own actions and behavior. I've consistently protested against wars. I participated in peace demonstrations, advocating for the cessation of armor and weapon production. I vividly recall an '80s demonstration when workers protested against the idea of including tank production in the industrial plan. If you continue to produce weapons . . .

Carpani: . . . you will have to use them.

Vezzoli: And these weapons will inevitably be traded. I'm wary of any false consciousness. We bear a responsibility for our actions, and though I'm content with having protested against this, we cannot evade our obligations.

Carpani: But ultimately, we return to Hillman's notion of a love unconquered by any other. My last question for you, quoting Ferrara again: He noted that the Iraq war had a profound impact on Europe. I vividly recall being in London in March 2003, participating in demonstrations against it. In the wake of these events, Ferrara highlights that Jacques Derrida and Jürgen Habermas pondered a crucial question: What binds Europeans together in terms of international order or in contrast to the USA? Their answer was a distinct political mentality founded on shared characteristics, differing from the American approach. This mentality emphasized aversion to the use of force, prioritizing law and respect for international legality, and endorsing a global system rooted in "liberal" multilateral institutions and human rights. Born from a history marked by opposing principles and centuries of violent conflicts, this mentality was meant to lead Europe to establish a foreign and security policy anchored in the EU, transcending the oversimplified dichotomy of war and peace. These traits are indeed significant.

My question for you is whether, in the years 2006, 2008, and 2014, Europe may have failed to recognize Ukraine as a genuinely European nation, possibly perceiving it as too Russian and consequently turning away. However, Ukraine was effectively signaling its desire to align with regions where these principles held paramount significance.

Vezzoli: My deep admiration for Derrida, who hailed from Algeria and was well-versed in colonialism, resonates with his words. Embracing multilateralism and fostering collaboration is of paramount importance, as there's but one planet that beseeches our collective action.

Carpani: Especially after two years of the pandemic. I fully agree with you. We need to progress towards co-constructing meaning with pluralism, embracing differences, and perhaps even addressing challenges in integrating these differences.

Vezzoli: Indeed, co-construction is pivotal, and my inclination toward delving into literature has been intertwined with this very principle – collaboration – and when need be, the willingness to stand up and fight.

Carpani: Yes. Thank you, Caterina.

Vezzoli: Thank you, too, Stefano.

Note

1 Hillman, J. (2005) *A Terrible Love of War*. Penguin Publishing, p. 214.

Chapter 5

Day 40 of War

Giuseppe Bettoni

Abstract

For Catholic priest Giuseppe Bettoni, this war is a blasphemous distortion of religion. He finds Vladimir Putin's use of God to justify the invasion of Ukraine intolerable, seeing it as a distortion of Jesus' message of life as a gift. Bettoni's conversation with Stefano Carpani revolves around the theme of religion. According to Bettoni, Patriarch Kirill of the Russian Orthodox Church aligning with Putin is opportunistic and blurs the boundaries between religious and political power, leading to a vassalization of power. Bettoni defines war as the absence of dialogue, emphasizing that without communication, conflict escalates into violence and deception, often characterized by false narratives. He advocates for peace education and dialogue to transform potential conflicts into opportunities for growth. The priest acknowledges that war is the easier path but believes in the importance of returning to dialogue. Bettoni highlights Kyiv as the cradle of Christianity, suggesting that Ukraine has a stronger identity in this regard than Russia. He personally traveled to Lviv during the war to provide essential supplies and witnessed Ukraine's desire to grow and align with European values. However, he laments that Europe has fallen short of creating greater cohesion among nations, with countries preferring individual organization under the umbrella of institutional charters. Bettoni also sees the energy crisis as an opportunity for the European Union to develop a unified yet diversified energy policy.

Carpani: *Father Giuseppe, as a Sacrament Priest and the founder of the Arché Foundation, [you] led a convoy of two vans to deliver humanitarian supplies to Lviv. What emotions did you experience during the journey, your time there, and upon returning?*

Bettoni: Let's say that the emotion was the driving force behind the departure, hearing what was happening to the people so violently invaded led us immediately to identify with the pain, with the suffering of these people. I heard from Father Igor, a priest in charge of the seminary in Lviv who, by the way, speaks Italian well – which is very important, because Ukrainian is really difficult, few know English. Speaking with him, he immediately launched this appeal to us to support him with an economic tribute. In Ukraine, you can find little or nothing in the supermarkets; things are rationed. He told us to bring food because they have refugees, women and children, people passing through, a series of needs of orphans that they take in and that they have to feed three times a day. It was beautiful when, on our return, they wrote to us: We had breakfast with your biscuits. It was a very long journey, 1,600 kilometers, to the Polish border and after the customs check to enter Ukrainian territory, emotions came up immediately. We perceived the people at war. I had already made a trip – more than one – to Croatia back in the day. Here it is even stronger, in the sense that here there are people with dignity, a cohesion that some even define as

DOI: 10.4324/9781003390039-7

nationalism, which I somehow understand because they are building their identity, which is more Western-oriented. They feel more European than Soviet, to be clear. The city of Lviv itself is a fascinating city, from what little we saw – a middle European city, Bulgarian architecture to be precise; even in the streets, there were people playing music, trying to exorcise this fear that, in any case, is perceived as violent. They asked us: "But why are you here? Why are you coming?" We expressed our position: Look, we are here, but in Italy, too, there are many Italians who are with you, who feel the injustice that you have suffered, that you are suffering. It is a feeling of sharing, of participation, of doing what we can, beyond what governments do, but as citizens to be in solidarity with the suffering that is inflicted on you and your people. We cannot watch this from an armchair in our living room, but we decided to make this journey for you. Now another convoy will leave, I will leave again at the end of the month, hoping that the war will not reach Lviv. In fact, the western part of Ukraine is quite less affected by the bombing, which seems to have been concentrated in the East. So, the emotion was to touch the pain of these women, of the children. . . . I saw the look in the children's eyes, the youngest, 2 or 3 years old, but the older ones make my heart squeeze. Their eyes have seen and felt the pain of separation from the father who accompanied them there and then went back to fight. This is the constant for everyone, and then the running away from the bombs, fleeing to shelters, spending the night there constantly woken up by the sirens. I was struck by a grandmother who was there with three grandchildren and I asked her: "But grandpa?" She replied: "Grandpa is there. It's sowing time, but the Russians won't let us sow." Cruelty, you don't sow because you have no future, you have no life here, this is the violence that accompanies the bombing. In these days, Sister Cristina, who I am in contact with because of the neonatal hospice, was telling me about Lviv, where women are going who have suffered violence. They have been beaten, suffered all kinds of torture, such an immense violence. We cannot give up the support right now, we can show them that we are close. Of course, we can do what little we can do, but at least they do not feel alone, they feel [as if] the international community is close.

Carpani: *You may know me as a Jungian psychoanalyst. Today, I want to talk about media voyeurism and media bulimia, inspired by Jungian reflections. When you switch on the radio, the content is often shallow and meaningless. Jung wrote in one of his most significant texts of 1936 examining [World War I]. Looking back to the period before 1914, we see that we lived in a world where war seemed impossible to us. By then, war between states, with civil action, was absurd, and it became increasingly unlikely in our rational, internationally organized world. However, there was [World War I], [World War II], and then another 70–80 years of peace. Jung stated that [World War I] wasn't the cause but instead a difference between the religious world of Western Europe and the secularized world following the Russian revolution that removed the Catholic Church. Putin used strong language, claiming that invading Ukraine was justified to protect persecuted compatriots. Mauro Magatti, writing for* Corriere della Sera, *noted that Putin even cited the Gospel: There is no greater love than someone who sacrifices their life for their friends. Some people claim it's propaganda or projections. What is your perspective on this, Giuseppe, as a Catholic priest?*

Bettoni: It sounded like a blasphemy. I have no other word for it. It means the misrepresentation of the words of Jesus, who makes his life a gift, in contrast to a criminal who uses lives to crush other lives because he uses the lives of his fellow citizens, of young men who are sent ignominiously to fight other citizens. I think this is a blasphemy. It always is a blasphemy when religion is used, when religious sentiment is manipulated for the purposes of power, and this is a constant in history. In my opinion, it is a constant that comes from afar, in mythology Mars and Aphrodite, Mars and Venus are the two divinities of love and war, who just happen to attract each other. Basically, [in] war, as you mentioned earlier, the morbidity of the media, fascinates, violence takes hold of man's heart; it interests him. There is little to say. We must not be hypocritical: war is instinctive, aggression . . . precisely – mythology teaches us this – when it is cloaked as mythology is by religious sentiment, or is manipulated by political power. It becomes a powder keg; it has devastating consequences that are not in control anymore, because it appeals to the deepest feelings.

Carpani: *While I've refrained from watching television lately because, for me, war is often presented as the latest gossip, drawing us to consume it, akin to bulimia.*

  Magatti also recalls that Putin's words were echoed in the speech of the Primate of the Russian Orthodox Church, who, in the early days of the war, went as far as to claim that the current conflict was not merely physical but metaphysical. According to this perspective, Putin's actions are deemed legitimate because they aim to counter the advance of a thoroughly secularized West. This notion is exemplified by events like the gay pride parade held in Kyiv in 2019, seen as an actual initiation ceremony.

  As a forward-thinking priest with 30–40 years of experience working alongside Arché, you don't just serve as a religious figure but also effectively function as a politician. Can you shed light on what Kirill's words imply in this context?

Bettoni: These words of Kirill really displeased and irritated me because they go in the line of what we were saying before, of the back bent to power, of those who in fact seize all the advantages of this throne/altar alliance. It is curiously called the symphony, the relationship between altar and parliament and government, political and religious power in the Orthodox Church is called a symphony. But in the history of the Orthodox Church, it is rooted to the point of identifying the church with the boundaries of a nation state this induces the church to vassalize power. The moment the same boundaries are those of a political government, at this point every form of prevarication and arrogance is identified with the action of a government. The church is reduced to this and indeed becomes a champion of values for their own sake. We were not so far from this logic 20 years ago, the so-called non-negotiable values – let us not forget, Luini's policy in the Italian church went in this direction, like the American church in the north now, the alliance with these values is instrumental to the alliance with power.

Carpani: *Similar to some Italian politicians, their comments seem outdated, displaying hypocrisy as they engage in multiple marriages and divorces or claim specific names, genders, and religions while remaining unmarried.*

Bettoni: The problem that upsets me, beyond personal consistency, is that they profess to be Christians, denying the Christian message, they put values – so-called values – before the Gospel, before Jesus Christ. Jesus Christ, if he had wanted to defend values, would not have ended up on the cross. He would not have proclaimed love for the poor. He would not have proclaimed forgiveness. He would not have done what he did. He would have organized a crusade against the adversaries, against the enemy. Jesus, the Gospel, stands on a totally different plane, what a metaphysical warfare, you have it in your brain: why don't you make peace with yourself, your conscience called by the Gospel, and the responsibility you have towards your people? You cannot contradict your membership in the name of political expediency. If you do this, you are a criminal, too.

Carpani: *You called a blasphemy. I'd like to share the words of James Hillman, an influential Jungian, from his book* A Terrible Love of War. *He asserts:*

There is no political solution to war, because war is not a problem that the practical mind can solve; it is better suited for its conduct than for avoidance or resolution. War is deeply ingrained in our collective soul as an archetypal truth of the cosmos. It is a human creation and an inhuman horror, a love unconquered by any other love.[1]

Hillman suggests that we must confront this harsh reality, become aware of it, and channel our passionate intensity into opposing the perpetuation of war. He believes that, despite the dark ages, culture has the resilience to persist and resist war while striving to better understand it and gradually detach it from the support of a hypocritical religion. However, war, in its essence, will persist until the gods themselves have left or are no more . . .

Bettoni: They are words of great intensity, very beautiful. There was a time when in the community where I live, the subject of how to talk to children about the war came up because usually mothers follow these events. One day I asked them how they live this: "We are worried. We want to hide this from the children, but perhaps we should take it as an opportunity. Let's not give in to pure morbidity. Let's not go on a lofty, philosophical rant, but it could be an opportunity." In fact, just before I had seen the gesture of a mother who intervenes in an argument between two children over a game. She takes the child, pulls him and takes him away: "Come away, you mustn't stay with that one!" I started from that example, but was it the only possible way? In a conflict, in a tension, in a misunderstanding, how much violence do we put into it? How do we resolve that situation? With words. When there are no words, there is war; if there is no dialogue, how could things change if that mother had knelt down, sat on the ground and said to her child: "Look, you can play this game a little and then he can play a little, or you can play together"? With dialogue and words, through mediation, she could cross tension, conflict, [and] different interests because it is the interests that conflict with each other. At a higher level, we call it diplomacy. Finding the word, because when the word is missing, bombs arrive, violence arrives, aggression arrives. Today, we are witnessing the triumph of lies, of falsehood. As someone says, the first victim of war is the truth. We are witnessing the continuous lie; hence, also the word that is used in falsehood, because sometimes the word itself is not enough. Many times, the word that becomes false – a lie – is a word that fuels conflict and violence. How much responsibility do we have as adults in educating our children to know how to manage conflict? Conflict is there; diversity is there; different interests are there. I am interested in one thing, you in the other, what do we do? Do we go to war? Let's sit there calmly. Let's try to understand. Let's try to find a way out. No one does this peace education anymore;

it becomes much easier to take one side or the other, instead of the effort of dialogue, of mediation, of transforming the potential conflict that could become war into an opportunity for growth, for regeneration; we easily give in to the easiest path, which is precisely that of making war.

Carpani: *While you were discussing the absence of peace education, two striking images came to mind. One is the old war school of the Nazi German Empire near my office, a building still bearing bullet holes from World War II. The other is the war school from my time in Madrid . . . however, I never came across a peace school, at least not an institutionalized one. This led me to contemplate the deeper implications of the ongoing war . . .*

Since 2006 and 2008, during the elections of Yanukovych and [Yulia] Tymoshenko, Ukrainians have been expressing their desire for independence from Russia and their aspiration to align with Europe. Their yearning is not for BMWs, McDonald's, Hollywood, or Armani; rather, they aim to establish a free, independent nation with robust institutions. They seek freedom from external influences that dictate from the top down and long for a democracy where they can elect and re-elect leaders every 4–5 years without tension or coups.

In this context, the U.S. stands as an example. Despite its crises, polarization, extreme left, and extreme right, it remains a democracy where leaders are elected and re-elected, showcasing the resilience of its institutions. Ukraine is clearly signaling its aspiration for stability and democratic governance akin to what the U.S. (and Europe) represents.

Bettoni: This I have touched with my own hands: Ukraine wants to grow, to rediscover its identity, its culture that was stifled during the years of the Soviet Empire. Even as far as the Christian tradition is concerned, Kyiv was the cradle of Christianity. The first person baptized was from Kyiv and it was from there that he left for Russia; therefore, the strong identity is not Moscow – the strong identity is Kyiv, Ukraine. It is a nation that has middle European traditions because it is between the most western part and the most eastern confinement. Therefore, the desire of those who come to Italy – of the numerous women caregivers or who do the humble services that we no longer do – do not want to stay in Italy. I have seen it with my own eyes – they have built very modern houses, as if we were in Milan. Around Lviv, there is very modern, avant-garde architecture; the city center is a historic center, but outside, the savings of these women who have worked here or who work here go to improve their people, their quality of life, their homes. This says a lot in this perspective of democracy, of growth in the participation of the people, of sharing; that there is a president like Zelenskyy, who has a history of his own. It is precisely the flourishing of that democracy that leaves the people the power to say that in this moment, I want to be governed by him rather than by others, a democracy that is still fragile, if you look at the last 30 years. Yes, but in the meantime, the Russians did not expect to meet such resistance.

Carpani: *Just two months ago, I questioned my friends about their willingness to defend their country in the event of war. Now, the understanding of one's relationship with their nation is evolving beyond rights and duties. An interesting case came up with a patient who, though born in America, has been living in Berlin for over a decade and recently obtained German citizenship. I pointed to a map of Europe in my office and asked him if he realized that he is now German. His response was, "What does that mean? Would I have to participate if something significant happened?" He firmly stated, "No, I wouldn't fight!"*

Around a month ago, Ferrara penned an article in Corriere della Sera discussing how the 2003 Iraq war deeply shook Europe, prompting widespread protests and demonstrations. He added that, in response to such war, Habermas and Derrida pondered what unified Europeans in terms of international order. Their conclusion was a distinct political mentality, differing from the American approach, grounded in common traits. These included a strong aversion to the use of force, an emphasis on law and respect for international legality, and support for a global system built on liberal multilateral institutions and human rights. This mentality, shaped by a history marked by centuries of bloody conflict, was meant to drive Europe toward establishing a foreign and security policy anchored in the EU, transcending the oversimplified division between war and peace.

Could it be that this reflection kept us Europeans from looking at Ukraine earlier? When, in 2006, 2008, and 2014, they were conveying their desire not to align with a particular bloc but to be part of a Europe defined by these values. They wished to belong to this kind of Europe.

Bettoni: They say it openly; the problem is that Europe is lagging far behind the appointments of history. We have created economic and commercial unity, we have dismantled customs, but at the appointments that history has given us – I am thinking of the reality of refugees arriving both from Libya/Africa and from the East – we have responded in a totally inadequate manner to what should be that charter of values that Europe still does not have. And this is another aspect that in my opinion is very serious: How can you not give yourself a constitutional charter, as the United Countries of Europe? Is there a design, is there a sharing of a future of defense, even of Europe? No, each country organizes itself. What we need is some European politician who has at heart to break down these indifferences, this superficiality, these closures, these particularisms in order to really create a more cohesive Europe with respect for diversity. Portugal, Norway, Italy, Ukraine are very different countries, but we cannot even imagine that each one goes its own way, trying to gain as much as possible for its own interests at the expense of others. The energy crisis speaks volumes – each has sought its own channels; Germany and Italy depend on Russia, France has made the nuclear choice. Is it possible that even there we cannot have a cohesive, united, diversified energy policy? We never talk about this. We always talk about abstract values. But those values must be embodied in political choices, of governments, of countries. Wars break out because they are the sign of this inability.

Carpani: *My call for an open dialogue between us is rooted in the understanding that, much like 30 or 40 years ago, when institutions were faltering, Europe today is sustained by civil society. In practical terms, civil society has stepped in to address the gaps left by these institutions. This is precisely what you've achieved with Arché. Currently, civil society isn't merely providing essential goods; it's also fostering thoughtful reflection. When we look at political Europe, it often appears preoccupied with financial statements and gas-related matters. Can't we, in simpler terms, decide that we've had enough and collectively manage our gas resources? Yes, it may mean enduring some cold winters, but at least we can put an end to certain things. Of course, I recognize that the situation is far more intricate, but sometimes it feels like we're needlessly complicating our lives.*

Last year, when we connected and organized a series of conversations about Lent, we explored various topics. Reflecting on the [COVID-19] pandemic, it seemed like an opportunity to decelerate, to restore a balance between our inner and outer selves, spirituality, and the creative depths of our souls. Yet, as I continue to contemplate this war, I can't help but feel incredibly naïve. It brings to mind Michel Houellebecq's assertion in 2020: "The world will stay the same, everything will stay the same, only a little worse."[2] Do you know why I say this? I genuinely hope, echoing the sociologist Ulrich Beck, that we can transition from declarations of independence to declarations of interdependence, where vulnerabilities become shared responsibilities and propel us towards a cosmopolitan perspective. Or maybe I'm being naïve in that hope too?

Bettoni: I remember what was being shouted from the balconies: Everything will go well! We will come out better! This also someone dared to say. But, as you said, at the end of the day, after two years, we are still here and even a little worse. I can hope that someone took the opportunity to rethink many things in life – that is to say, a deceleration, a greater interiority, a greater possibility of reading, of study, of knowledge, because in the end, the real problem is that people no longer read, no longer study, no longer know, [there is] no deepening, and even these opportunities slip away. Even war will leave an emotional aftermath. But if there were a true reflection, on non-violence, on peace, on pacifism, educating for peace, where are these opportunities to be found today? They are to be found in private citizens, in some organizations because it is not certain that they all have this willingness to get involved, to think. The human soul is bizarre, because at the same time what I see here in Milan, in a neighborhood context such as that of Quarto Oggiaro, even in a city context, is the growth of an impressive pain of the soul. There is a psychic, moral, spiritual suffering – let's choose all the adjectives we want. I like to call it pain of the soul because people find it hard to give it a name, so this dissatisfaction either becomes explicit, verbalized. It is given a name; otherwise, it turns into violence, aggression, indifference, closure. So many adolescents are really sick – so many young people are really sick – [that] they look at the adult world that finds it hard to listen, continues to do the things it has always done. One of the few voices left in the field today to call this out is Pope Francis, the only voice that says something different, something that opens the soul and opens up high horizons, but he is alone.

Carpani: *No, he's not alone. That's why I value our conversations. I've had patients who hold a negative view of the Church and its priests, often generalizing them as all being the same and tainted by the actions of a few. But when this comes up, I mention your name and point out that priests like you, from Bergamo, have achieved remarkable things, countering such stereotypes.*

Bettoni: I thank you. For goodness' sake, I have done my part. I do these things and continue to do them because they are the salt of my life. I like doing them. I am happy to do them. I do the same things I did when we were all a bit younger. Of course, mistakes are made; there are mistakes, and when they tell me the church is like this, I reply that even with Jesus in the end he chose 12 – one betrayed him, the other denied him. What do we expect? Perfection in others? Let's look at our own house, each one of us has his own shortcomings, his own things that they hide, that they don't show, we are all poor Christs. Many times, it is an alibi to shift the blame on others, the usual scapegoat mechanism. I do not think that this also helps people. Let's stop with this mechanism, look inside yourself, look in the face of reality – what do you expect from others, from the Church, the state, the government? For charity, it is necessary to be critical. You know that I am very critical, if the Eternal Father gave us a head to think. Faith is not blind; it is anything but blind: it is enlightenment, the ability to read, to understand, to make the effort to think with one's head, to confront oneself. A faith that is not like this is a faith that then calls everyone to war. I do not believe much in an ideal church. It frightens me.

Carpani: *Because it does not exist – never existed.*

Bettoni: It becomes something else that is not what Christ wanted. He would have chosen the best of his time, the wisest. He would have held a competition, sent out resumes he would have chosen the best – but no, he took what was there, the human being for what he is.

Carpani: *He accepted responsibility instead of evading it, a crucial message he wishes to convey. This sense of soulful pain and purpose revolves around taking responsibility for our actions. It reminds me of someone important to both of us, my first analyst and colleague, Alessandro Albizzati, who, during Arché's 30th anniversary, described volunteering as a self-cleansing (sanificante) duty. Recently, I've encountered patients feeling lost and empty. I've recommended volunteering at Berlin's central station, where trains arrive.*

Bettoni: Which is the most subversive gesture one can make in these times. You have done a subversive action in full coherence with our history. It is this because if we don't do these things, nothing changes. Instead, take your daughters there, send your patients there – that is where you start again, right from the depths, from the abyss, precisely from that emptiness that is full of humanity. Albizzati said he was waiting – to do politics, the interest of the polis, of the city against the closure on the private, out of reality, the political choice is to get one's hands dirty, to get involved in life, to be touched, to be questioned, to get involved . . . it is politics.

Carpani: *Giuseppe, thank you very much.*

Bettoni: Thank you.

Notes

1 Hillman, J. (2005) *A Terrible Love of War*. Penguin Publishing, p. 214.
2 The quote is from *Corriere della sera*, 4th of May 2020: www.corriere.it/esteri/20_maggio_04/houellebecq-cari-amici-mondo-sara-uguale-solo-po-peggiore-e512c852-8e40-11ea-b08e-d2743999949b.shtml

Chapter 6

Day 42 of War

Joseph Cambray

Abstract

Psychoanalyst Joseph Cambray notes the activation of the collective unconscious, causing a shift in archetypal patterns, with a preference for restoring old empires leading to tension. He stresses the importance of
understanding societal interdependence. While Vladimir Putin justifies actions using the Bible, Cambray
questions our connection to the numinous, suggesting a potential re-enchantment of the world in response
to current challenges. Cambray shares a personal experience of a bodily reaction to the Ukraine invasion,
emphasizing the collective forgetfulness that can occur if we lose touch with nature. He contrasts human
organized wars with the cooperative side of nature. Introducing the concept of "adjacent possibility,"
Cambray explains how leaps of imagination often collapse without adequate resources, citing examples
from history, including Mikhail Gorbachev's vision for Russia, which, having gone too far, led to Putin's
return to Cold War dynamics.

Carpani: *Joe, what is currently happening?*

Cambray: We are currently experiencing activations of the collective unconscious, with a
noticeable cyclical pattern. It seems that we're in the midst of a significant shift,
marked by changing archetypal patterns and a clash of opposing forces, akin to tec-
tonic plates in motion. The old heroic egoic states that once defined our individual
and cultural identities are no longer relevant. Instead, there's a growing need for
a broader ecological understanding of our interdependence. This shift is causing
substantial friction and giving rise to regressive efforts to resurrect old empires,
evident in the imperialism of the U.S., Russia, and China. These major powers are
caught in a polarizing dynamic.

Carpani: *How does religion and spirituality address this issue? According to Jung, significant changes are occurring in the religious realm. It's not surprising that in Russia, the vibrant Eastern Orthodox Church has been overshadowed by a godless
movement. Jung continues to state that despite the scientific enlightenment of the
19th century being at a low spiritual level, it was inevitable for it to eventually
dawn in Russia.*

*It's worth noting that Jung's reference to Russia pertains to its loss of religion
and spirituality after the revolution. Presently, Putin is quoting the Bible, emphasizing the value of laying down one's life for friends. Meanwhile, Kirill, the Moscow leader [of the Russian Orthodox Church], argues that Russia is engaged in a
battle against the secularized Western world, particularly opposing the LGBTQ
community. This seems contradictory to his stance.*

DOI: 10.4324/9781003390039-8

Cambray: The relationship with the numinous is a crucial consideration. When the numinous is forcefully removed from a society, it tends to re-emerge in repressed and unsettling forms. This prompts us to question our connection to the numinous and its role in these conflicts. From an academic standpoint, there's a movement towards re-enchanting the world, countering the Enlightenment's disenchantment, as described by Max Weber in 1917/18, when a scientific view led to the loss of the divine. Simultaneously, figures like Jung underwent experiences that unleashed the numinous energies, not only in them but also in other artists and individuals. This poses a significant question for our future: What is our relationship with the numinous? Figures like Putin and Trump attempt to co-opt it for political purposes, as seen in the "Make America Great Again" slogan, which can be viewed as a secular misuse of the numinous, aiming to regressively restore something rather than embracing it as a transformational force leading to something new. This dilemma stands as a significant challenge we must confront.

Carpani: *Can you clarify the meaning of "numinous" and define "disenchantment" and "re-enchantment"?*

Cambray: The term "numinous," coined by historian of religions Rudolf Otto, was adopted by Jung during his time. It represents the profound sense of awe and emotional overwhelm we experience when encountering something vastly greater than ourselves, whether positive or terrifying. These encounters take us beyond ourselves and connect us with something we may not have acknowledged previously, making them central to religious and mystical experiences, as exemplified in William James' "Varieties of Religious Experience." The shift from disenchantment to re-enchantment was a sociological theory, particularly associated with sociologist Max Weber. It traced the progression from a magical worldview in "primitive" societies, gradually replaced by religious systems, and ultimately supplanted by a scientific perspective, with the scientific revolution in the West reaching its zenith in the 17th century. By the late 19th century, there was a belief that everything was understood, leading to scientific arrogance. However, the advent of quantum mechanics, relativity, and the discovery of the unconscious challenged this assumption, overturning the paradigm. This transition prompts us to reconsider how indigenous cultures' views of spirits, demons, and animate forms in the world might not be mere superstitious misreadings but could reflect a genuine dynamism in the world itself. Jung's concept of synchronicity explores the porous relationship between the psyche and the external world, leading to a re-enchantment of the world. It offers a sense that the world is reanimated, not solely through our psyche's projection, but by something intrinsic to the world – a world soul.

Carpani: *I wrote a paper about the fall of the Berlin Wall, focusing on Günter Schabowski's response at a press conference on 9 November 1989. When asked when people could leave East Germany, he answered "sofort," meaning "right now" in German. I propose that a numinous or transcendent experience influenced his response and that experienced politicians would typically not react in that way. Instead, it was the collective unconscious and the mounting tension over several months that led him to utter those words.*

Cambray: Reconnecting with altered or non-ordinary states of consciousness is crucial. I've developed an interest in oracles, especially as I've delved into altered states of consciousness and revisited historical figures like the priestess of Delphi, who served as the oracle of Apollo. Modern classicists have shed new light on this, revealing that Mediterranean civilization was sustained for over two millennia by a group of middle-aged women with no political power. They had a direct connection with the numinous and delivered often cryptic messages that compelled people to pause and reflect on their actions. The rich layers of meaning in these messages required individuals to contemplate their deeds and motivations. Our modern approach of linear problem-solving and seeking single, definitive conclusions tends to foreclose on such reflection. Jung's pursuit, as seen in his *Red Book*, was to make sense of visionary experiences and their relevance. In current times, there is a growing need for visionary guidance in the political realm. We must consider where we derive our visionary insights and information for shaping our political vision.

Carpani: *Let's consider the concept of the masculine and the feminine. It might seem simple, but it raises the idea that as we gain enlightenment, acquire more scientific knowledge, and adopt a more linear language, the world becomes more masculine, potentially losing its enchantment and becoming less feminine. Is this a simplistic view?*

Cambray: One way to address this challenge is through embracing complexity. Linear thinking is potent for solving specific problems, but it breaks down when confronted with increasingly intricate issues that don't neatly fit into predefined categories. Complexity demands a more intuitive and holistic approach that recognizes and navigates patterns. Traditionally, this approach has been associated with the feminine side of consciousness, although it's important to recognize that these binary distinctions can be limiting. Embracing more permeable and flexible forms of consciousness can help us move beyond the polarized thinking that patriarchal systems have often encouraged.

Carpani: *Is war masculine?*

Cambray: War often embodies masculine desire, and desire itself is a fascinating aspect of this phenomenon. The connection between desire and the violence of war underscores the use of power and sexuality. There is a captivating element to this dance, with war often framed as an attempt to win feminine attention. This erotic dimension is a potent instinctual energy, making it challenging to find a "moral equivalent of war," as William James suggested, because it requires reckoning with the depths of desire. While women can unquestionably be warriors, exemplified by the Amazons and contemporary female combatants, it's not solely a feature of the masculine psyche. Nevertheless, there's an element within the masculine psyche that taps into the irrational in a particularly powerful manner. Understanding and navigating this aspect is a complex challenge we must confront.

Carpani: *In my introduction, I mentioned that we are drawn to war, that war holds a peculiar attraction. I believe war is the antithesis of what Truman Capote referred to as "Vanity Fair" – it's more like an "Atrocity Fair." However, when portrayed through the media and TV, it often takes on the guise of a "Vanity Fair." What, then, truly defines war?*

Cambray: The age-old human question revolves around the extremes of polarization, where opposites cannot find a middle ground. When war erupts, it signifies the collapse of tension and a failure of peaceful efforts. This presents a significant challenge to our rational thinking, as we often believe we can transcend the power of these emotions. However, it becomes evident that war emerges from a cultural, paranoid, schizoid state where[by] negativity and evil are projected onto the other, serving as a justification for the subsequent atrocities committed against them. The inability to maintain a relationship and work with this tension dissipates in such states.

Carpani: *In a conversation with our colleague Caterina Vezzoli, war was described as a failure. An Italian priest, Padre Bettoni, equated war with blasphemy, in reference to statements by Putin and Kirill. I'd like to share the words of someone who's dear to you and whom you've worked with closely – James Hillman. In his book* A Terrible Love for War, *he suggests:*

There is no practical solution to war because war is not a problem solvable by the practical mind, which is better equipped for its conduct than for its avoidance or conclusion. War belongs to our soul as an archetypal truth of the cosmos. It is a human work and an inhumane horror, a and a love that no other love has been able to overcome. We can open our eyes to this terrible truth and becoming aware of it, devote all our passionate intensity to undermining the enactment of war, strengthened by the courage that culture possesses even in the dark ages to continue to sink as it resists war. We can understand it better, postpone it longer, work to gradually remove it from the support of a hypocritical religion. But the war as such will remain until the gods themselves leave.[1]

Joe, recall Nietzsche and Freud's notion of religion as an illusion and Jung's revival of the gods. Have the gods already departed?

Cambray: I certainly resonate with some of that perspective. On the morning of the Ukraine invasion, there was an almost tangible presence in my room. I felt a sudden chill, as if the numinous – the gods of war – had awakened. Although not visible entities, their presence was felt, not just in my room but in the world at large. It's reminiscent of narratives like *[The] Lord of the Rings*, where a dark force reanimates and returns. This resurgence seems to occur when we forget the lessons of history. The cyclical nature of history is striking; as we age and the living memory of such events fades, we become more vulnerable to their recurrence. Many World War II veterans have passed away, and we are increasingly removed from those experiences. Collective forgetting seems to play a significant role, contributing to the allure of television's portrayal of war, particularly for those who have never witnessed its horrors firsthand. The war metaphor that Hillman uses is rooted in an evolutionary model of competition, particularly the Darwinian concept of competition that dominates Western thought. This model may be inadequate. Animals may kill each other, but they don't wage organized wars; they know when to stop. We've overlooked the cooperative side of nature, which we are rediscovering through indigenous knowledge and a deeper understanding of interconnected systems like forests. Nature maintains a balance between cooperation and conflict, but our focus has been on competition, fueling paranoia about being overpowered. We must shift our perspective to consider how we can engage in competition differently.

Carpani: *That's why European Union countries are making substantial military investments. My perspective is less psychological and more psycho-political. I believe Ukraine, starting from 2006, 2008, and the Maidan protests in 2014, is signaling a different path. It's as if Ukraine is saying, "Hey Europe, United States, we aspire to build a democracy with strong institutions." What's significant here is that in the United States, despite the political tensions and changes, the institutions remained robust enough to handle these challenges and transition to a new leader without descending into authoritarianism or dictatorship. To me, this is the most critical lesson from this war.*

Cambray: Comparing 1920s/30s Germany to the United States and drawing parallels between Hitler and Trump, a crucial distinction lies in the United States' robust constitutional history and entrenched governmental mechanisms. In Germany, these mechanisms were swept aside at the end of the Weimar Republic, leaving little room for challenge. In the United States, however, it felt perilously close. We teetered on the brink of a potentially catastrophic situation, and although we might wish to believe we've moved past it, that sense of safety remains elusive. The struggle is ongoing, marked by a regressive longing for a fantasy of greatness, a narcissistic yearning for grandiosity.

Carpani: *When discussing figures like Putin, Orban, Trump, [and] Obama, we're delving into the theme of the father and possibly the fatherland. As Jungians, we understand the concept of compensation, and it could be argued that this war is a form of compensation. Just as World War II compensated for the first one, this war might be a way for Russia, the fatherland, to seek compensation for the collapse of the Soviet Union. I sense that when one becomes ashamed of their father, like Freud and Jung, or when individuals like Hitler were embarrassed by their predecessors, such as Putin's discomfort with Gorbachev's actions, they tend to look for someone to blame, as Lenin was in Putin's case.*

Cambray: Yeah, I remember. Strange moment.

Carpani: *Freud and Jung believed that their fathers were heroic, but in reality, they fell short. Perhaps Jung and Freud felt compelled to become heroic themselves to make up for this. Is this also true of Putin, and of modern nations that must act when their fathers were unable to demonstrate heroism and confirm their status as heroes?*

Cambray: The psychodynamics of the individual may indeed play a part in these dynamics. It's essential to consider why this holds true for an entire society. The situation is not solely attributed to Putin's individual psychopathology, as there are levels of societal buy-in. Exploring the concept of the "adjacent possible," discussed by Vittorio Loreto, sheds light on how culture evolves and changes. Typically, this evolution occurs step by step. Imagine the "adjacent possible" as discovering a new room in your house within a dream, and as you explore it, you find another door leading to another exciting possibility. This concept reflects one path of development. On occasion, individuals attempt significant leaps. Consider Leonardo Da Vinci, who designed advanced machines like tanks and submarines using rope and wood. His imagination leaped into what they termed the "far possible." Without the necessary resources to support these ideas, they tend to collapse. A similar scenario unfolded with Gorbachev, who envisioned a new political direction for Russia but lacked a society ready to work out the steps to achieve it. In this context, Putin's actions appear to reset things, moving back towards a Cold War–like scenario to create an alternative pathway. He seeks to establish a connection with the heroic fatherland, but this regression is a result of the earlier leap that couldn't be sustained due to its excessive ambition.

Carpani: *How would Russia and Europe be different today if Gorbachev had been allowed to carry on with his work? He was removed from office while he was in Crimea, which is a disputed area. It's noteworthy as a symbol. He was too far ahead in time for what people and the mind could handle. Shall we explore psychopathology for a moment? Is it helpful, or is it improper to psychopathologize?*

Cambray: As a clinician, it's almost impossible not to observe behavior and consider what it may indicate. The challenge lies in the rapid conclusions one might draw. While it's possible to make a diagnosis based on behavior and reactions, understanding this within the broader societal and cultural context is more complex. Trump's presidency posed a significant challenge, especially with the "Goldberg rule," which discouraged psychiatrists from diagnosing a public figure without direct, in-person interviews, citing ethical concerns. While there's merit in this approach, it's only natural for individuals to have gut feelings and concerns about the behavior they witness. The key question, though, is what to do with this understanding. It's easy to become self-satisfied or smug with a diagnosis, but it had limited impact on changing public perception or the trajectory of Trump's leadership. The value of diagnosis lies in its effectiveness and how it can be used to address the issues at hand, and this is the challenge [of] clinicians.

Carpani: *Jung's portrayal of Hitler, as seen in "After the Catastrophe," seems to echo in contemporary populists. For instance, Putin denies the events in Bucha, labels Ukrainians as Nazis, and alleges genocide in Donbas. These claims appear to be projections of his own actions, even though many historians hesitate to label them as genocide. It's as if his actions serve as a form of compensation for this projection.*

Cambray: This situation could be described as a form of projective identification, where you attribute to the other what you're unwilling to acknowledge about yourself. By labeling someone as a genocidal murderer, it can rationalize treating them inhumanely. The act of projecting and denying these shadow elements of one's own or the collective psyche can be part of a strategy to trap someone within a paranoid structure. As a clinician dealing with projective identification, it's essential to work on countertransference, to process and understand these dynamics thoroughly, and create a meaningful engagement. When both parties become ensnared in a mutual projective identification, there's a powerful annihilation energy at play. Recognizing this and not merely absorbing it is crucial. Instead, it's about listening to what the person is attempting to convey and understanding their underlying paranoia and concerns.

Carpani: *What do you think is his paranoia about?*

Cambray: Putin appears to have a complex perception of Russia. He may view it as having shrunk from its Soviet days, leading to a sense of deflation and a loss of a grander identity. Simultaneously, there's a deep-seated fear of invasion and encroachment, fostering paranoid feelings and a belief that others are intent on annihilating Russia. These feelings serve as justifications for taking actions against others, often leaving little room for negotiation. It's notable how the Ukrainians are seeking middle ground, but on the other side, it appears that such a middle ground may not even be considered possible.

Carpani: *I'd like to pose a challenging question without oversimplifying. We frequently place responsibility on our leaders. Jung's "After the Catastrophe" suggests that a leader is essentially the product of the collective unconscious, reflecting the collective psyche. The complexity arises from the possibility that this reflection could ultimately be turned back upon us.*

Cambray: The issue of systemic racism and historical injustices has become a pressing concern. These deeply ingrained cultural patterns persist, even if individuals personally reject the practices or thoughts associated with them. The challenge lies in understanding our relationship to this cultural inheritance. It's not about succumbing to guilt but rather embracing a broader, ecological perspective on our cultures without allowing shame to create divisions. Leaders often manipulate the anxiety stemming from collective shame, as Hitler did regarding Germany's humiliation from the [Treaty of Versailles]. Addressing our collective histories within each country is essential, focusing on how we relate to them rather than pointing fingers at others. The real question is how we navigate this relationship. Initiatives like the diversity, equity, and inclusion task force at Pacifica, which has been re-evaluating core theories, exemplify the necessary efforts to bring about meaningful change. This is a substantial challenge for all of us, and collaborative thinking harnesses collective wisdom that surpasses individual perspectives. It raises the question of what the international community's collective wisdom has to say on these matters, and how we can tap into it.

Carpani: *And perhaps is the answer to my question: War as reset, but reset of what?*

Cambray: There's a looming threat of resetting us back to a state reminiscent of 75 years ago. This appears to be an international trend. The crucial question is whether we passively accept this reset or adopt a meta-level perspective to understand the psychological implications of this reset and consider what truly needs to be reset. While I don't have a definitive answer, your question is pivotal. It's ironic that the Clinton–Lavrov reset button was inaccurately named, considering the current situation.

Carpani: *I think it was a "Freudian wish" by the Obama administration . . .*

Cambray: Obama's presidency in the United States may have arrived prematurely, making him one of the youngest presidents in office. The weight of being the first African American in the highest office, combined with his relatively limited experience in the shadowy aspects of Washington, posed significant challenges. He struggled to fill positions at the outset of his term, which may have led to a suppression of the energy he represented. The reset, in this context, might have been more of a wish and longing, like a Freudian desire.

Carpani: *In 2008, he stood in Berlin, addressing an audience of three million, offering Europe a ray of hope. It was the era of [Spain's José Luis Rodríguez] Zapatero and [France's Nicolas] Sarkozy, both harboring their own hopes, which ultimately waned. On March 1st, Mauro Magatti, expressed in the Catholic newspaper* Avvenire: *"By bringing war back to Europe, Putin has shattered the global liberal order that emerged after the Berlin Wall's fall." Magatti highlights Putin's intent to test the West's resilience. Now, the words of Ulrich Beck, the German sociologist, seem distant. He once advocated a shift from a declaration of independence to a declaration of interdependence, advocating global cooperation due to shared vulnerability and collective responsibility, including our survival. He emphasizes the toxicity of nationalism and proposes a cosmopolitan perspective.*

Cambray: I wholeheartedly agree. It's heartening to witness the countries of the EU attempting to present a unified voice in response to the atrocities. This collaboration is emerging under immense pressure from the ongoing conflict. This is what I refer to as an emergent form, where[by] multiple agents interact to create something greater, possessing properties not found in any single individual. Achieving interdependence is the path to fostering emergence, but it can be fragile and short-lived. The question is how to trigger a lasting chemical transformation, a shift towards a broader inter-dependent awareness that becomes an integral part of people's identities, guiding them towards a new way of being beyond just crisis solutions in the moment. This transformation points the way to a more sustainable future, and that's the challenge we face.

Carpani: *My previous series,* Lockdown Therapy, *was driven by a hope that the pandemic would prompt us to slow down, seek equilibrium between our inner and outer selves, and move away from a world overly reliant on aspirations and expectations, placing more emphasis on inner reflection and spirituality. In agreement with the French writer Michel Houellebecq, I now recognize that the world may not undergo significant changes after the pandemic, but it might take a turn for the worse. In hindsight, my initial optimism was somewhat naïve.*

Cambray: I have to agree with this perspective. When discussing this topic, I often refer to the transformation in airport security after 9/11, which fundamentally altered the way we experience air travel. The ease and freedom of walking up to the gate were replaced by thorough security screenings, and we've adjusted to these changes. With a background in science, I believe that the coronavirus is just the tip of the ice-berg. Research from Harvard's [Chan National Institute of Environmental Health Sciences {NIEHS} Center for Environmental Health] suggests that the coronavirus may be a direct consequence of global warming. As the planet warms, animals that used to reside further south in the northern hemisphere migrate to higher latitudes, bringing them into contact with birds and bats they wouldn't typically encounter. This increased contact leads to opportunistic cross-species transfers and contrib-utes to the emergence of diseases like COVID-19. If this holds true, shifts in our ecology are likely to bring about more pandemics and disasters. Without taking the time to reflect on the lessons learned and make meaningful changes, we may find ourselves wearing masks for an extended period, even for the rest of our lives.

Carpani: *Let me present a challenging question to an American perspective: How do we dif-ferentiate this conflict from past actions of the U.S. in wars like Vietnam, Iraq, and Afghanistan? Also, how should we evaluate the U.S.'s history of invasions, occupa-tions, and attempts at political stabilization since 1946, framed as endeavors to bring liberation, and since 1991, under the banner of spreading liberal democracy?*

Maurizio Ferrar, an Italian journalist, noted on March 7th (Corriere della Sera) that the U.S. invasion of Iraq was a premeditated carnage, similar to the invasion of Ukraine. He references Nobel Prize laureate Svetlana Alexievich, who painted a clear picture of an imperial reality not only linked to Putin but also to a significant portion of the Russian population, stemming from the enduring legacy of the Homo Sovieticus and the tsarist tradition. Yet, when considering the imperialistic actions of the U.S., China, and the British crown, with their history of invasions and occu-pations resulting in problems and civil wars, it becomes vital to scrutinize whether the actions of these powers, along with the associated loss of lives, significantly dif-fer from those attributed to Putin. This question is crucial not to establish a simplis-tic dichotomy, but to engage in a nuanced exploration of meaning co-construction and our interconnectedness. What are your thoughts on this matter?

Cambray: Human tendencies to preserve the status quo, assert one's dominance, and exhibit xenophobia are pervasive traits that operate in societies, and U.S. policies are not immune to these forces. Despite the rhetoric used to justify actions, systemic issues may persist. The rationalizations surrounding the weapons of mass destruction in the lead-up to the Iraq war were exposed as a farce, leading to a profound deconstruction of the heroic individualism narrative in the United States. Vietnam was another turning point, challenging the prevailing narrative. Political historian Gabriel Kolko's work delves into the history of post-World War II U.S. diplomacy, highlighting a series of disasters and the underlying anxieties, fears, and motivations behind these actions. While there have been protests against such policies, a crucial difference is that in Putin's Russia, there is no room for public dialogue or protests. In contrast, during the Vietnam era, people in the United States could voice their dissent in the streets and oppose actions they deemed unjust.

Carpani: *Reaching out to my Russian colleagues is challenging due to safety concerns, much like conducting a webinar in China with potential uninvited listeners. We must exercise caution to avoid putting our friends at risk.*

In conclusion, I'd like to underscore the concept of cosmopolitanism and the co-construction of meaning. Ferrara, on Corriere della Sera, highlights that the Iraq war had a profound impact on both Europe and the U.S. In the aftermath of extensive street protests, European intellectuals Habermas and Derrida pondered what unites Europeans in terms of the international order. They identified a distinct political mentality characterized by an aversion to the use of force, a commitment to law, respect for international legality, and support for a global system built on multilateral institutions and human rights. Rooted in a history marked by opposing principles and centuries of conflicts, this mentality led Europe to develop a foreign and security policy within the EU, transcending the simplistic dichotomy of war and peace.

I believe that Europe's essence isn't solely defined by its politics or institutions like the European Union, but by its people, philosophy, and what sets it apart. This is what Ukrainians aspire to join – these circles of shared values and a future founded on a collective sense of purpose.

Cambray: Indeed, there is an appreciation for the depth of history and the hard-won progress that has been made over time. From a Jungian perspective, there is a gradual shift in dominant archetypal patterns and a growing recognition of the failures of nationalism. The 20th century was marked by a fierce struggle with nationalism, and its resurgence isn't surprising. The challenge lies in forming a transnational consciousness, particularly in the face of global issues like climate change and pandemics. Recognizing that we must consider everyone as part of the solution is crucial. If we don't inoculate everyone against a virus, we can't eliminate it. The mentality of sealing off a small, isolated vessel and protecting it has become obsolete. The United States is still relatively young in this process, and it's early in their history of learning. Their suffering hasn't been as extensive as in other regions.

Carpani: *The Ukrainians' resilience and determination for a brighter future remind me of Rosa Parks, and Mary Watkins' work on liberation underscores the importance of individual growth.*

Cambray: The concept is that one cannot achieve liberation if they don't invest in the liberation of their fellow humans, regardless of gender. This isn't a purely individualistic endeavor. It revolves around an idealized image that serves as a beacon of light and hope in the midst of complex challenges.

Carpani:　Let's aspire to a collective liberation, echoing the wisdom of Mary Watkins. Thank you, Joe.

Cambray:　Wonderful to talk with you, Stefano.

Note

1　Hillman, J. (2005) *A Terrible Love of War*. Penguin Publishing, p. 214.

Chapter 7

Day 52 of War

Dmytro Zaleskyi

Abstract

The conversation with Dmytro Zaleskyi delves into his unique situation as a Ukrainian psychoanalyst serving as a medic in a battalion near Kyiv. His team cares for injured and traumatized individuals amidst the war, and Zaleskyi himself grapples with post-traumatic stress. He believes that resilience is vital for survival in such circumstances. Despite the ongoing conflict, Zaleskyi continues his work as a Jungian analyst. Reflecting on a recent visit to a massacre site near Bucha, he uses the metaphor of a pneumothorax to describe the suffocating feeling of gradually closing in on oneself. This parallels his experience as a psychoanalyst, as he now struggles in sessions with clients undergoing similar traumas. He acknowledges a profound shift in his analytical approach, moving away from transference and countertransference, now providing more of a supportive role. Zaleskyi also expresses a complex mix of emotions towards the Russian people and uses the metaphor of master and apprentice to describe the historical relationship between Ukraine and Russia. He emphasizes Ukraine's advantage as a democratic nation, considering Ukrainians to be more European than Europeans themselves. When asked about the possibility of reconciliation with Russia, he believes it is currently too early for such a prospect.

Carpani: *Good evening, Dmytro – not Dmitry, and that distinction is very important. Today you're in Kyiv. Unfortunately, Kyiv has once again been heavily bombed by the Russian army. This may be in retaliation for the sinking of the most important Russian warship, the* Moskva. *Before we move onto my questions, let's discuss today's events. Today is a significant day for Christians as they recall Jesus Christ's resting in his tomb after his crucifixion and the day preceding his resurrection. Additionally, today marks the commencement of Ramadan. Moreover, tonight is the second night of Pesach, during which Jewish people observe the exodus from Egypt. It's about being free from slavery, but there is once again a lot of tension in the world, whether in Palestine, Jerusalem, Yemen, or Afghanistan. Thank you for speaking with me, especially given your unique circumstances. I initiated a series of discussions entitled* War as Reset, *where I assert that being neutral is not an option. There are many issues where neutrality is not an option. I support Ukraine as it is a part of Europe. And this war is not like the previous ones since it has a distinctly European character. But firstly, how are you?*

Zaleskyi: Thank you, Stefano. I appreciate talking to you, it's a big pleasure. And I think you do a very important job. I will try to give you some answers. Yes, our orthodox Easter is coming next weekend. What I'm doing now: I'm a veteran of war and I was called into the army in 2015. And I served one year. I served as a doctor, a battalion doctor. And now I also proposed my service to the army. It's a big pleasure

DOI: 10.4324/9781003390039-9

that a lot of young doctors now are in our medical service. I'm just serving in the territorial defense system. It's also some part of our system of defense. But this is just a territory south of Kyiv. I'm staying 30 kilometers out of Kyiv. We have an office in Kyiv – my wife Oksana and I. On ordinary days, we just travel to our job. But now I'm staying here and my wife's staying in Ljubljana with Tine Papič's family. I'm very thankful to him. He offered his help. Here I'm working in a mobile hospital. We have a car and voluntary activity. I train people who are going to serve in the army and the defense system of the first aid – tactical medicine. I'm very good trainer, I think. That's just one thing I'm good at: I teach people. I also keep my practice but reduce hours for different reasons and work online. At least I keep my practice. I have no offline meetings, of course. But it's OK. That's my activity now.

Carpani: *I met your family in Ljubljana three weeks ago and wanted to convey my admiration to them, especially your eldest child, for your courage and dedication in fighting for Ukraine. In a matter of weeks, ordinary individuals with regular jobs like doctors and analysts have transformed into soldiers. One of my Ukrainian patients expressed, "Last month, killing was illegal; now it is a must." This resonates with a point you made during our recent Jungian meeting about the intersection of analysis and activism.*

     Could you share your daily routine? What's your typical morning like, and do you head out with your weapon? Given the deluge of misinformation in the media, could you provide insights into what's transpiring 30 kilometers from Kyiv? It's crucial to hear from someone on the ground like you to cut through the sensationalism and overindulgence in media coverage.

Zaleskyi: Just to add to the Catholic Easter: I'm working with my neighbor. My neighbor is a Polish man; his wife is Ukrainian. She is in Poland now and him and I, we are working together because he has a van and is working as a paramedic. My neighbor is a religious man and that is why tomorrow I also have a celebration. We have a Roman Catholic monastery here, and he has been to the service yesterday and tomorrow and celebrated that. We have an international team. I have one person who is from my village. Another person is from Azerbaijan, a very young boy but very talented and I teach him. What we are doing most of the time is to provide evacuation. That's our task. To provide evacuation for people who have injuries and trauma. We follow the Hippocratic rules; we help everybody – civilians and soldiers. Thank God, our area is quieter than in the west of Kyiv. But it is very close and it is a terrible place, Bucha. I have been there three days before, visiting my father; he is staying in a village over there. I've seen it my eyes: There is a lot of destruction and you see a lot of black metal, you know, it's a special color. And I could see this civilian's car with many bullet holes and the burning technique from Russia. Burning tanks, burning machines, many. I think we live in a special kind of psychological state. I have an idea but I'm not ready to elaborate. I think it's a good and Jungian idea if we talk about my state, about the state of my mate, not about term[s] like post-traumatic stress disorder.

Carpani: *What is your mood? How do you feel? What is the morale of your brigade?*

Zaleskyi: I think all of us – not only me, my colleagues who are staying in Kyiv for example, who are just under attack in the left bank of Kyiv, they are attacked and have alarms very often now, especially after the event of the sunken ship of the Russian fleet, it's a very symbolical event; also, it's very good for our defense because this ship was a real threat – but I come back to the idea about post-traumatic stress disorder, which is a very typical approach. I understand when we talk about psychosis, for example, Jungians do not talk about the psychosis in a diagnostic way. We talk about psychosis like some inner part of our wholeness. Everybody has a psychotic part, that is what I mean. And we talk about some psychotic state more than about a psychotic structure. In this situation, I prefer to speak about a kind of post-traumatic stress state. It doesn't mean that this is now. Now I think we are in the trauma, we are getting traumatized now and that is a special state. I think in other situations, we can resolve and maybe digest what's going on now, and it looks like a post-traumatic state. It's a very special state and it's not a disorder, but sometime it looks like a disorder. But the meaning of the state is positive; it's our way of getting resilient. The difference is that this state – if I have some space for that and people who could be resilient for me and could be a container for my state, that could help me prevent this situation to become a disorder. It stays a normal situation. It's like a crossing, like in a fairy tale where one street goes to the left, the other goes to the right.

Carpani: *You're saying that being resilient is helping you keep going. I have a simple question for you that you don't need to answer, but as Jungians, we always ask: How do you feel? What emotions did the dream bring up for you? How did you feel when you visited Bucha and witnessed the actions of the Russians?*

Zaleskyi: I have just been near Bucha. I have a metaphor about that emotion. I am training in tactical medicine and I tell people that when bullets hit the chest, we have a state called pneumothorax. That is when air accumulates in the pleural space between the lungs and the chest, and if you try to breathe, your pleural space gets inflamed, and every breath causes tensions – every time you breathe, you have less and less space for it. It's very dangerous. You survive for five or seven minutes if nobody comes to help you. This is something that I'm feeling in a psychological way because I have a very strong feeling when I'm working with my client. I understand that I do not survive more than 35, 40, 45 minutes. I have 45-minute-sessions with clients, and I feel that it is physically difficult to get through the last maybe 30 minutes. For the last 15 minutes, I feel tension. Every minute, I look at the clock. All of us look at the clock, but this is a different situation. I understand that I did not connect with the situation, the client, the patient. It's my containing function that is reduced in this moment. And it provoked me also to change my style of practice because my patients are mostly also traumatized, but in a different way. And I try to change my practice individually. I change my analytical attitude. I can't analyze it; it's not connected to the dreams. I never analyze transference, countertransference, dynamics like that. I lost the possibility to do this analytical stuff. My analytical style is usually very [oriented] towards transference and countertransference. But that is absolutely nonexisten[t] now. It's a more supportive way to just be with my patient. Some of them just keep me at distance, but all time checking where I am. I understand for them it's also important, the time they are staying. Of course, I have different difficulties working with client in this way. But this is a real deep change. I have lost my analytical position. It is connected to my feelings, also. It is a challenge to my professional position. Am I an analyst, or what am I doing? That is a big question now. I hope our interview will be watched by some important people for me like Francois Martin-Vallas. Because he writes a paper and tries to discuss some positions about Russians with our colleagues. You talk about that, too. That's so funny,

I don't know, it's a very special feeling because I respect Francois very much and love him like friend, like a teacher, like a master, in many ways, and I love his paradox way of thinking and his ability for deep feelings. And you know, now it's a very strange feeling when I read what he writes at some point. I feel myself, like my pneumothorax became narrow because he tried to analyze it, us. But my colleagues tried to write the ordinary letter. But it's not an ordinary letter, it's a scream that you should listen to.

Carpani: *To provide context for our readers, there have been discussions among Jungians, including you, and the president of the Ukrainian development group, Iryna. She penned a comprehensive letter addressing current affairs, and she's not alone in her perspective. Many of your colleagues have expressed their views on the matter, and Francois's response aligns with expectations, prompting this conversation with you. I also encourage other Ukrainians to engage in conversations about recent developments.*

In 2006, I compiled a collection of poems about Russia and Ukraine, including the Maidan protests of 2014, symbolizing a quest for liberation. The Maidan represented Ukraine's aspiration for self-governance and a desire to be part of Europe, driven by a determination to choose their own destiny. It's about embodying European traits, as outlined by Habermas and Derrida during the Iraq invasion. They identified the common thread that unites Europeans: a political mentality distinct from the American one, rooted in shared characteristics. This mentality encompasses an aversion to the use of force, a commitment to law, respect for international legality, and support for a global system built on "liberal" multilateral institutions and human rights. This mentality, shaped by a history marked by opposing principles and centuries of conflict, impelled Europe to construct a foreign and security policy within the EU, transcending the simplistic duality of war and peace. I believe this is what fuels your enduring fight – a commitment to these values.

Zaleskyi: This war is not just happening now. Ukraine never stopped war with Russia. Every generation has been fighting with Russia in a different way. That's a huge part of our history. I think you have the same situation between French and Germans in the West, or the French and British, where there is a long history of war between countries. We have the same long history of war. The big difference for us is that Europe is not imperial. For the USA, it's a big question. Of course, it's a democracy, but there is all this conversation that the USA has some imperial trades maybe. But Russia is the last huge empire with empire thinking. In our relation with Russia, we do not have a subjective position. We are an object. They couldn't understand us. Russians talk about us like we are the same nation. But we have a border with Russia, and this border is not only physical, it is also between people, between organizations, between language and culture. The Russian invasion is a hybrid invasion on a different level. And that's sometimes that happens absolutely not consciously, [but] unconsciously. We had the same situation in our society; after the beginning of the war in 2014, we created the new society, a Ukrainian society. And in this society, there were many Russians, a big percentage of Russians.

Carpani: *As far as [I] understand, the official language was Russian, not Ukrainian.*

Zaleskyi: Not really, our developing group (UDG)[1] mostly uses Ukrainian language. Russian is not prohibited, but 80% of the time, we are using Ukrainian language; that's our mainstream – it's our strategy. And a lot of young people came to us because this is a huge value for them now. The Ukrainian Jungian Association (UJA)[2] have rarely a chance to use Ukrainian because people do not understand Ukrainian language. It's part of them. And that's a real big problem. This is one of my biggest mistakes in my professional life because in 2015, in Avignon, I made a mistake because I've agreed with IAAP (International Association for Analytical Psychology) officers. They persuaded me to take part in and support this organization. For me, it was really huge and bad compromise. And I think IAAP officers also made a big mistake because now we have a very strange hybrid organization. Maybe you understand why I'm using this word hybrid; it's a strong word, I think. We want to create the second union of analytical society in Ukraine. And now we should ask for permission from our Russian colleagues.

Carpani: *This is the coming hybrid future where there will be divisions and polarization among Russians and Ukrainians. What are your thoughts on Russia and Russians these days?*

Zaleskyi: We didn't think about them, truly. We didn't care about our Russian colleagues.

Carpani: *I'm not referring to your Russian colleagues specifically, but rather, I'm interested in your perspective on Russia as a nation and its people.*

Zaleskyi: Look, of course, it is very difficult to feel this tension because there is a lot of anger in us, a lot of hate. You can imagine. If we come back to the situation with my colleagues and Russian colleagues – that's my very personal feeling and attitude. I never told anybody publicly – I want to tell you a analytical secret. Now we have a situation that we can talk about it after many years. What do you think? How many people in this organization UJA are my students or my clients?

Carpani: *I have no idea, but I think many. You are a very busy person.*

Zaleskyi: [One hundred] percent of them. Maybe just two or three people haven't experienced long-term work with me. But all of them have some part of identity from me. I never talked about that. It's a very special relation. That's why I want to present a paper about the master and apprentice relation in Argentina. Because I think it's a big mistake if we are using this metaphor of Abel and Cain. We heard that all time about the relation between Russia and Ukraine. It's like Cain and Abel, the Bible legend. But it is not – we are not brothers. We are absolutely not brothers. Our relation is much more complicated. Maybe it's much better to use the metaphor of master and apprentice.

Carpani: *Or, and it's also a passage in the Bible: master and slave.*

Zaleskyi: You mean Russia as the master? No, because the master is Ukraine, as we are an ancient nation – older than Russia. That is a system mistake because Russians never think about Ukraine like a more ancient nation, like Ukrainians did not exist. All of us Soviet people, in my childhood, all of us, live in this mythology. The same situation is with my colleague here, because my Russian colleagues, they are also my students and I'm a teacher for them. And I have a good relation to them really. We are not in contact and I think now our relation is broken. But it's okay, we are not personally hating each other. But I couldn't understand why they keep their positions there because you know, we quarrel and fight a lot. But they just keep silent and watch what's going on around them. They take no position, give no statement, nothing. There is no position about the future, what's going on, about the future.

I don't know. I really appreciate that you try to talk about that. For me, it's a personal and heavy burden.

Carpani: *Your frustration is evident in your words, particularly since you're currently experiencing bombings. What is your outlook on how this situation will develop? What are your thoughts?*

Zaleskyi: To build what we just did. To develop our Jungian society, just our own – with our culture, with our history, with a connection to the Jungians of the world. That's our way.

Carpani: *And on the ground? You have a military life now. The Russians that are repositioning, you are defending, what do you think is going to happen?*

Zaleskyi: The situation currently develops very slowly, but the change is more towards our victory. I hope so very much. But there are this crazy government and the many crazy people in Russia, and they have nuclear weapons. This is a special state of mind because all of us think about that in a different way. We feel that if we started to win, there would be a greater possibility of nuclear attacks.

Carpani: *How do you feel about Europe? Do you think Europe is doing enough?*

Zaleskyi: Now it's better. We were very angry about the position of before. We also understand that nobody believes that we could stand against Russia. In the beginning, we only had weapons for a partisan war; now we have better weapons and a huge resource of very motivated soldiers. Being a democratic country is our huge advantage. Putin and government couldn't understand that. They couldn't understand how our government just believes in people and gave weapon to people for self-organization. That was a really good decision. Of course, we had some friendly fire in the beginning, but this group of people is the most important resource for our professional military army. Now we have support from the West. That's very useful, and we have enough motivated soldiers and a much better situation than in the beginning. I hope we will win.

Carpani: *I'd like to conclude with a poignant moment from Pope Francis, who dedicated the Stations of the Cross to this ongoing struggle. At the 13th station, where the body of Christ is returned to his mother, two remarkable women stood together. Iryna, a Ukrainian nurse, and Albina, a Russian doctor practicing in Rome, share a deep friendship and held the cross in unison. Typically, during the Stations, a narrator provides explanations of the events. However, on this occasion, silence prevailed. The speaker wisely recognized that, in the face of death, silence often conveys more than words ever could. The image of two women holding a cross together is a profoundly powerful one. Do you see any potential for healing or continuing to stand side by side in the future?*

Zaleskyi: Not this way.

Carpani: *It's too much, too early, right?*

Zaleskyi: For me, it's very difficult to explain. Especially in this situation, I feel this pneumothorax. For me, it's very difficult to explain that Russians with a Western or liberal approach want to hold us back. For us, they are cowards, because we know what it means to go to the Maidan, to stand against a government, against the police, against everything. We know that very well. We know how terrible and frightening it is. I understand that – that it's not good to blame people this way. I understand the position of Francois Martin-Vallas and Andrew Samuels maybe, but for us, it's too early.

Carpani: *Well, it's too late for the Maidan, but it's too early for everything that comes.*
Zaleskyi: Yes.
Carpani: *I'd like to conclude by expressing my deep appreciation. I wanted to relay this message to your son – that he should take immense pride in his father. In contrast, if we, as Western Europeans, were ever called upon to defend our nations, we might have found the notion laughable. However, your courage and unwavering dedication convey a crucial lesson: that, unfortunately, it's a time to safeguard our freedom and not take it for granted. In 1941, Erich Fromm wisely noted that when we relinquish our freedom, those who receive it from us may ultimately evolve into authoritarian leaders.*
Zaleskyi: It's true, yes. Even for the simple people in Ukraine, it's very clear that we are now more democratic and more European than Europeans. I have been to the Museum of the French Revolution in Grenoble and I understand that there is a big distance between revolutionary France and modern France. Of course, the French have this culture, but we are closer to the real fighting for democracy. If you ask [if] me or any one of us could, we would die for democracy. Many of my friends that are on the frontline would say yes. Thank you, Stefano. I very appreciate your interview, and I want to share my feeling and my heart with all Jungians who watch this interview. EAP also needs some changes because the world around us changes very fast, and I think we should think about all the conflicts on the planet. Our organizations and institutions have become old. We should think about some new way or transformation. You with your interviews are representing this approach.
Carpani: *I'm not sure about that. I simply believe we have a moral obligation. I'd like to end with a poem by an Italian poet who passed away 19 years ago. This poem is called Foreign Policy, by Giovanni Raboni. I am translating from Italian: "If you do not wish for others to suffer the same fate as you, then speak up. Standing up in your situation entails fighting." Thank you, Dmytro, and good night.*
Zaleskyi: Thank you, Stefano. Good night.

Notes

1 UDG means Ukranian Development Group.
2 The IAAP – the umbrella organization for all Jungian societies around the world which supports high standards of education and provides the compliance with the ethical norms and develops the interests towards analytical psychology – accepted the "Kyiv Development Group (KDG of IAAP) by IAAP in 2004 as the official group supported by IAAP with the purpose of the creation of the professional analytical society in Ukraine. The members of the group were citizens of different countries such as Ukraine, Russia, Belarus, and Moldova. The group was later named Ukrainian Development Group (UDG). The second development stage was marked by the availability of the Router program in Ukraine. The Router status is given to the participants of the international analytical training of IAAP in the country which does not have its own professional training society. The first group recruitment took place in 2008, and the last Router recruitment took place in 2016. The following meaningful development stage of analytical psychology was the appearance of certified analysts in Ukraine, the Individual Members of IAAP (IM of IAAP).The first analyst was Dmytro Zalessky in 2010, as he had received and completed the Russian Router training of IAAP in Moscow. After that, three analysts finished their training in Kyiv and were certified in 2013: Inna Kyryliuk, Elizaveta Molostova and Konstantin Slepak. Six more analysts finished their education and were certified in 2016: Iuliia Sharipova (Kobzieva), Olga Sikorskaya, Tatyana Pervouchina, Sergii Tekliuk, Mark Kriukov, Yuri Danko. After this, the last but one stage of support of IAAP in Ukraine became accessible – the creation of the professional society without the training status. In 2017, six analysts, after finishing their training in Ukraine, took upon themselves the

commitment and the responsibility to unite and register the professional society in Ukraine, the Ukrainian Jungian Association (UJA). UJA was accepted as the Group Member in the International Association of Analytical Psychology (UJA, GM of IAAP) in 2019 at the XXIst Congress of IAAP in Vienna, Austria. The separate status of all UJA members has changed from individual membership after the certification (IM of IAAP) to the group membership of everyone in the IAAP (UJA, GM of IAAP). All the UJA members started their way from the membership in the Kyiv Development Group. And also, all of us have earlier been Routers in the IAAP training in Kyiv, Ukraine.

Chapter 8

Day 55 of War

Dmitry Kotenko

Abstract

Dmitry Kotenko, the sole Russian guest in Stefano Carpani's series, is motivated by a desire to reveal the truth amidst the fearful atmosphere in Moscow. He predicts that this war will inflict more damage on Russia than Ukraine. The conversation centers on uncovering Russia's motives for invading Ukraine – is it spiritual or imperial? Kotenko emphasizes the importance of the cultural complex and the Russian soul in comprehending the conflict. He believes that Russia's history is marked by aggression and that the trauma of the Soviet Union's collapse in the 20th century still affects the nation. Therapists in Russia are just beginning to address this pain, he notes. Kotenko sees Ukraine as a heterogeneous society in the process of forging its national identity since the early 1990s. Putin's actions, he argues, have accelerated the unification of Ukrainian society. On a spiritual level, Kotenko highlights the strong historical connection between Ukraine and Russia through Kievan Rus. Ukraine's quest for its identity, he suggests, symbolizes the potential demise of Russia. The bond with Kyiv holds great significance, and Putin will go to great lengths to protect it. Kotenko views Putin as embodying an archetype, and his nuclear threats serve to demonstrate his power. Ultimately, Kotenko believes Putin raises a global question: whether this planet is worth saving or not.

Carpani: *Dmitry, good afternoon and thank you for agreeing to speak with me in this web series I call* War as a Reset. *The title may make you hesitant, but I appreciate that you decided to participate. I must ask, are you certain about showing your face, giving your name and surname? Are you aware of the possible consequences?*

Kotenko: Yes, sure, I'm a little bit afraid of the consequences, but I think it's the right time to speak truth and to take responsibility when you're speaking truth. So, this is a decision, yes.

Carpani: *Thank you very much for your bravery. You are currently in Moscow and your name is Dmitry. A couple of days ago I talked to our colleague Dmytro, a Ukrainian colleague. Dmytro is spelled differently according to one or the other language. You know I'm not impartial, I support Ukraine because I believe this is a barbaric act by the Russian government against a self-governing nation. Can you inform us about the current situation in Moscow and throughout Russia? What is the overall feeling like?*

Kotenko: There is a like a general feeling and atmosphere of illness in Moscow. Many people run away from Moscow, especially very active people and those who had a political opinion on this difficult situation. Now in the center of Moscow there are so much less people to meet and many abandoned places like shops and restaurants now look different.

DOI: 10.4324/9781003390039-10

Carpani: *What's going on? Are people finding shelter, hiding, or getting ready for an attack, or is something else happening?*

Kotenko: People are trying to understand the new reality, but it's rather difficult because it seems like nothing has changed in Moscow. That's like a regular life but yes, it's like a sheltering in memories about the reality. And now most of the people are not ready to face the new world, the new reality which is true for Russia and Moscow.

Carpani: *Can you clarify what you mean by "the new reality?" You mentioned it twice. What are you referring to?*

Kotenko: There is no future if you live in Russia. This is a new reality. There is no future for your children, for your family, for yourself. You cannot think about making any plans in Russia now. You only live in the present. There is no past, there is no future – only the present. It's like a spiritual practice, like yoga.

Carpani: *Many people are denying the events currently unfolding. President Putin announced his intention at the end of February to conduct a military operation in Ukraine aimed at removing all traces of Nazism and overhauling the leadership. Was his statement taken seriously?*

Kotenko: Russia is a very big country. There are so many different people, and you should understand that the Russian government uses a very strong weapon: propaganda. It's used against Ukraine and against the people in Russia. This is the real weapon. Our television is a 23-hour propaganda channel; they use it just all day.

Carpani: *What do you mean by propaganda? Can you explain what is happening clearly?*

Kotenko: The propaganda is a way of telling a reality of what is going on now. If you want to know something about reality and if you need to feel the earth under your feet, you need something. And if you do not have the time for research, you take the propaganda they give to you and their explanation and way of thinking about what is going on. It's like producing boots: You can do it by yourself if you are master in shoemaking. But if you're not, you go to the shop. But in Russia, there is only one shop where you can buy only one type of boot. If you are not ready enough to produce your own opinion, to produce your own boot, you go to the shop. This is now happening. This phenomenon is not new in Russia, it [has] roots in the Soviet era. Some people have an antidote for propaganda – some not.

Carpani: *Those who don't have it, believe the propaganda. This is not a recent phenomenon and is not only prevalent in Russia. These discussions aim to delve deeply into the current situation, instead of solely labeling Russians as antagonists as portrayed in Hollywood movies. The purpose is to understand the reason why war has resurfaced in Europe. My previous conversations underlined that war is the opposite of the "Vanity Fair" – the "Atrocity Fair." What is war?*

Kotenko: I think war is the only one way that people know how to react to aggression. This is very common for Europe, and it was very common for Europeans to be involved in a war. I think this is like a common reaction to it. We have been in a in a very peaceful situation but now we need it to react to aggression. Thinking about Jung that Russia is a shadow of the West is very useful in this situation. Where is the European aggression? If you do not act aggressively, aggressions come to your house. I think, from the depth level, this is very helpful to understand what is going on.

Carpani: *Can you explain further for those who may not be familiar with Jungian language? Your previous statement could be misinterpreted as being supportive of Russia or Putin.*

Kotenko: To go through this very awful situation it's very important to take in the mind the two polarities, not only one. So, if you're saying that you are pro-Ukraine, that you are on this side of the conflict, then your shadow is Putin, and you should take care of the Putin inside you. It is very important to not forget that it influenced you. This is something you put in the basement, and you should check what it is in your basement that you should take care of.

Carpani: To explain in simple terms for those who are unfamiliar with Jung's theory: You state that you are Russian and your shadow lies in the West, which means Europe. Your shadow is the part of yourself that you deny or reject. And our – the West's – shadow is in Putin. As a Jungian, I understand this concept, but what about Ukraine? Was Ukraine attacked without reason as a sovereign nation towards the end of February? Do you believe there was a reason for the attack?

Kotenko: Ukraine wasn't attacked in February this year. Ukraine was attacked in 2014; it was much earlier.

Carpani: Correct, one can say that. However, the scale of this invasion is massive.

Kotenko: In 2014, it was a shadow operation, a covered invasion of Russia. But it was still an invasion of Russia at that time, because Russia said that there were no Russian soldiers in Ukraine. The reaction of the West was very interesting. The West said: OK, it's something Ukrainian going on there. And now the February situation is the consequence of ignoring the Russian aggression in 2014.

Carpani: Do you mean the impact on the West or on Russia?

Kotenko: The consequence for everybody. Because this war is more dramatic for Russia, this war is destroying Russia much more than Ukraine now.

Carpani: Let's emphasize the importance of respect, considering the tragic loss of lives on Ukrainian soil, with people enduring unspeakable horrors. I've often noted in our discussions that Ukraine – notably during the 2006 and 2008 elections with figures like Yanukovych and Tymoshenko, and during the 2014 Maidan protests – has unequivocally expressed its desire to align with nations where the people can choose their leaders and hold elected officials accoun[table]. The rejection of authoritarian rule and the pursuit of democracy have been consistent themes. I share your view that Europe and the U.S. largely disregarded these pleas. It's only now that Europe is paying attention, driven by genuine concerns over the severity of the military intervention, air raids, loss of life, and the heart-wrenching images emerging from Ukraine. Yet, amidst all this, we must contemplate the deeper significance of this war.

Kotenko: There are so many levels of meaning, and so many deeper levels. First, you said that the Ukrainians do something or want something. But you should understand that Ukraine is not so homogeneous. There [are] very important reasons for today's situation. The Ukrainian society had severe split problems as a consequence of the Soviet Union. The basis of the process that Ukraine is going through now is the process of obtaining the identification of the nation. It's like a very painful birth of the nation – and indeed it's awful, sure, but in this case, we can say thanks to Putin for allowing Ukraine to be united now and quickly obtain this identification and to understand that this is Ukrainian language, this is Ukraine, and I will pay my life to save my country. I understand this is my country.

Carpani: Allow me to clarify that I never suggested Ukraine is a homogeneous country. Drawing a parallel, as an Italian, I can empathize with the historical complexities. Just

a little over a century ago, the northeast of Italy, parts of Poland, Galicia, and the
Czech Republic were part of the Austro-Hungarian Empire. Consider this: What if
the Austrians were to attack Italy today, or if the French attacked Germany, prompt-
ing a German response over Alsace? The point is, Ukraine is diverse, with strong
Russian connections in the East, which may also have Polish or Hungarian ties.

Even before 2014, Putin couldn't accept Ukraine's aspiration to align with the
West, based on values that encompass a unique political mentality distinct from
the American one. This perspective, rooted in European thought, was outlined in
a study by Habermas and Derrida. It includes a collective aversion to the use of
force, a resolute commitment to law and international legality, and support for a
global system anchored in legal multilateral institutions and human rights. Shaped
by a history marked by opposing principles and centuries of conflict, this mentality
propelled Europe to develop a foreign security policy under the EU, transcending
the oversimplified choice between war and peace. It's this shared vision that draws
Ukraine towards the West.

Kotenko: If you ask about the depth level, you need to investigate the symbols that appear
now or had before. Ukraine and Russia are so closely connected, and they have such
a long story. In the beginning there was only Kyiv, then Moscow. Kyiv and Moscow are just like mother and child. And you know, communication could be like a
separation process between mother and child, but the child would like to just eat the
mother. Kyiv is a sacred place for Russians, because the baptism of Saint Vladimir
happened in Kyiv; hence, it established the Kyiv Rus. So, it was the roots of the
Russian identity. And if the Ukrainians say: OK, we have our own identity, we'd
like to have our own state and to have our own language and politics – this means
death for the Russians, because they lose a sacred wheel of the Russian civilization
or for Russian ideology in general. It is very important for Russia to be connected
with Kyiv. The main sign was that Putin put the monument to the Saint Vladimir
close to the Kremlin in Moscow. This has no sense if you can look at this because
Moscow wasn't there at that time when the Saint Vladimir was in Kyiv. There was
only forest at that time, but Putin said: Yes, we have the right to have Kyiv; we need
it. Without Kyiv, there is no Russia. This war is so dramatic because Putin will do
everything to take Kyiv. He will kill everybody, he will destroy everything, only to
take a sacred place. If you want to understand what is going on, you need to go to
the spiritual level.

Carpani: *Yes, but wars were waged for religious or spiritual reasons for centuries. And they*
were wrong.

Kotenko: Wrong from what point of view? It is wrong from the human ego, yes. But if you
look at it from the Self – yes, this is a reason. We need it. We need this war all
together: Europeans, Russians, Ukrainians. Everybody needs it, and everybody benefits from it. Yes, in a very bad way, but we do.

Carpani: *True, this is why I called my conversations* War as Reset.

Kotenko: Yes, when was Europe ever so united before this crisis?

Carpani: *Yeah, you're right. Europe wasn't united and neither was Ukraine.*

Kotenko: Let's look at the relationship between Poland and Ukraine. How else was a real reset
of this relationship possible?

Carpani: *You're aware that Poland had a negative reputation within the EU up until three*
months ago. [Hungarian Prime Minister Viktor] Orban was also viewed as a

negative figure, and he still is. However, my main point is regarding the homogeneity of Ukraine. Although it was not initially homogeneous, even the pro-Russian Ukrainians are now in support of Ukraine. This was unexpected. But do we need to go to war like this? This is the question I ask you: Is it necessary?

Kotenko: You know, there is no logic. You are trying to use logic to answer these questions, but there is no logic. Within everything happening now, there is no logic from the ego level. If you'd like to have a new territory, you shouldn't destroy it. You need to have infrastructure; you need to have hospitals, houses where people can live. You need this territory in a good condition, so you shouldn't bomb it.

Carpani: *Do you realize that your message might be misinterpreted if not expressed clearly and firmly? Some might associate your stance with Putin's spiritual motives regarding Kyiv. However, this conflict goes beyond mere warfare. In my view, it's a genocide, akin to the Holocaust, as you've pointed out before. Putin's intent seems to be the destruction of everything and everyone. Pursuing a spiritual objective, or even a political one, like the U.S. and the UK had in their invasions of Iraq, Vietnam, or Korea, doesn't seem fitting. There's another crucial aspect I want to address: soft power. The United States has employed soft power in Europe and worldwide since 1945, building strong relationships through collaboration with people and cultures, rather than top-down approaches involving bombings, as witnessed in Hiroshima and Nagasaki. Scandinavian nations, in particular, excel in wielding soft power. Since the fall of the Berlin Wall, Russia has struggled to exercise (any) power and integrate into the community of nations without resorting to aggression and violence. In our childhood, we watched the Rocky movies on TV, featuring an Italian-American boxing hero and his adversary, Drago, depicted as a cold-blooded Russian. I always disliked this stereotype, but it appears to hold some truth.*

Kotenko: You know the story of Russia is a story of constant aggression. There were so many wars all over the borders of Russia. It was very common for Russia to be involved in different kinds of aggression. You can see on the map how big is Russia. It is still the last empire on our planet and all this territory was united not solely in peaceful way. As a rule, it was by Russian aggression to take a new territory. It's a kind of identity of Russians to be aggressive.

Carpani: *The pattern of empires exerting aggression and asserting territorial dominance is an age-old narrative. In our past conversations, I've probed whether the current war and aggression significantly differ from historical actions undertaken by the UK and the United States, citing instances in Chechnya, Georgia, Kazakhstan, and now Ukraine. What emerges is a stark absence of democratic resolutions. You've mentioned concerns about potential penalties or even incarceration due to our open dialogue. This raises a fundamental question: Why is it so challenging to voice discontent, label our leaders as dishonest, without the looming threat of fines or arrests? This is my fundamental concern, and I sense that our conversation might be veering off course.*

Kotenko: You don't understand it because your live in the West. You don't understand what is going on with the Russian soul. This is one of the cultural complexes. I read about the lawlessness complex; this is a very important point of view on the world what it means to be Russian. It's just to behave illegally; law is nothing in Russia – you should understand [this]. It's not possible to talk about democracy because it's like there is no place for it. You shouldn't make the mistake to think about the Russians from the European point of view, but you can think about the Russians like Christians. If you talk about war as a reset, I think this is the answer for the problem of Western civilization. This is like a new chance for the West to overcome the limits that this civilization already reached. I think Putin said exactly these things about the war against the West, against the Western values, against the value of Western civilization. When in this case I say Putin, I don't mean the person but an archetype, the manifestation of the collective unconscious in Russia now. This is very important to understand, that we live in an apocalypse scenario and Putin has the ambition to be like the Antichrist. He talked about nuclear weapons for this reason – to show that he has the power to change this world dramatically, fully destroy it, or bring it to a new order, to bring a new reality to the world. So, he has a power.

Carpani: *If I grasp your point correctly, it suggests that following Putin's lead could lead to a continuous state of ungrounded expansion until a breaking point is reached, possibly resulting in catastrophe. Consequently, it's evident that merely condemning Putin's actions is insufficient. Instead, we must remain vigilant not only against potential threats from Putin but also from other aggressive empires with a history of violence and death.*

I often question the distinction between the ongoing war and the actions undertaken by the United States since World War II, such as in Vietnam, Iraq, Afghanistan, and Korea. Additionally, I inquire about the fine line between invasions, occupations, and political destabilization carried out since 1946 under the guise of liberation and, more recently, since 1991 under the pretext of spreading liberal democracy. Maurizio Ferrara (Corriere della Sera) on March 7th, discusses the U.S. invasion of Iraq, likening it to a premeditated carnage, much like the situation in Ukraine over the last decade, which unfolded after months of preparation. He also draws attention to Nobel Prize laureate Svetlana Alexievich, who has painted a clear picture of imperialist impulses, not just in Putin but in a significant part of the Russian population, referring to it as the enduring legacy of the Homo Sovieticus and, before that, the tsarist tradition. I then delve into the imperialistic actions of the USA and the British, who, after invading and occupying countries, often left a trail of problems and civil wars in their wake, citing examples in Kenya, Cyprus, and Palestine. The consequences of these actions, resulting in loss of life, appear to parallel the actions of the U.S., Putin, and the British crown. While colonial Britain pursued occupation for resource acquisition, the U.S. often invaded countries for geopolitical motives. Putin's spiritual motivations for invading make the situation even more complex and ominous.

Kotenko: Yes. I think this is a very important moment to stop and understand what the message is that Russia and Putin send out to [all] of us. That is the question we should find an answer for. This is a spiritual question. We need to think about our planet, our lives, about everything that is going on. Is this planet OK to be saved, or not? This is the question. This is a global question from Putin to [all] of us.

Carpani: *You've mentioned the notion of a spiritual war, but is this truly a spiritual conflict, or is it primarily about power, an imperialistic struggle? To illustrate, we can consider the Orthodox Church's involvement in Africa, which seems more about seeking influence and power rather than a spiritual endeavor. What has prevented Russia from achieving transformation and developing a unique identity through peaceful means over the past three decades, a path that some other Slavic-rooted countries have successfully pursued? During my interviews, I consistently pose the question: What would have been the fate of Russia and Europe if Gorbachev had continued his policies of Perestroika and Glasnost?*

Kotenko: I'm sorry, Stefano, but you know the tragedy of Russia is that it was too traumatized in 20th century. [There] was so little time to heal the trauma that Russia suffered in 20th century, so many millions of people were just killed. Unfortunately, we are still going through re-traumatization. I feel we're just trying to heal in this awful way, but we are trying to do it. As a therapist, I can say that now many of my clients and my colleagues' clients are feeling relieved because now they can talk and feel the traumata of their families, of their grandparents. Only now we are ready to talk about it. Only now we can touch it – in this awful situation, yes, but it seems like there is only one way. This is the reason why Russia wasn't able just to go through the transformation, unfortunately. You know that in the center of each trauma there is a psychotic content. Eventually, you can observe it.

Carpani: *This is a pivotal point, Dmitry, that the Western world needs to grasp. The 20th century was marked by widespread violence, and it's challenging to ascertain whether Russians suffered more than Western, Northern, or Southern Europeans because suffering was widespread. The foundational traits of Europe that I mentioned earlier should serve as the basis for the reset that we, as Europeans, require. I also ponder if your remarks tie into the betrayal of King George [V of England] and his cousin Nikolai [II of Russia], particularly the moment they exchanged uniforms, until King George disassociated himself, leaving Nikolai to confront his own revolutionaries. It seems like a lingering resentment between Russia and Europe.*

Kotenko: Yes, and I like to talk about the war not just a reset, but about a way of transformation of the relationship, which is not good.

Carpani: *It's a long way, isn't it?*

Kotenko: Yeah, but I hope there is a time for this now.

Carpani: *We can engage in dialogue, despite our differing perspectives, stories, and histories. I believe our grandfathers fought in the same war but for different sides. What matters most to me is that if we embrace what you've described, this reset, and let it reach its final conclusion, it would be a truly grim outcome, as it implies a loss of the opportunity for repair. Thank you, Dmitry.*

Kotenko: Thank you, Stefano.

Chapter 9

Day 59 of War

Iryna Semkiv

Abstract

Iryna Semkiv's story is directly connected with the war: A Ukrainian citizen, she fled her hometown of Lviv immediately after the war began. Just two days of warning sirens and constant running to shelters affected her son so much that for some time afterwards, he was still frightened by ordinary car horns while being in safety, she says, explaining her family's war experience. In her work with patients, she notices a big change. None of the participants – including her – would have experienced war before, and each person would try to adapt in his or her own way. She found some striking patterns: young women would suffer from their husbands putting all their libido into the war. She even recognizes a love triangle whereby the war is another woman in the relationship. Semkiv highlights the feminine article in the Ukrainian word for war, "viyny," and notes the recurring dream image of a divided man, part savior and part perpetrator, revealing women's suffering from male inattention and their desire to support and fight alongside men. The psychoanalyst does not try to predict how the war will develop, and continues to work and lecture from Poland, where she has found a temporary home. When asked if it would be possible to repair the relationship between Ukrainians and Russians in the future, she strictly denies it – to even think of making amends with Russia would be tantamount to identifying with the aggressor. Her thoughts would revolve around how to survive and keep Ukrainian culture alive.

Carpani: Good morning, Iryna, not Irina – and that is very important.

Semkiv: Right! It is nice to hear my name and pronunciation in Ukrainian language (Iryna) instead of Russian (Irina).

Carpani: *Let's focus on your story. I want to emphasize that I stand in support of Ukraine, and I view Putin and Russia's aggression as highly concerning. As Jungians, it's essential to begin any analysis by delving into the symbolism and archetypes at play. We must unequivocally condemn the ongoing violence. It's vital not to repeat Jung's mistake of approaching his analysis of National Socialism and Hitler from an intellectual standpoint and being misconstrued as sympathetic to their cause, which was not the case. Iryna, could you please share your personal journey? Where are you from, and where are you currently?*

Semkiv: Now I'm in Poland, in Toruń, with my son and with my colleague, my friend, and her children. We live together now. We left Ukraine right after the invasion started. We hardly experienced the circumstances, but I'm still there all the time. In Toruń, I work as an analyst, keeping my practice online. We have a great support in Poland from our colleagues at university because I'm not only analyst and psychotherapist, I'm also an academic teacher at university. Our colleagues were very supportive. They gave us a flat to live in for this time. And now we are trying to adapt.

DOI: 10.4324/9781003390039-11

The university here supports academics. We can do research regarding the war and get financial support. What is good is that our children can go to kindergarten and school. They have the opportunity to integrate and communicate with people. This is the nice part of all this; we meet nice people here.

Carpani: *It seems you've managed to find a secure location, and those around you are supportive and empathetic. Let's take this one step at a time. The conflict started on February 24th. Can you tell us where you were at the time?*

Semkiv: I'm from Lviv in the western part of Ukraine, very close to Poland. That's why it is easier for us to come here.

Carpani: *Lviv hasn't faced as many attacks as other cities, but you chose to leave as soon as the war began. Why did you make that decision?*

Semkiv: For me, it was difficult because of the alarms that happened all the time. When they started, we would have had to go to shelters and hide. When you are an adult, you can do it whenever it happens. But when you have a child, it is very stressful. My son was stressed. He was afraid of those alarms, even after two days. It was enough for him to feel anxious. When we came here to Poland, our kids were reacting to all kinds of noises. When cars were honking, they were reacting, looking for something in the sky. This was after two days. I can't even imagine what is going on with children who are still in Ukraine and under attack and have to go to shelters all the time. It's difficult.

Carpani: *How did you feel when the war began and the alarms sounded? I understand it's a challenging question.*

Semkiv: It is a difficult question. I will give you, I think, a very strange answer because my first reaction was that I was not reacting. It was a very strong defense because when I heard those alarms, I thought that it is not alarms because rockets were shot. I thought that these alarms were because war started and people would have had to know that war had started. In my imagination it was an informing alarm, but they were not. They were real alarms. And when my son woke up because of those alarms, he ran to me and asked why we were not going to the bombproof shelters, I told him because it is an informative alarm. It was my explanation, very defensive, very rejecting of the situation. And then he told me that at school, they said that whenever there is an alarm, people would have to hide. Only then it came to my mind that it is a rule that we have to hide.

Carpani: *The attack was almost unbelievable. Can you describe what you did after the assault began? I believe it's important to hear your firsthand account, rather than relying solely on the often banal news reports.*

Semkiv: One morning, we were at home – me, my husband, and my son. My husband went to fill his car. It was half past 5 in the morning, and there already was a very long queue at the gas station. He was in the queue for three hours. People were reacting to the start of war by filling their cars with gasoline. That day, I still had several online consultations with patients. We all were shocked. The next day, I asked my patients to postpone and we were preparing to go to Poland. It was not an easy decision, you know? When I found myself hesitating, I was surprised. I thought: "How can I go when everyone is staying?" But on the other side, I had to protect my son. So, it was a very strong inner conflict, which was very painful in that moment. When I found myself in this situation, hesitating, I thought that when in peaceful times, I would not have hesitated to go to Poland because I was studying there. I'm still in the training in Poland, I'm used to go[ing] to. But at that moment, I was so hesitant and I told myself that I can go to Poland – and if I feel like, I can come back. This thought helped me to take my son and go. Only this thought that in [any] moment of the future, I can come back if I feel like that.

Carpani: *Your husband went to fill the tank of the car. There was long queue. What happened then?*

Semkiv: My husband wanted to send us to his relatives in Poland. So, we get in the car and drive towards the border, together with another befriended family. And as we are in the queue at the border, our friends decide not to go. We also decide to return. But then I called another friend with a car. And then we decided to go together – two women and three kids.

Carpani: *Why did they decide to go back?*

Semkiv: The plan was that the mother of the other family and I get all our children in one car and drive to Poland while our husbands go back to Lviv. It was our plan, but they decided not to go. And then I was alone. I thought that it would be difficult to reach relatives after the border, you know? And when we are going, we thought that in a – in a week, we will come back. We didn't think that we were going for two months. But I called my friend Oksana and we decided to go together with our children, using her car.

Carpani: *How long was the queue?*

Semkiv: We were in queue with three children for 36 hours: two days and one night. We arrived at our relatives in Poland in the evening of the second day.

Carpani: *Why did it take so long? We know that there were loads of people at the border. What happened there? Did the military or police stop you?*

Semkiv: Nothing like that. It was very organized, it was just a lot of people – 30 kilometers of cars queuing. I thought that we would have waited even longer. But my friend Oksana has a small child, less than 3 years old. That is why it was easier to cross the border. It still took long.

Carpani: *Can you tell me how you felt during the 36 hours?*

Semkiv: I think that in that moment, we all were in mania. It was very exciting. Just after crossing the border, I touched my depressive feelings and I felt how maniac it was before crossing the border. I think that these primitive defenses helped us to do everything we could.

Carpani: *When did you say goodbye to your husband?*

Semkiv: One day before crossing the border, but we are in touch all the time, talking.

Carpani: *You said earlier that you planned to be away for one week. But it has been almost two months now. Is this also a way of protecting yourself? Are you hoping that this situation will end soon? What is your plan for the future? How do you think this will develop?*

Semkiv: I try not to predict, because it is very difficult to set deadlines for being somewhere. I think that I will see how it'll be. I decided to keep my life going from where I am. I still work online at the university in Lviv. I lecture, and I am still the director of the master program of clinical psychology. We organized support for people, crisis counseling where our master students – those with a certificate – are counseling.

Carpani: *You're still very busy. Maybe this is also a part of the manic resistance.*

Semkiv: Yes, of course.

Carpani: *It's really helpful. What do you observe in your patients? You said your son feels anxious, which is understandable. How are your patients?*

Semkiv: Oh, it's very different. Sometimes I'm very surprised about what I see and observe. Everything is new. We are not used to liv[ing] in a war; we are used to liv[ing] in peace. Every person is trying to adapt in their own way. What is interesting for me, when I was looking at previous records and conversations of your web series, when you were asking your guests about whether masculine or feminine sides of wars, I observe some interesting patterns. My clients, who are young girls or women, are suffering that their husbands give all libido, all their energy, to the war. They react with very huge envy. It's like this war is like another woman for my clients' partners – as if the husbands betray them with the war. Men are giving a lot of energy to the war, also psychic energy. Some of my clients even have some image that maybe they really find something else. I think that it is also defense.

Carpani: *That's logical. What you're saying is because the goddess of war in Greek mythology was female.*

Semkiv: And war is female in Ukrainian language.

Carpani: *Also, in Italian! You're only the second woman I've had the chance to speak with about this. I reached out to many men and women, and it seemed women may find it more challenging. The women I contacted expressed that they might not be ready to talk for another month or two, while the men were more eager to participate. I appreciate your willingness to engage in this conversation. I recently read an article in the Italian newspaper* Repubblica *by Arianna Farinelli on April 16th. She wrote about women challenging regimes, including figures from global history like Aung San Suu Kyi [of Myanmar], who won the Nobel Peace Prize, endured imprisonment, became a prime minister, and was imprisoned once more. We can also contemplate the struggles of women like Svitlana Tsikhanouskaya and the [feminist punk rock band] Pussy Riot, who are fighting in their own unique ways. Additionally, we can consider the female prime ministers who led their countries during the onset of the [COVID-19] pandemic and how their nations and citizens seemed to fare better. I'm curious to hear your perspective on what sets male war apart from female war.*

Semkiv: One thing is the supportive position of women. Females feel like they have to give a lot of energy to be close to their men. Another point I want to make stems from my clients' dreams. Some dreams now show images of split men. There are some features who show men as saviors, other men who tried to abuse. For three weeks, I get such kind of dreams by my clients now. I think that shows that on one side, women are suffering because they lack attention and investment from their husbands as their minds are at war, and on the other side, they try to support with whatever men need from them. I also observe that women try to enter the military system to fight alongside men. There are many thoughts in my head of the gender of war, as you see. Some women are volunteering and supporting, others are trying to protect children, some take care of the protection for woman. One of my clients is supporting like that: She is volunteering to go to Poland and look for what is useful for the army and get it. People are trying to do what they can. Psychologists are giving consultations. At some point, there were more people willing to help than things to do. People even started to suffer when they could not invest as much as they want during the first month.

Carpani: *It's indeed intriguing, especially considering our earlier discussion about the emotional conflict of whether to stay or leave, and the jealousy some Ukrainian women experience when their men go off to war. I recently stumbled upon an article by an investigative journalist from Radio Svoboda that left me utterly shocked. He uncovered conversations between a soldier and his wife, Roman and Olga, who had moved to Crimea a couple of years ago. In these conversations, Roman, a young man in his 20s, asked his wife for permission to sexually assault Ukrainian women. Astonishingly, she agreed, with the condition that he use a condom and not inform her about it. When he inquired again, she insisted he shouldn't tell her. This exchange underscores how war affects women as well. So, my question to you is: Why does war transform ordinary people into beasts? This seems to be another illustration of that phenomenon.*

Semkiv: I remember that this was after Bucha. And it reminds me of custom of tribes. When a tribe was trying to get the land of another tribe, the symbol of conquering this land was to rape all women there. It was like a symbol of occupation and power – this land is now our land. I think that this is what was going on in Bucha, something like that. It was very primitive and very symbolic: It is close to Kyiv, the center of Ukraine – the main castle, so to speak.

Carpani: *Discussing basic concepts is compelling. Aristophanes wrote about Lysistrata around 400 BC. She was a remarkable woman who excelled in strategy. Athens and Sparta were at war for 20 years, and the bloodshed seemed unending. Lysistrata had a plan to persuade women from Athens and Sparta to end war by halting sexual relations with their husbands. While the story is fictional, it demonstrates that women possess significant power and are capable of influencing men. The tale prompts us to consider the reason behind the conflict and its underlying purpose. What is this war about? What is the ultimate meaning?*

Semkiv: It is a very difficult question. I don't know how to answer it. What I can I say is that we have [had] a very difficult relationship with Russia for several centuries. It is about differentiation. Many Europeans call Ukrainians and Russians brothers. I think that with this war, we all can see that we are not – that we have a border, different cultures, different views, and that we are different. In 2014, all [of the] world thought that it was our inner war, an internal political issue.

Carpani: *It's almost akin to a civil war situation. Europe and the West initially didn't grasp the gravity of the situation or the depth of Ukraine's desire for independence. The Ukrainian aspiration for self-determination, beyond merely relocating to Western Europe, has become abundantly clear. If we look at countries like Switzerland, Finland, and Sweden, they are actively considering NATO membership for security reasons. I believe that the events on Maidan in early 2014 signified Ukraine's wish to align with a part of Europe where the political mindset differs from the American one, emphasizing common traits (as underline by Habermas and Derrida): an aversion to the use of force and a strong commitment to law and international legality, support for a global system rooted in liberal multilateral institutions, and human rights. This mindset, forged through a history marked by opposing principles and centuries of devastating conflict, was what compelled Europe to develop a unified security policy under the EU, moving beyond the simplistic dichotomy of war and peace. You, too, aim to be part of this vision. Ukraine boasts a rich history. Looking back to the 19th century, the northeastern region was under the sway of the*

Russian Empire, while the southern part fell under the influence of the Hungarian Empire, featuring diverse languages, religions, and a vibrant Jewish community. The Austro-Hungarian Empire's management of such diversity serves as a parallel to modern Europe's efforts to maintain peace. However, the absence of alternatives has again pushed nations to war, viewing it as the sole means to safeguard their sovereignty.

Semkiv: When you are talking about this . . . : In the western part [of Ukraine], we have different nationalities: Austrians, Polish, Hungarians, Romanians. This western part is still using Ukrainian language as everyday language. And in the eastern part of Ukraine, Russian was spoken for only for 40 years, but the majority speaks Russian. In those 40 years, Russia occupied our culture more than those other countries for many more years.

Carpani: How do you feel now about Russia and the Russians?

Semkiv: It's very difficult to talk about feelings. There is anger and hate – and a lot of disaster, even. I don't even want to talk about them. I don't want to even think about them. It is like that, but I try to keep those emotions on a cognitive level. But I don't know if it is good idea. Maybe it is better to keep this emotional level.

Carpani: I sense reparation is far from being possible, right?

Semkiv: To think about reparation now is like identify[ing] with the aggressor. I'm not thinking in this category. I am thinking about how to survive, how to keep our culture alive. I think about how to protect our society from death and keep teaching and talking with people – but not about cooperation. I don't think that there is [any] time for that now. I also notice in conversations with my clients that there are those who are more used to identify with the aggressor are trying think to do some cooperation – to write to their friends, for example.

Carpani: When journalists discuss Ukraine and Russia trying to mend their relationship, it reminds me of a metaphor about this theme by Yaryna Grusha Possamai: a woman being forced onto a stage, her hair pulled, to confront the man who abused her just the day before. The audience demands a kiss and a hug from her, wanting her to make peace with her attacker. Western viewers often desire to return to normal lives swiftly. The continuous coverage of wars on television and the recurring stories of girls being raped and abused have drained us. However, we must remember that the girl who was assaulted yesterday is still in pain. Her wounds are still fresh, and her lips remain swollen. She stands on that stage, wondering how much more she must endure before anyone takes her suffering seriously. Possamai suggests it's time for psychologists, psychoanalysts, and psychotherapists to step in and explain what is happening to this woman and her country. This is why I wanted to speak with you. She believes dialogue is impossible for now, as you mentioned. Perhaps, one day it will be possible, but only when our wounds have transformed into scars of the past.

Semkiv: Yes.

Carpani: I thank you very, very much for our conversation.

Semkiv: Thank you very much for your invitation and for your work. What you are doing, looking for the meaning of what is going on, is very important.

Chapter 10

Day 69 of War

Vickie Sims

Abstract

Vickie Sims is a priest of the Church of England and gives a definition of war in which she quotes the Old Testament, in which war is seen as part of the human condition, the routine of life. It is also, she says, a product of the inability to negotiate needs as well as the ability to go to war. Sims herself sees it as a result of insecurity and fear of not having enough. War, she says, is never between two individuals, but on a larger scale between leaders who would then draw their people into the conflict. Religion cannot provide an answer to the bestial behavior of people in war, she says. Yet, he said, atrocities can be seen in wars around the world. Even in Europe, where 8,000 Muslim men and boys were killed by Serb units in Srebrenica during the Bosnian War. Sims thus states that the beast must be innate in our human nature. With Stefano Carpani, she links the COVID-19 pandemic to current events and shares her concern that society has not yet learned to heal the wounds caused by the pandemic. The impact on the collective is still immense, she says, and the experience of the invasion of Ukraine now adds to that wounding. She felt the threat of social breakdown. The concept of co-construction of meaning brings the discussion to the topic of love, which Sims believes can serve as a response to the conflict; love, she argues, is rooted in respect for the other person. To quote the Old Testament again, according to Sims, love shows itself in the law that God gave to people so that they could live in community. The idea of love is not only emotional, but it is about how we structure our society.

Carpani: *Vickie, you are a female priest with the Church of England based in Milan. As Italians, we usually associate priests with being male, which makes your perspective particularly interesting. Could you provide a definition for war?*

Sims: It's interesting that you mentioned that some people claim that war is in our nature. And it brought to mind for me a line from the Old Testament, which is in the Book of Kings; it said something like: In the spring of the year when kings go out to war, and it just took for granted that war was part of human existence and that it was almost part of the routine of life that kings would need to go out to war for whatever reason. War is a product of the incapacity to negotiate conflicting needs of various peoples. It's the product of [and] also the capacity to go to war. So, if you are able to exercise power and think you can take what you want by force, then you'll do so. It's also the product of insecurity and fear of not having enough or not being able to meet the needs of the people that you're responsible for. If it's just a fight with your neighbor, then that's not a war. A war actually is something a bit larger scale and a bit more official than that. It does tend to break out between leadership groups, which then bring their people into the conflict in some way.

Carpani: *What are the masculine and feminine elements in warfare?*

DOI: 10.4324/9781003390039-12

Sims: That's a very psychoanalytic question. I think a lot of women in casual conversa-
            tions like to say: "If we were running the world, we wouldn't have all these wars."
            So, there may be intuitively within us this sense that war has an inherently mas-
            culine kind of power demonstrating function in some way. And it's interesting to
            speculate: if women were running the world, would there be fewer wars? But as
            a woman speaking, it would be really kind of displacing the problem of anger and
            violence onto men simply to say: "Oh, well, if women were running the world, there
            would be no war." The feminine aspect of war in some cases seems to be also about
            how the vulnerable are affected by war. The feminine aspect can be mothers trying
            to protect their children. It can be mothers fleeing to safety. It can be mothers whose
            sons are sent off to be fighters. It can be women who – in some cases, in a more
            traditional society – lose the breadwinner in their family. That's looking at it from a
            kind of traditional point of view. I know that now many women are also part of the
            armed forces of countries. But my intuition and my feeling is that there still is kind
            of a very strong masculine element to warfare.

Carpani: *Why do wars turn ordinary people into animals? I read a newspaper article about a
            young Russian soldier at war who asked his wife for permission to rape Ukrainian
            women. She responded by saying, "Do it, but use a condom." This is evidence that
            war can change people into beasts.*

Sims: The thing that you've just recounted really provokes in me a kind of visceral
            response. It's – it's a sickening story. It's sickening on every level and it's kind of
            hard to react to what you've just told me in a rational and receptive way. I'm left
            without words. One could almost imagine without justifying that in the heat of a
            moment in battle when soldiers attacked, women got in the way and then a kind of
            gang rape would happen and people afterwards would say: "Oh, I've – I lost my
            mind. What have I done?" with a sense of remorse. To have asked your wife permis-
            sion to do it and to decide you're going to be beastly in a rational and predetermined
            and premeditated way is so appalling. [If] you are the psychoanalyst, you say some-
            thing about it.

Carpani: *Your emotions are essential, and I believe they are shared by many others. The
            news was difficult to accept, and at first, I wondered if it was fake. However, upon
            double-checking the sources, it became evident that the news was genuine. During
            World War II, this type of warfare occurred, but I do not want to vilify the Russians.
            Instead, I aim to comprehend why war can bring about the worst in people. As a
            religious individual, I would like to know: Does this notion reflect in the Bible or
            in the Gospels? Is it also observable in literature that humans can act barbarically
            when they are desperate?*

Sims: I was thinking if we do get pushed as a human being to a point – maybe in the
            fight-and-flight kind of reaction – where we do choose to push back and to fight
            back in ways that then become beastly. And that's why I mentioned the thing about
            the premeditated aspect of deciding to be beastly, which was so shocking to me. But
            we've seen the war in Bosnia and the Balkans, we saw the beastliness of that, of
            neighbor turning against neighbor. We saw it in Rwanda with this incredible explo-
            sion of violence. If the media reports are true, we see potentially systematic rape
            also in China with the Uyghurs in an attempt to kind of force a kind of integration.
            The scripture does not teach us that war is the answer. Certainly, the teachings of
            Jesus do not say to fight back. It doesn't say to rise up against the oppressor. And so

there may be something more innate in our human nature, which is an aspect of our failings as human beings to be the full human beings in its fuller sense.

Carpani: *Could war be a form of compensation? Consider World War II, a response to Germany's defeat in World War I. It also contributed to the fall of the Austro-Hungarian Empire. This war may essentially continue the longstanding hostilities between Austria and Russia prior to World War I when two distinct worlds coexisted. Some argue that it's a consequence of the Soviet Union's collapse, as Putin mourned its fall. Alternatively, could war, including this one, act as a reset? It's possible to argue that Christ's crucifixion, though barbaric, was ultimately a step forward, leading to salvation for all.*

Sims: I was thinking of the issues around honor and shame and this desire to defend honor and the way in which in many respects, the crucifixion of Christ tried to undo that set of values in which I need to defend my honor, which is part of the compensation aspect: I can't bear that my homeland looks humiliated or it looks smaller or it looks less grand than I had put my hope in. In theory, Christ's refusal to play the game of the political authorities and his willingness to go to a completely shameful death undoes some aspects of this need to defend honor. And yet as human beings, we find it very difficult to avoid doing. In terms of a reset . . . that's a pretty political concept in a sense. There may be elements of it. I could imagine countries that go to war wanting a reset. If there is that sense of not having a rightful place in the world or the rightful respect and power that one should have, one will demonstrate power and one will reset the table. But it's a failure to use the means of dialogue, negotiation, respect for peoples, because one could argue that there was a kind of reset when the Soviet Union split up. Agreements were reached, borders were set, people signed up for a reset and for people to say: "I'm just not going to accept that, so I'm going to go and just sort it out myself" – I suppose that happens often in the world, but I don't think that we can use that to justify that way of going about business.

Carpani: *My view of the reset pertains to Ukraine's desire to move westward, not merely for our way of life, but because they seek a nation with resilient institutions that can withstand attacks from authoritarian rulers or nations. The American experience demonstrated that even with a flawed democracy, the United States endured the attack on the Capitol on January 6th, 2020. What role can the Church – including the Roman Catholic Church or the Church of England – and other religions play, given NATO and the West's support for Ukraine? In the Italian context, pacifists advocate for negotiations. While I believe negotiations are essential, they have been feeble thus far. However, someone seems uninterested in discussions and seeks destruction. How do we negotiate with someone who shows no interest in it? Putin won't halt until he achieves his goal of taking over the entire country and specific regions. What's your perspective and that of the Church? People are divided between sending weapons to Ukraine and advocating for peace negotiations. Both sides overlook the fact that there's an aggressor and a country that has been invaded. Lately, I've been reading a book by Stefan Zweig that delves into the question of why we go to war, especially when considering the tragedies of [World War I and World War II]. It's a solemn consideration, raising the question of how to prevent such catastrophes from recurring.*

Sims: If the Pope, who is a great man of tireless energy, is struggling to find a way forward, you know, it's not as if a little person like me or any other person professing a religion will suddenly come with a breakthrough because you ask this key question: "How can you begin to sit down and talk, call it, negotiate with a person who doesn't want to?" When the perception is that the vision of the world and the vision of right and wrong is so vastly different, that we actually don't seem to have common ground in our vision, that makes it immensely difficult. One of the things that has depressed me looking at things and seeing the media is an excessive desire to just find the white cowboy hats and the black cowboy hats. We divide it into good and bad because we like things to be clean and neat. That doesn't take us very far. The other thing that I found disturbing is what I would call a kind of armchair diplomacy, when in the comfort of my Western democracy and my freedom of speech and freedom of expression, I decide for the Ukrainian people that they shouldn't fight for the borders of their country and their freedom because it doesn't suit me because my cost of living will go up. It offends my sense of values as someone who doesn't like war and violence. And I found that also distasteful – that other powers would presume to decide for the people of Ukraine whether they should want a kind of freedom that they desire or not. And so, I'm not answering your question, in a sense, because I don't have an answer. The role that the church can play is perhaps the role that Pope Francis is trying to play, which is simply a continual invitation to those in power to talk, to begin to speak, to begin to negotiate. And I don't know if he will make any headway with that, but I think it's also important to stay in touch with the people of Ukraine as much as it's possible so that they don't become the person in the middle that other people are deciding for: "No, you need to fight this to the death, to the total defeat of Russia" or "No, you need to give up now. Set it down, give away part of your territory and – and accept what Russia has decided for you."

Carpani: *What you just mentioned aligns with our shared goal of ending the conflict to return to peaceful living. Ukrainian writer Yaryna Grusha Possamai illustrated the situation in Ukraine when the West encourages Ukrainians to converse with Russians. She likened it to a scenario where[by] a woman is forced on stage to face her assailant, and the audience chants "kiss, kiss, kiss" as if she must reconcile with her attacker. But she's still wounded, trembling, and swollen. She invites psychologists to discuss the trauma and the difficulty in consciously processing ongoing events, emphasizing how it [affects] the unconscious mind. According to Grusha Possamai, a meaningful conversation is currently unattainable, and it might not be possible until the wounds have healed.*

I recently spoke with Ukrainian colleagues, a woman who recently fled Ukraine. They were part of the same psychoanalytical society as Russians. They recounted that just two months ago, killing was illegal, but now it's seen as necessary. While we in Europe observe conflicts in places like Syria, Yemen, and Afghanistan, which seem distant, we still react impulsively and yearn for their resolution, as you astutely pointed out.

When the [COVID-19] pandemic began, I initiated a series called Lockdown Therapy with the aim of providing mental support during the lockdown when we were all confined to our homes. I hoped that [COVID-19] would allow for a slower pace, a better balance between our inner and outer lives, and a rekindling of our spiritual and creative aspects. However, I was mistaken. That was my desire, and I've come to realize that Michel Houellebecq was correct in 2020 when he said that the world would remain unchanged but worse. The question is: can we strive for an improved sense of balance between the inner and outer to gain a better understanding of the world around us?

Sims: As a Christian, I want to say we never lose hope – but I, too, have been disappointed to think that it seems we haven't learned very much collectively from [COVID-19]. Maybe some individuals have learned some things about themselves and have made some decisions about how they want to live their lives. But I don't see a big collective learning or any kind of shift. I see a lot of anger, and as you spoke about the media bulimia, what I can think of that talk show entertainment – it was not about educating and informing: Let's create entertainment based on tragedy, stir up extra anger, because in some way this will get us good audience ratings. That may be excessively cynical on my part about the media, but it's what I felt was happening with [COVID-19]. It's what I felt is happening with the situation with the war in Ukraine. It does make the world a little bit worse, because if we're using tragedy of any sort – the tragedy of [COVID-19], the tragedy of war – for audience ratings and entertainment, rather than to build up the common good, then that does leave us a little bit worse off in order to learn. It's what you were saying about having to heal the wounds. And I have a hunch that we're not even very sure yet what the wounds are that we have experienced during [COVID-19]. We probably are carrying some wounds that we are not really even personally aware of – that maybe we're aware that we've changed slightly or that something isn't quite right, but we're not maybe so easily able to name it, to explore it, to see where we take it. And it could be that it's kind of the buildup of this kind of woundedness of a whole series of individuals because we each experienced [COVID-19] in a different way; that somehow, the effect on the collective is such that we were still in quite a model and – and quite a mess with it all, really. For this invasion to come on top of that is particularly difficult, and one wonders whether society has any resilience left at all to experience any hardship that might come about because of this war. If it does mean that prices rise or we can't have our air conditioning the way we want it, or that we find a lot of foreigners in our cities and need to accommodate that in some way, I don't know how resilient our collective will be in being able to absorb that – and that's very worrying, because that's where you get a kind of social breakdown and inner social conflict within your own society.

Carpani: *How does the Russian invasion of Ukraine differ from the post-World War II conflicts involving the United States, such as Korea, Vietnam, Iraq, Afghanistan, and the subsequent invasion, occupation, and political stabilization efforts since 1946? In a recent article, Italian journalist Maurizio Ferrara (*Corriere della Sera*) noted that the U.S. invasion of Iraq was a premeditated carnage, much like what we've witnessed in Ukraine over the last ten days, following months of preparation. Ferrara also drew attention to the insights of literature Nobel laureate Svetlana Alexievich, who depicted the imperialistic impulses not only of Putin but also of a significant portion of the Russian population, stemming from the legacy of the Homo Sovieticus and tsarist tradition. This prompts us to consider the imperialistic tendencies of the U.S. and the UK, [which] invaded and occupied countries. The U.S., driven by capitalism, invades nations for geopolitical reasons. Can we then argue that Russia's actions share similarities with this pattern?*

Sims: I do think that colonialism in the 19th century, the 20th century, and even in recent times – there is a lot to condemn in it. It's not for nothing that Britain is giving up and has been giving up its colonies, that even countries in the Commonwealth are voting to leave the Commonwealth as they gain a new sense of kind of strength and capacity to run their own affairs [and] their desire to exercise their agency in the world, freed from what looks to be a kind of imperial pressure on top of them. The U.S. is in an awkward situation. Having been taken from an isolationist view of the world, which is what they held really up until the times of the World Wars and ending up with this kind of hegemony, which you can say they want but you can also say is thrust upon them slightly. If people or nations say: "Gosh, democracy and capitalism look pretty good to me because I can say what I want, I can do what I want, [and] I can make money and have a good lifestyle." If that looks attractive and other countries say: "Well we'd kind of like that, too," it's not the U.S.'s fault that it has an attractiveness. Someone – I don't remember the name – wrote about this idea of the difference between hegemony and imperialist ambitions. I think that everyone has been pretty convinced that the war in Iraq was wrong. It was a war in a country that was not responsible for what they were blaming it for. It was something which, as you say, had been planned and I think probably manipulated by both the UK and U.S. government[s]. The question is how much it was perceived that Iraq was actually a threat to the Western world. There is this difficult analysis of at what point do you seek to reduce a threat that exists in a nation that's not your own. If you are perceiving a lot of terrorist activity coming out of a place, do you seek to stop it before it reaches your home shores? That's not to justify the war in Iraq because I do think that it was completely ill thought out. There's so much that America needs to be sorry for having done. But I do think also that to say that the invasion of Ukraine is simply equivalent to the mistakes made by the U.S. and Britain. It's overly simplistic and to claim that NATO is an imperialist power in its own right. When the borders of Ukraine were drawn up from 30 years ago and signed off and agreed on, what is the reason for this invasion? Ukraine was not threatening Russia; from what I know, Ukraine was not sending terrorists into Russia to destabilize Russia. Ukraine wasn't seeking to take over other bits of territory that it didn't already have. It's just very difficult to see a justification for this invasion. It just doesn't add up to being an equivalent of something else.

Carpani: *One perspective on the reasons behind these conflicts comes from a Russian colleague who sees justification in Russia's actions due to a revival of traditions, perhaps exemplified by figures like Kirill. An alternative viewpoint, inspired by Ferrara's analysis, suggests that the memory of the Iraq war has had a significant impact on Europe. This event led to street demonstrations and prompted two eminent European intellectuals, Habermas and Derrida, to define what unites Europe in terms of the international order. They highlighted a distinct European political mentality, different from the American one, rooted in specific common traits. This mentality emphasizes an aversion to the use of force, a commitment to law, respect for international legality, and support for a global system based on liberal multilateral institutions and human rights. Shaped by centuries of historical conflicts with opposing principles, this mentality has driven Europe to establish a foreign and security policy anchored in the EU, transcending the simplistic dichotomy between war and peace.*

So, Ukraine's desire to align itself with Western values and embrace these principles is a key reason behind Russia's invasion. When Ukraine adopts these values, it becomes a challenge, not only due to its geographical proximity but also because it sets a precedent for other nations in the region, like Belarus, and even potentially Russia itself, to embrace these values, as we saw with Gorbachev in 1989. This shift is perceived as a betrayal by many in Russia. A Russian colleague once told me, "You don't understand. You're a Westerner. Russia's different." What's your take on this perspective?

Sims: I've seen it stated elsewhere that the fear in Russia or amongst certain of the Russian leadership is the fear of the liberal values. They are the values of human dignity, of human agency, of individual liberty of self-expression to the extent that you're not interfering with somebody else's self-expression, the rule of law, and a desire in most of the European countries to create the grounds for a kind of equality – not just equality of opportunity, but a sense that there isn't, you know, the kind of the class of the hyper-rich and then the everybody else who's just the kind of the servant or the slave of that system. I can understand that that could be very frightening for Russia, and you can see then also why Ukraine would like to move west if that's what they're aiming towards. Having grown up in the generation that I did in the country I did – growing up in America and then growing up in Italy, both countries very committed to the rule of law, to democracy, with all of the flaws in those systems. But the commitment is to that and to the individual liberties and rights. I take it so much as a given that it's very hard for me to imagine that that is just of equal value to another system based on something else. And that we can just negotiate it away either for ourselves or give it away for somebody else who's striving and wanting to have that as the way that their government is set up.

Carpani: This is why the reaction from European countries was particularly strong. Co-constructing meaning is crucial in our societies, especially in Europe and the U.S., where it's built on the foundation of co-constructed meaning and love. Speaking of love, I'd like to introduce the thoughts of theater actors and directors Scimone and Sframeli. They emphasize the idea that "love for others saves them, saves us, and saves the world. The world can only be saved by love. It's not a rhetorical act. Only love can conquer violence."[1] As a priest, how do you respond to this perspective?

Sims: Yes, the problem with the word love is that it's become so overlaying with sentimentalism, but the real content of love is immensely rooted in dignity and respect for the other person – and from looking at it from a kind of religious point of view, if you go right back to the Old Testament and the construction of the law, which is this gift from God to the Hebrew people. It is the law not because it's supposed to be annoying or restrict our freedom or make life difficult. It's seen as a gift because it begins to enable people to live in community with respect for one another and to recognize that society needs to be based on a respect for others, not based on my personal capacity to exhibit power over you, but by agreed communal agreement to respect a structure which is bigger than we are, a kind of judicial structure. When Jesus then begins to speak about love in a stronger and deeper way, it also has to do with the kind of love which is rooted in – in our capacity to set aside self and desire and greed and personal exercise of power in order to offer space, kindness, respect, [and] mercy to those around us. So, this idea of love is extremely deep and it's not just an emotional thing, but it also has to do with how we structure our life as a society. Our supposedly secular European society in effect is very rooted in a religious tradition, which is a tradition going right back to this first giving of a law, which was about how we govern ourselves – how we find a way to live in communities and in cities and societies and nations with a dignity and capacity to hold our common life together.

Carpani: *You mentioned the importance of putting aside the self. I agree, as failure to do so can hinder collaboration. Thank you for your time, Vickie.*

Sims: Thank you for the conversation.

Note

1 It was an article in the magazine *Doppiozero*, 23rd of December 2022: www.doppiozero.com/scimone-e-sframeli-una-grazia-filosofica

Chapter 11

Day 70 of War

Verena Kast

Abstract

Verena Kast challenges the notion of war as an archetype, citing cave paintings that lack battle scenes as evidence. She rejects the idea that war is inevitable and emphasizes the need to combat its negative bias. Even in the midst of conflict, Kast believes in the potential for creativity and rebuilding. Discussing Vladimir Putin, she and Stefano Carpani explore the concept of the absent father, likening Putin to an authoritarian figure. Kast criticizes Putin's narrative, labeling it *pseudologia phantastica*, a term from Carl Gustav Jung. She notes the difficulty in establishing a well-founded pathology of Putin's personality due to the unpredictable nature of events in the war. Kast briefly touches on Ukrainian President Volodymyr Zelenskyy, highlighting him as a counterpoint to Putin but questioning his role post-war, when heroism may no longer be necessary.

Carpani: *Verena, could you explain what exactly war is?*

Kast: It's a barbaric regression to archaic times – senseless violence and ordered destruction, driven by a grandiose idea.

Carpani: *James Hillman wrote in* A Terrible Love for War*:*

There is no practical solution to war because war is not a problem solvable by the practical mind, which is better equipped for its conduct then for its avoidance or conclusion. War belongs to our soul as an archetypal truth of the cosmos. It is a human work and an inhumane horror, a and a love that no other love has been able to overcome. We can open our eyes to this terrible truth and becoming aware of it, devote all our passionate intensity to undermining the enactment of war strengthened by the courage the culture possesses, even in the dark ages to continue to sing as it resists war. We can understand it better, postpone it longer work to gradually remove it from the support of hypocritical religion. But war as such will remain until the gods themselves leave.[1]

What comes to your mind?

Kast: We can't avoid conflict, but war is a different matter. I'm not convinced it's an archetype. It's a human possibility, but I resist the idea that nothing can change. We should strive to humanize our culture, rather than returning to traditional gender roles. Perhaps Hillman is right and I am naïve, but I wouldn't like to accept it that it is just there and it is coming from God.

Carpani: *"Love, that no other love has been able to overcome." Could this imply – from a Freudian point of view – that war is inherent in our nature, in our instincts and our desires?*

DOI: 10.4324/9781003390039-13

Kast: This is how I understand him, but our instincts encompass more than just love. They include aggression, fighting, and a caring, seeking system. Conflict doesn't always have to escalate into war; it can manifest on different levels.

Carpani: *Hillman's perspective suggests that war will persist until the gods withdraw, echoing Nietzsche's declaration that God is dead and Freud's assertion that religion is an illusion. This leads to a profound philosophical issue. However, Jung reintroduced spirituality to the discourse. In his essay "Wotan," Jung begins with a reflection (paragraph 371): "When we look back to the time before 1914, we find ourselves living in a world of events, which would've been inconceivable before the war. We were even beginning to regard war between civilized nation as a fable." The same happened to us after World War II, right? And then Jung continues thinking that such an absurdity would become less and less possible in our rational internationally organized world.*

            Isn't this a pattern that has repeated itself over the past 78 years?

Kast: Yes, but this is really frightening because then we have the idea nothing will change.

Carpani: *Paragraph 372 continues: "But in the sphere of religion, we can see that once that some very significant things have been happening. We need to feel no surprise that in Russia, the colorful splendor of the Eastern Orthodox Church has been superseded by the Movement of the Godless." I guess he's talking about the revolution, Lenin and communism. "And however deplorable the low spiritual level of the scientific reaction, it was inevitable that 19th century scientific enlightenment should one day dawn in Russia." Did he suggest that the decline or cessation of spirituality was the cause of the Russian Revolution of 1916?*

Kast: Jung might suggest that the Soviet Union's problem is the loss of spirituality, leading to revolution and mass movements. However, the meaning of a revolution is complex, and it often arises when people need transformation urgently. It's not always a gradual process but can occur when a critical mass decides action is necessary. While Jung's idea connects war with spirituality through Wotan, it can be limiting to focus solely on male archetypes; there are other aspects to consider.

Carpani: *Of course, when we consider the gods associated with conflict, Athena is merely one amongst many, and also serves as a goddess of other domains.*

Kast: She is the goddess of creativity. And I think this would be the remedy against war – to be more creative.

Carpani: *This aligns with what we discussed about [COVID-19] two years ago, where[by] the impact was similar. What's intriguing is that these paragraphs suggest that the Orthodox Church in Russia lost its spirituality and connection with the people, eventually leading to the revolution. Both Putin and Kirill echo this sentiment, framing the conflict as religious in nature. Putin defended the invasion of Ukraine, citing the Gospel: "There is no bigger love than to give your life for your friends." However, Kirill, the head of the Russian Orthodox Church, anticipated Putin's words by stating on the first day of the war that this is not a physical war but a metaphysical one. Putin believes this war is lawful and justified because it's against the secularized West with its events like gay pride parades and similar things. Over the past century, we've seen a swap of ideals: Russia lost its spirituality but is now attempting to reintroduce spirituality from the top down. Do you find this attitude to be somewhat paranoid?*

Kast: Some of Kirill's ideas come through here. The Russians have created various fantasies in this war, including feeling threatened and trying to restore the nation's spirituality in Ukraine. At its core, it seems like a revenge for the Soviet Union's dissolution, an attempt to make Russia great again. The bigger issue is their aim to destabilize Western democracy, our freedom, and our values, which is why we feel threatened. To claim that this is done in the name of God feels like blasphemy to me.

Carpani: *This echoes what Father Bettoni stated: War is blasphemy, and using God for such reasons is even more blasphemous. Regarding values, in March, Italian journalist Maurizio Ferrara referenced Habermas and Derrida in* Corriere della Sera, *raising questions about what unites Europeans. Their response highlighted a different political mentality from the American one, grounded in certain shared characteristics. These include a strong aversion to the use of force, an emphasis on law and respect for international legality, support for a global system based on "liberal" multilateral institutions and human rights. This mentality has its roots in a history of opposing principles and centuries of bloody conflicts. It's the very mentality that pushed Europe to establish a foreign and security policy anchored to the EU, moving beyond the simplistic divide between war and peace. Putin seems to be challenging this, while Ukraine strives to embrace it.*

Kast: Yes, it's related to democracy, which is under threat globally. Putin aims to combat this creative freedom, but I've witnessed the West uniting more than ever. We are fighting for these values.

Carpani: *I've recently been revisiting my old poems, written before 2006. What struck me was the number of poems I'd penned about the tensions between Russia and Ukraine, with Yanukovych's role in those conflicts. Ukrainians were genuinely fighting for their independence, no longer wanting to be part of an authoritarian system. They aspired to become an independent country capable of making its own decisions – a place where an event like the storming of the United States Capitol would be a significant issue, but democracy would ultimately prevail. It's as if Ukrainians are sometimes excluded from the ideas of Habermas and Derrida because they aren't seen as fully European.*

Kast: We made a crucial mistake by not taking Putin's announcements seriously; he signaled his intentions. We wanted to believe in peace and cooperate, which is the right approach, but we lacked skepticism. While cooperation is valuable, we must balance it with a healthy dose of skepticism. We were blinded by the idea of peace, which brought prosperity and creativity. We often forget that within us, there's both an aggressive and destructive side, as well as a creative side. The West leaned towards cooperation and creativity, but overlooked the potential for conflict. Ukraine has a rich history, but many in the West view it simply as "the East." I consider Ukrainians as Westerners, but not everyone shares this perspective.

Carpani: *Do we in the West take peace for granted? Could this be why we are so troubled by these events?*

Kast: While we may not take peace for granted, we've made substantial efforts toward peace in recent centuries. We still believe in the possibility of global peace, and we must remember this, especially considering Germany's history of two devastating wars. Germany has actively invested in peace efforts. At the individual level, we all encounter conflicts and strive to resolve them, which is a deeply human quality. However, we can't always guarantee that conflicts will be peacefully resolved.

Carpani: *It's evident that a conflict occurred at the border between Ukraine and Russia.*

Kast: The problem is that Putin instigated this conflict or asserted his power. If there were a genuine conflict, it would be easier to deescalate and engage in dialogues to find solutions.

Carpani: *This is why it's so tricky to negotiate presently. You can't negotiate if one side is not willing to participate. The only solution is to resolve the conflict in his way.*

Kast: Yes, it's truly depressing. Many are seeking a solution, but there are those who believe it will continue to be a battlefield. This is terrible because it entails immense human suffering. People are not allowed to live in peace; women and even men are subjected to horrific ordeals. This isn't merely an intellectual debate; it's about the devastation of lives and homes, and it's unbearable. To say there's no solution is incredibly disheartening.

Carpani: *You're touching so many points: whether war is masculine or feminine, [if] the concept of the hero and the fatherland applies, and whether it will be total war in Europe, and if it's appropriate or not to psychopathologize Putin. But let's go one by one: Is war masculine? Is war feminine? Is war something else? It may seem simplistic, but I'd like to know what you think.*

Kast: If we view it in terms of principles rather than gender, fighting may be associated with masculine principles, while caring could be considered more feminine. Women can engage in combat, as well, but nurturing is often seen as a more feminine trait. In this war, we witness both aspects. However, the media tends to emphasize negativity by showing destruction. Yet, there are moments when we see people caring for and saving others. War encompasses not only devastation but also acts of compassion. Labeling war as inherently male or female doesn't necessarily aid our understanding. War can bring out the worst and the best in humanity, presenting an ultimate challenge. Both aspects are present.

Carpani: *The emphasis on care is important to me, as traditionally soldiers have been men, but now there are also women involved in combat. Examples include Anna Politkovskaya, a slain journalist, Pussy Riot, Sviatlana Tsikhnouskaya, and recently Marina Ovsyannikova. This demonstrates that war can also be seen as a feminine or female experience, but in our view, the warrior or gladiator is typically a man. However, I once met a woman who had served with the American military, and she told me that she was the first female soldier allowed to press the missile firing button. The Army had previously believed that women were unsuitable for such duties, but in reality, it was simply a matter of pressing a button. Who knows if the Army is correct? – but I believe that this woman was severely traumatized by it.*

Kast: In a war, all individuals – and especially soldiers – are subjected to trauma. We've observed this phenomenon during World War II. The theory suggests that when you kill someone, you kill a part of yourself. Research has also indicated that many soldiers find it extremely challenging to take another person's life. This is why technologies like remote-controlled drones, tanks, and missiles are employed, making it easier to kill from a distance. However, in the Ukraine conflict, we've witnessed many women learning to use firearms. This phenomenon is influenced by societal gender norms, but it shouldn't be this way. I can empathize with them and think, "I'd also want to be able to defend myself before being harmed." It's a complex issue.

Carpani: *You know, a Ukrainian patient I've been working with for two years said this: "until two months ago, killing was against the law; now we must kill before being killed ourselves." This is a tragedy not just for Europe, but for the world. Some critics wonder why Ukraine is getting more attention than other countries. It's because of the values and history we mentioned earlier, but what does this lead to, Verena? It's already a huge wound across Europe, maybe even worldwide. The country was ruined in a mere two months, and it's unknown how much longer it will last. Both psychologically and in terms of infrastructure, we're still recovering from the aftermath of World War II.*

Kast: Are we currently experiencing a negativity bias? I've encountered people who fled to Switzerland and have chosen to return to Ukraine. When asked why, they express their desire to help rebuild the country. They might be traumatized to some extent, but they exhibit remarkable courage. We can anticipate an increased prevalence of post-traumatic stress disorder, especially among the young girls who have suffered sexual violence. I believe Europe will provide support in terms of psychological and psychosocial assistance. I maintain hope that even when something is destroyed, it can be rebuilt. It would be challenging for me if this war persists, but when I speak to these individuals, and they display the courage to say that life must go on, it signifies a new normalcy. Nobody claims it's easy, but I hope that the West will provide aid. This crisis also affects our convictions and beliefs. It's not solely about cooperation; we need to exercise caution regarding interdependence and dependency. In this context, we can leverage the ecological crisis to transition away from fossil fuels and embrace renewable energy sources. It's not all negative.

Carpani: *Yesterday, I spoke with Vickie Sims, an Anglican priest, who, like you, emphasized the importance of hope. Can you explain why war transforms ordinary people into savages? Recently, there was a news report about a Russian soldier who was recorded on a phone call with his wife. In the call, Olga gave her husband approval to sexually assault women in Ukraine. This is unimaginable.*

Kast: It's unimaginable. Raping women is like killing them without physically taking their lives, as it shatters something within them. This brutality also erodes the moral compass of men, as husbands and fathers endeavor to protect women, and when they can't, it profoundly [affects] their self-perception. Raping women essentially weaponizes sexuality.

Carpani: *I wanted to provide an example of the lowest point. You highlighted the importance of creativity and hope or positivity, as war can make people more connected. Could you give an example? Can you expand on what we may miss due to our focus on negativity?*

Kast: I think this is archetypal: In moments of apparent despair, hope emerges. It's not mere expectation, but a profound human sentiment that says, "I'm alive, and as long as I am, I'll fight for life." Hope is the companion of existence. When you feel hope, you act. It's the time when people generate numerous ideas and show inventiveness. In times of scarcity, individuals organize and rely on their intelligence, working together to create positive solutions. We must remember that people are resourceful and capable of cooperation, which can foster creativity and energy.

Carpani: *Let's return to Russia. You mentioned this war is for revenge. I wonder if it could also be seen as compensation from an Adlerian perspective. I think that if someone feels embarrassed by their father – like Freud or Jung, whose father was a priest but lost his religion – then the individual may feel compelled to become a hero to make up for it. If we consider Putin, who claimed that the day the Soviet Union ceased to exist was the most terrible day of his life, we may question who his true father figure could be. Was it Yeltsin or those who placed him in the KGB? Perhaps Lenin or Stalin? It is uncertain. However, we must ask ourselves if this conflict is related to seeking reimbursement or a sense of ownership to one's homeland. What are your thoughts?*

Kast: We must remember that Putin is not a young man. I hope today's fathers don't have to be heroes in the traditional sense. Fatherhood in our time is more about nurturing and fostering connections, not heroism. Young men might seek to be heroes in moments of crisis. I see a compensatory element in the younger generation. Putin is older, and he had a challenging childhood. His father was a war veteran. Putin carries multiple traumas and has tried to compensate throughout his life, accumulating immense wealth and power. He feels most empowered when he believes he can destroy another country. It's a form of compensation, but not just that. What's remarkable is Putin's sense of empowerment. He felt threatened by NATO and the West, despite their arguments that they weren't the ones invading Ukraine. It's a kind of paranoia, fueled by a lack of mirroring, combined with a compulsive fear of being threatened. This mixture is deadly.

Carpani: *Fear of being threatened compulsively. Ultimately, this fear is also a fear of losing one's identity, whether it be the collective identity of the Soviet Union.*

Kast: Putin's leadership has seen an authoritarian shift in Russia. Many Russian people look up to him – although not all, as many suffer greatly. He has managed to convey the idea that they are good and great under his leadership.

Carpani: *That's why I remember the notion of the far-off hero. The hero who is distant in his castle, who has no relationships, and thinks he's better than everyone else. We'll discuss another hero or potential hero shortly. However, may I ask you: Is it fitting to attribute psychopathological traits to Putin? The media did it, and we just did it, too.*

Kast: It's not appropriate to diagnose someone like Putin without real contact. We can discuss narcissism and power as normal people in a conversation, but making a formal diagnosis isn't proper. It only shows our uncertainty about his intentions. Putin appears to be unwell. When confronted with death, the question of what is important comes to the forefront. In his perspective, it's bringing Ukraine back home, while in ours, it's causing harm to much of the world.

Carpani: *What do you think of Zelenskyy?*

Kast: Zelenskyy is brave and represents a new kind of politician with a different communication style. I appreciate him, but I'm concerned about the pressure he puts on the West, making demands and judgments. I understand his situation, but I wonder about his role after the war ends. Heroes are needed during a crisis, but their role changes when the crisis is over. Zelenskyy is a hero for a specific time.

Carpani: *A bit like Churchill: He was good during the war, but after, he couldn't govern the country.*

Kast: I sometimes wonder if it might have been better to seek a diplomatic solution early on instead of allowing destruction to continue. It's challenging for us to make such judgments from an external perspective.

Carpani: *This is the topic that my interviewee, the priest Vickie Sims, spoke about yesterday. She is from America and said: Who do we think we are to tell Ukrainians not to fight? While we sit back and enjoy our 78 years of peace and continue to watch*

Netflix, Ukrainians are showing us the meaning of freedom in 21st century Europe. We're coming to the end of this chat. Two years ago, we spoke about [COVID-19] and I want to apologize because I believed it was a chance to slow down and keep a healthy balance. To move from a world based on hopes and expectation, the linear world, to one where interiority and spirituality can be contemplated, the creativity of the soul. You said many times in that conversation [that] it's about creativity. I wonder whether I was naïve. We were naïve. I'm unsure if Michelle Houllebeqc's prediction that the world will remain similar, but slightly worse in 2020, was accurate. Are we also naïve talking about this war?

Kast: Maybe we were both naïve, but I won't abandon the path of hope. It's true that during the pandemic, people found solace in their inner world and memories of better times. They believed those memories would become a reality again, helping them overcome traumatic experiences. Maybe we were naïve in thinking this would be the case for everyone, but it has changed the way we view our own invincibility. Life always involves seeing both sides. When someone says it'll be the same world, just a bit worse, I'm not so sure if that's accurate. Some things will undoubtedly be worse, but perhaps some things will also improve. I'm committed to fighting against this negativity bias.

Carpani: Maybe there's another way to view war as a reset, a transition from negativity to hope, and the chance for Europeans to build and continue creating a shared understanding, as Habermas and Derrida suggest. We are Europeans; you are Swiss, I'm Italian. We've been talking about Putin and Russia – but what about the U.S.? What is the difference between this war and U.S. military actions since the end of the World War II, the invasion and occupation or political de-stabilization since 1946 under the guise of liberating them, under the pretext of spreading liberal democracy. According to Ferrara, during the U.S. invasion of Iraq, some commentators back then observed to be a determined carnage just like the one of the last two months. Ferrara also reminds us that the Nobel Prize for literature winner Svetlana Alexievich has drawn a very clear picture of the imperialistic process, not only of Putin, but also for large part of the Russian population, the long wake of the Homo Sovieticus, and before that of the tsarist tradition. But what about the imperialistic tendencies of the UK or the USA? They have invaded and occupied other countries, leading to problems and fatalities. One might argue that the deaths caused by these actions are no different from those caused by Stalin or Putin. Colonial Britain took everything from the countries it occupied, whereas the capitalist USA invades countries for their geopolitical interests and resources. And we cannot overlook the situation in South America. It's controversial, but I believe it's necessary to ask this question. Otherwise, we'll [be labeling only] Putin and the Russians as bad, which isn't accurate.

Kast: The Russians are not inherently bad, and we should acknowledge this. What's happening in Ukraine is not new, as similar conflicts occur in various countries around the world. We react more strongly to Ukraine because it's closer to us and aligned with our values, and it [affects] our economy, but starting a war is inherently wrong. What sets this situation apart is Putin's narrative, where he projects his fears onto others. I didn't catch such a narrative with Iraq or Afghanistan. However, this issue isn't exclusive to Russia; all wars need to stop.

Carpani: After the Catastrophe by Jung comes to mind, where he examines Germany, National Socialism, and Hitler, and suggests that Hitler suffered from a unique form of hysteria that causes one to lie. This reminds me that war should cease, and we should listen to each other. You believe your lies, and your followers believe them, too, creating a cycle of deception. Is this not what is occurring presently?

Kast: It's almost like *Pseudologia Fantastica* or something similar, but not just that. Putin has significantly controlled the media in Russia over the years. Media freedom has declined, and those who didn't align with Putin's agenda faced imprisonment or even death. He's worked hard to establish the idea that "I am the truth."

Carpani: *Yes, "I am the only one who speaks the truth." I have asked you many questions and allowed you to answer them fully. If there is anything else you wish to share that I have not yet asked, please feel free to do so now.*

Kast: I'm deeply impressed by the concept of negativity bias and positivity bias. It's vital to remember that life encompasses both positive and negative aspects, possibly even more than just these two. We believe in the positive – creativity, imagination, and using them for addressing the ecological crisis. However, we also have destructive tendencies within us and in the world. To prevent war, we must find a balance within our psyche that acknowledges and holds both of these tendencies.

Carpani: *Thank you so much, Verena.*

Kast: Thank you.

Note

1 Hillman, J. (2005) *A Terrible Love of War*. Penguin Publishing, p. 214.

Chapter 12

Day 84 of War

Luigi Zoja

Abstract

Luigi Zoja analyzes war from a psychological perspective, drawing parallels with animal behavior during mating season. He discusses the Russian invasion of Ukraine, highlighting the cultural impact on innate instincts, especially male aggressiveness. Zoja delves into media influence, emphasizing the drawbacks of consuming short, fast, and spatially limited news, which he believes leads to paranoia. He explores the role of personification in crises, contrasting the COVID-19 pandemic's inability to be personified with Vladimir Putin's role in the war. Zoja contends that Putin's perceived need to compensate for Russia's historical loss makes him dangerous. Using the analogy of a cat stuck in a tree, Zoja illustrates Putin's blindness to alternative solutions. The conversation with Stefano Carpani touches on the concept of millennial culture, noting the Russian tendency for patience and long-term thinking, as evident in Putin's speeches.

Carpani: *Professor Zoja, thank you so much for accepting my invitation to this new web series* War as Reset.

Luigi Zoja: Thank you.

Carpani: *Luigi, we're almost three months into this chaotic war. You recently released a book in Italy, currently titled* Dialogues on Evil, *or* Conversations on Evil. *I started a web series called* War as Reset *because I wanted to understand more about the current situation in Russia and Ukraine. However, with the recent events in Palestine, not to mention Yemen and Afghanistan, everything seems even more complicated. To comprehend war better, and specifically because I believe that media outlets and newspapers have lost the ability to thoroughly analyze events beyond advertisement sales or even populism, I'll begin with a basic yet challenging question: What is war?*

Zoja: Clausewitz's famous definition likely stems from an evolutionary perspective, resembling animal fights and mating season duels among males. This instinctual aspect, as per Konrad Lorenz, is ritually regulated in nature. However, at the cultural level, it becomes complex and amplifies the instinct of masculine aggressiveness.

Carpani: *In February and March, we became aware of the threat of war. This was due to the situation in the Balkans and events that occurred in 2014. It was the first time in a long while that Europe was facing the possibility of a powerful conflict. I think that war is the opposite: And the "Vanity Fair" – Truman Capote's "Vanity Fair" – actually it is the "Atrocity Fair." But thanks to TV and social media, we're constantly bombarded with information. It's like a media bulimia, making us crave more and more until we reach saturation point. We don't have time for the important things anymore. What are your thoughts on the media and*

DOI: 10.4324/9781003390039-14

television? What are your thoughts on social media content created by ordinary citizens through images, rather than professionals?

Zoja: I'd like to mention two references, one being the Vietnam War, often called the first war in real-time television. The other is my study on paranoia and history, viewing paranoia not as an individual ailment but as an archetypal trend in history. In the evolving landscape of civilization and progress, the media exacerbates the situation by emphasizing shorter, more aggressive, and paranoid messages, especially through social media. This instant and concentrated communication favors simplistic scapegoating, inhibiting in-depth psychological analysis. The media, once a helpful tool, now contributes to a psychic infection, amplifying our instinctive self-defense reactions and perpetuating a cycle of aggression.

Carpani: *Do you believe we are living in a time filled with fear? Allow me to clarify: Over the past two years, starting in early 2020, we grew accustomed to news coverage solely about [COVID-19]. Suddenly, [COVID-19]–related headlines vanished and were replaced by news of war. What can we expect in the coming months and year? I think we live in a nervous society that's the opposite of an intuitive one, where we require this information to satisfy an inner emptiness.*

Zoja: The pandemic's chronological connection has been widely acknowledged, and I've previously delved into this topic in video lectures. A crucial distinction with [COVID-19] is its non-human nature, preventing personification. Scapegoating, an archetypal instinct, arises as a defense mechanism when confronting our own imperfections. Historically, this ritual served to expel evil in small and tribal societies. But even Adolf Hitler did the same with the Jews, a collective. Germany was suffering and obviously for the humiliation, the economic loss – whom can you scapegoat? Let's scapegoat somebody! However, in the modern context, personifying the adversary is essential for scapegoating to occur. [COVID-19] posed a unique challenge as an invisible enemy, triggering a pervasive sense of paranoia. I explored this issue in my book on paranoia, drawing on experiences from discussions on terrorism in Europe several years ago. Despite statistically low risks, the fear and need for a scapegoat persisted. Interestingly, the ease of scapegoating is now exemplified by Putin, offering psychological relief by fulfilling our deep-seated needs for a target.

Carpani: *I want to end the discussion on the meaning of war. Some authors, particularly feminists and women, debate whether war is a male or female characteristic, such as Nobel laureate Svetlana Alexievich.*

Zoja: As a Jungian, I discuss psychological figures and share the belief, consistent with Jung's ideas, in the broader concept of evolution. The observed evolution tends toward a negative manifestation of the animal drive, particularly in its masculine aspect, rooted in the historical behavior of male combativeness. My writings have explored the masculine identity, emphasizing figures like the good enough father, irrespective of gender. While masculinity remains an archetype and instinct, its contemporary expression can be embodied by both men and women.

Carpani: *I need to connect this query with another: why does war transform ordinary individuals into savages? We observed Marina Ovsyannikova's circumstance as an example. She revealed her disdain for war on TV with a large sign. However, now she has had to escape, and her husband forbids her from seeing or conversing with their children. We know Anna Politkowskaja, Pussy Riot, Femen, [and] also*

Aung San Suu Kyi. There is a story that I read a month ago in La Republica, *one of the most important Italian newspapers: the story of Olga and Roman. Roman is a Russian soldier deployed in Ukraine [and] Olga is his wife; they're both 25. Roman wrote to his wife the following: Do I do I have your permission to rape Ukrainian women? She answers: "Of course, my love, break them, use the condom." Roman's message to his wife is deeply disturbing, and her response is equally troubling. They discuss the possibility of sexual assault with a shocking lack of empathy or concern for the victims. This dialogue highlights the historical issue of sexual violence perpetrated by Russian forces during wartime, a pattern that seems to persist today. It underscores that the dynamics of war involve not just male and female roles but also masculine and feminine aspects, revealing a broader dimension to the problem of sexual violence in conflicts.*

Zoja: This is an isolated case, and such unfortunate incidents can occur. Generalizing from Roman and Olga's actions may not be valid, as their behavior appears to involve a sadistic perversion and a lack of awareness. They exhibit fragility akin to paranoia, projecting a shadow that necessitates splitting. While the use of a condom is officially a hygienic measure, it also carries symbolic weight, signifying a desire to avoid contamination from collective aggressiveness. However, the individual involved is already affected by such aggression and is personally vulnerable. It's crucial to note that occurrences like these typically happen in wartime. The key distinction lies in whether a country, even during war, upholds the rule of law, such as having military courts, to address such issues. In the case of the Soviet Union in 1945, military courts were effectively halted upon entering Germany, underscoring the importance of a nation's self-respect in punishing misconduct.

Carpani: *You wrote a masterpiece about the father:* Ettore. *I have a fantasy that this war is very much about the concept of father, fatherland and compensation. Let me provide an example: when someone feels ashamed of their father. As teenagers, Jung and Freud idolized their fathers, but later realized they were flawed. They became fixated on the concept of the heroic archetype and felt the need to become heroes themselves, potentially to overcome their fathers' fate. Putin stated that the collapse of the Soviet Union was the most difficult day in his life. Putin was not a senior KGB agent, but rather a young one with a lower rank. However, a series of traumatic events may have led to his compensation. Who was Putin's father? Yeltsin was a power-hungry drunk. It is unclear if Gorbachev played a role. Is this war like compensation for World War I, and does it mark the end of a century-long rivalry that dates back to before World War I, when Franz-Josef opposed the tsar?*

Zoja: Avoiding a personalistic approach, I'd discuss compensation and rebounds in terms of archetypes, specifically how leaders activate these archetypes in the population. Referring to Jung's Knickerbocker interview before World War II, the attempt by leaders like Mussolini and Hitler to personify the nation or Volk can be analyzed. While Jung's insights seem generally accurate, his judgment on Italian identification with Mussolini might be questioned. The complex interplay of national psychology, identification with the nation, and state allegiance varies across countries. The discussion transitions to the current context with Putin in Russia, where there seems to be identification among the common people but less so among the more cultivated or educated population, including colleagues. This raises the question of the risks individuals take in expressing a critical attitude in such situations.

In the 1930s, Putin coincidentally aligns with Jung's interpretation, feeling that a historical mission can compensate for Russia's significant loss. A key difference with Hitler is that the latter faced a military loss in World War I and attempted compensation through World War II. Hitler's inner struggles, including inflation and paranoia, activated a *self-fulfilling* prophecy of encirclement. Despite being the master of a vast country, Hitler felt surrounded by enemies. His claim of the 20th century's greatest tragedy, the Soviet Union's disappearance, was not a military defeat but an inner and economic collapse. Putin, facing a Russia composed mostly of the former Soviet Union, lacks the means to compensate for this loss, with fringe countries becoming critical and projected enemies, albeit with limited economic strength beyond being a gas station for Europe. Russia's GDP [gross domestic product] is lower than Italy's, and comparing it to China is misguided. While China aspires to global economic dominance, Russia relies on passive wealth from resource exploitation, making it unsustainable. Putin's lament about the Soviet Union's disappearance might indicate a loss of empire, not the egalitarian aspects. This concern is troubling because the true tragedies of the 20th century were two World Wars and the Holocaust rather than the loss of an empire.

Great Britain and France lost their empires through decisions made in London and Paris, demonstrating that democratic systems eventually curbed colonial violence. Russia's economic power is only a tenth of China's, making Putin's unrealistic aspirations dangerous. The situation is reminiscent of Hitler's ambitions, which led to *self-defeat* and suffering for normal people in Russia. There is concern about why Russia, a regular country with normal citizens, should endure such consequences.

Carpani: *You might think about words such as failure or ruin. You aspire to create something brilliant, but, in reality, you do the opposite. Why is this the case? This is why I referred to the period preceding World War I. You were discussing empires, including the Russian Empire and the Austro-Hungarian Empire – regarded as a distinguished empire[s] within Europe. Europe begins and ends in Vienna. But if we consider Wotan and Jung and the reason why the Russians and the Communists were alarming is because Lenin's Red Revolution of 1917 rejected industrialization and opposed the bourgeoisie of Western Europe. Therefore, it was an intellectual, advanced and direct revolution against spirituality. Jung at the beginning of* Wotan *writes*[1]:

When we look back to the time before 1914, we find ourselves living in a world of events which would've been unconceivable before the war. We were even beginning to regard a war between civilized nations as a fable, thinking that such an absurdity would become less and less possible in our rational, internationally organized world. And what came after the war was a veritable witches sabbath.

Didn't we believe the same until a couple of months ago, a couple of years ago? Even when somehow Russia joined externally NATO. But in the sphere of religion, we can see it [at] once that some very significant things have been happening. We need to feel no surprise that in Russia, the colorful splendor of the Eastern Orthodox Church has been superseded by the movements of the godless. And however deplorable the low spiritual level of the scientific reaction, it was inevitable that 19th century scientific enlightenment should prevail one day. In Russia now, the opposite is happening. Someone suggested that the alliance between Putin and Kirill is based on spiritualism. I disagree. The notion of returning to Kirill or Vladimir – as the Nazis did with their gods and beliefs – is alarming and frightening. There is a political aspect and a false spiritual aspect that could cause significant harm.

Zoja: Putin's psychic inflation is highly dangerous, akin to what you'd identify as a psychological analysis of dictators. His demeanor, characterized by a lack of smiles and a sense of possession, aligns with this analysis. A recent comment compared Putin to a cat that climbs too high on a tree, highlighting the clever ascent but the lack of readiness for the descent. This analogy emphasizes the potential danger of inflation, cautioning against underestimating the risks, much like handling a frightened cat on a tree.

The recent parade for the 9th of *May* commemorated the Great Patriotic War, an event over *three-quarters* of a century ago. While it was a significant victory over Hitler and the Nazis, insisting on that identity as the defining aspect risks dwelling in a negative past. The continuous infection, described in Jungian terms as omnipotence, ties the identity to the Red Army's triumph in Berlin, but this historical feat is increasingly disconnected from the present reality. Attempting to impose an identity linked to a bygone era, especially with the fall of Russia, poses serious challenges. It's akin to being possessed by the past, a situation not reflective of current circumstances. The analogy of the cat or the potential use of nuclear weapons as a last resort reflects the difficulty of navigating a situation when there's a lack of clarity on how to move forward.

Carpani: *I enjoy discussing with you and Jungian psychoanalysts in general because we can use psychosocial comparisons to comprehend Putin's stance. However, I want to know what sets this war apart from the unpleasant events the U.S. has caused since the end of World War II. We tend to focus on Putin, the Russian government, and the Russians. Let's examine the opposing forces in the war instead. You mentioned Vietnam, Iraq, and Afghanistan – what about the U.S. invasion and occupation, political destabilization since 1946 under the guise of liberating them and since 1991, under the pretext of spreading liberal democracy? I believe that pandemics and wars cause empires to change. This has happened in the past with the European empires, now potentially with the U.S. empire, and possibly in the future with China. Maurizio Ferrara in early March [in]* Corriere della Sera *looked at the U.S. invasion of Iraq and underlined that some commentators back then observed that it was a predetermined carnage just like that of the last 90 days, which took place after month and month of preparation. Ferrara reminds us that Nobel Prize for literature winner Svetlana Alexievich has drawn a very clear picture of the imperialistic impulses, not only of Putin but also of the large part of the Russian population, the long wait of the Homo Sovieticus, and before that of the tsarist tradition. What about the imperialistic impulses of the United States? Apart from the United States, every time Britain has invaded and occupied a country, it has caused problems and civil wars upon leaving, such as in Kenya, Cyprus, and Palestine. As seen currently, the conflict in Palestine has been ongoing for the past 70 years. Are the actions of the British crown – the death it causes – different than any of those the U.S. or Putin does right now? When*

Britain colonized other countries, they took everything. Similarly, the capitalist U.S. invaded other countries for geopolitical reasons and their goods. Although we are Westerners with certain values, it is important to acknowledge that we are not innocent, either. Instead of solely focusing on the "bad guys," we must also examine our own actions.

Zoja: In the 1970s, I left Italy to participate in the anti-Vietnam War movement. Later, during the invasion of Afghanistan and the rise of an extreme right-wing faction within the Republican Party in the U.S., I expressed my opposition. Despite these complexities, there is a nuanced truth in the U.S. position – they have been involved in conflicts like Vietnam and Afghanistan but haven't sought to annex these regions. On a symbolic archetypal level, the U.S. can be seen as an empire with a continental reach, having acquired a significant portion of North America from the British and native tribes. Unlike some European empires, decisions related to U.S. actions are not just military but also involve political considerations. The difference lies in a commitment to what we may term European or Western democratic values.

Carpani: *And we are back in Europe. Ferrara also wrote in the same article that the war in Iraq gave an important shock to Europe in the wake of massive street demonstration. Two great European intellectuals, Habermas and Derrida, asked themselves the question what binds Europeans together with respect to the international order. Their answer was: a political mentality, different from the American one, and based on some common traits – an aversion to the use of force, first of all, and therefore an insistence on law and respect for international legality; [and] support for a global system based on "liberal" multilateral institutions and on human rights.*

Fruit of a past inspired by opposite principles and characterized by centuries of bloody carnage, precisely, this mentality had to push Europe to build a foreign and security policy anchored to the EU, overcoming the "stupid" and simplistic opposition between war and peace.

Italian Prime Minister Mario Draghi and many others have emphasized that Europe's core values, as articulated by Habermas and Derrida, are the true driving force behind Ukraine's desire to join Europe, rather than materialistic factors like McDonald's or BMW. They see Russia's authoritarian approach, which opposes self-determination, as something Ukraine can no longer tolerate. To uphold these values, there's a growing need for a stronger Europe, not just in terms of military funding, but also in terms of cultural strength.

Zoja: In contrast to the confusion and slowness of the 27 European countries, there is a distinction from Anglo-imperialism, where[by] translation is often necessary. Europe seeks to level the playing field by offering translation assistance, countering undemocratic practices that disadvantage millions. Despite financial challenges, Europe maintains a minimum level of socialization and equalization, fostering a sense of solidarity. This commitment extends to cultural events, exemplified by assistance in translating works into less widely spoken languages. Italy, facing financial disorder, still accommodates 115,000 Ukrainians and upholds the European tradition of providing a minimum income and addressing homelessness. These values, rooted in decency, are likely to endure into the 22nd century, regardless of military uncertainties.

Carpani:	*You previously used the German word "Volk." Emperor Franz-Joseph used to say it: "my people, mein Volk." Europe is comprised of 27 countries, but also contains many different groups of people. This variety is our strength, but can also cause issues.*
Zoja:	In Russian culture, there is a notable absence of belief in the state; instead, there's a historical pattern of leaders like Stalin embodying the role of the tsar. This lack of faith in the state contributes to a high level of corruption. However, Germany, with its complex relationship with Russia, grapples with discussions on supporting Ukraine due to a romanticized view of Russia. Russian culture, characterized by enormous patience and a focus on quality, contrasts with the shorter-term thinking prevalent in the United States. Despite Putin exploiting these cultural aspects, there's admiration for Russian culture, especially in Germany, where the love for it allows politicians to express sentiments that might be less common in other countries. The appreciation for Russian literature and theater, known for its slow pace, is a testament to this admiration.
Carpani:	*The narrative unfolds at a deliberate and uncomplicated pace, devoid of any sense of deprivation. This is why I often find myself revisiting the era preceding World War I, particularly the Austro-Hungarian Empire, which stands as one of the largest empires of its time. Within the Austrian domain, there existed a stark contrast – a sense of moral decay prevailed. The army, once poised for war, eventually succumbed to a lifestyle marked by incessant drinking, encounters with prostitutes, and indulgent gambling. Isn't this war just about values? As the West becomes more prosperous, there seems to be a rise in vices such as gambling, prostitution, and a decline in morality. This creates a need for someone to step in and reset the balance. In this case, it was the Russians who introduced communism as a solution.*
Zoja:	In Russia's centuries-old culture, the 80 years of communism appear almost inconsequential, with little observable change. Reflecting on Austria, there's a complex view – the country, despite corruption, contributed significantly to culture, as depicted in Stefan Zweig's *The World of Yesterday*. However, the declaration of war on Austria played a role in its collapse, leading to the destruction of culture and multiculturalism, giving way to fascism and nationalism. George Steiner's essay on Bluebeard's Castle captures Austria's paradoxical blend of creativity and the anticipation of war. Despite the richness of culture, the aftermath of conflict prompts a rhetorical question from the "History of Paranoia": who has truly won? Despite democracy theoretically prevailing in World War II, the conflicts of both World War I and [World War] II witnessed widespread ethnic cleansing. Criticism of Israel notwithstanding, it remains a democracy. Until the [19]50s, Egypt was a multicultural haven, home to influential Italian literates like Marinetti and Ungaretti, as well as the Greek poet Cavafis. The fringes of the Soviet Empire experienced ethnic cleansing, a trend reflected in the current Russia-Ukraine conflict. In the best-case scenario, Russians might need to withdraw, engaging in land exchanges or prisoner swaps, potentially exacerbating nationalism in both nations.
Carpani:	*One final idea that aids me in exploring the concept of war is reset, or as you might call it: culture as a reset. The period spanning from the mid-18th century to the early 20th century, marking 70 years of peace, was a time of cultural prosperity. The notion is that war is a consequence of culture, whereas culture becomes*

a reaction to war when it transforms into pure hedonism – culture as a coun-
terbalance to war and war as compensation for culture's displacement towards
hedonism.

Zoja: It's a bit too soon to see that in the conflict between Russia and Ukraine – if you
           can call it war in spite of not having Putin's permission to call it war.

Carpani: *I am reflecting on the past, specifically the exciting period leading up to the end of*
           the 19th century and the beginning of the 20th century, filled with hope even after
           war. However, when culture devolves into hedonism, leisure time loses its value
           and becomes solely about indulgence. This ultimately leads to conflict.

Zoja: You're correct. On a macroeconomic level, our unconscious complicity is evident.
           To secure cheap gas, we shook hands with the devil, Putin, turning a blind eye to
           his previous attacks, as Ukrainian colleagues had warned. Now, with hindsight,
           our foreign minister, Luigi Di Maio, advocates helping Ukrainians with weapons,
           yet he was in a position to scrutinize those gas contracts when he held the role of
           foreign minister. Our collective consumption of this cheap gas implicates us all,
           revealing a hidden corruption.

Carpani: *Luigi, thank you very much for your time. It's always a pleasure.*

Zoja: Thank you, Stefano.

Note

1 Jung's CW 10, para 371.

Chapter 13

Day 92 of War

Elana Lakh

Abstract

The interplay of good and evil in wars and how their boundaries blur is a topic of discussion with Elana Lakh, a psychotherapist and political activist born in Ukraine and raised in Israel. She lives in Tel Aviv, where she observes violence against Palestinians on a daily basis. At the same time, she says, the common view – strongly nurtured by the Jewish population as victims of World War II – that the country faces a constant existential threat dominates the discourse. Today, that threat would be caused by the Palestinians. For Israelis, the role of Israeli perpetrator would still not fit this view. Lakh transfers this to the blindness of Europe and the U.S., which would feel a superiority to cultures on the margins: But the good West and the bad others would not be a valid concept. So, the war in Ukraine would remind us of what we don't like to hear and see: The West can kill, too. For the Israeli, the war in Ukraine, on European territory, is an opportunity to bring this to consciousness. Deepening the motif of the beast in war, Lakh draws a parallel with Israel: in public opinion, she says, Palestinians are portrayed as inhuman beings. This practice of seeing the opposing party in a war or conflict as monsters is a common practice of dehumanization, Lakh says. While Russian soldiers are derided as orcs by Ukrainians, Vladimir has called Ukraine a state of Nazis from which he must rid it. Calling someone a Nazi, he said, is the ultimate justification for doing whatever you want to that person. In this psychosocial and political discussion, Stefano Carpani and Elana Lakh talk about the Nazi narrative that serves Putin well and venture back into the history of Ukraine and Russia.

Carpani: *Good afternoon, Elana.*

Lakh: Good afternoon, Stefano.

Carpani: *Thank you for agreeing to join the new web series* War as Reset. *I am pleased to have you participate for several reasons – you are in Jerusalem, [an] Israeli, and an activist working in the Israeli–Palestinian conflict field. The recent Russian invasion of Ukraine urged me to discuss war and the concept of resetting through it. Many wish the pandemic was nothing more than a dream, a nightmare, but the truth is that even still now, [COVID-19] is a reality. Many people wish [that this war was nothing more than a nightmare – and as we know, psychoanalyst know very well that nightmares can only disappear when their meaning is represented. And this is what I'm trying to do with this series. Also, because – following Andrew Samuels – I truly believe that these conversations in depth can be an antithesis to what I call media bulimia. Media outlets, TV, even radio, newspaper[s] – they share news that [is] really empty and [they] do not help to understand the situation. War is the*

DOI: 10.4324/9781003390039-15

antithesis of the "Vanity Fair"; war is the "Atrocity Fair." But when it comes to TV media outlet, it becomes again the "Vanity Fair" when we are attracted by war. We eat war on a daily basis, yet nothing alters. Recently, I conversed with Luigi Zoja about the anxiety during the [COVID-19] era and the consequences of [COVID-19] news. What will happen next? You reside in a constantly threatened country. Elana, define war.

Lakh: It's a big question. But I think that basically war is in the basis of human nature. And it's sad to think about that it's an archetype and this is something that people have been doing since primordial times. Maybe in the past few decades, we grew up thinking that war has ended. War is conducted by various means, but not in the way that it was conducted before. At least for the past 20 years, I learned to look at war and at acts of war from the perspective of who profits from that and how acts of war influence the minds of people and how we are taught to think about it. But this invasion of Russia in Ukraine struck me totally shocked because it's so violent and it's so aggressive and it's so savage. We had hoped these kinds of sights [were] not here anymore – but it's an illusion, because they are here. They are all over. It's just that the Western world doesn't look at it. And as you said, I live in a country that is at war. The question is: What kind of war we are in? We are used to think[ing] about Israel being under existential threat. But that's not true. Israel [has not been] under existential threat for long time, but that's the notion that we are taught: We are under threat and we have to be strong so we will not [become] extinct. I want to question that, because that notion is dangerous in my eyes. And there are attacks on Israeli civilians, and there are missile attacks in Jerusalem. As I have lived here for the past 30 years, I've been witnessing attacks. When I hear more than one siren – as every Jerusalemite – I jump, because we are programmed like that. When I hear more than one siren, I immediately think that something happened. Living in the middle of Western Jerusalem during the mid-[19]90s and the beginning of the 2000s, more than one siren meant that there was some kind of a bombing. That is different today. But nowadays, the violence that we perpetuate on Palestinians is much greater than the violence that we suffered here in Israeli areas. And that I'm trying to do something about and that I worry about because the way that things are presented in the media, we as Israelis are protecting ourselves against the great violence. We are the powerful side here. And as a powerful side, for me, we have the responsibility to think about it and to stop it – and that's not happening. So, war for me is big suffering caused to human beings. That's one thing. And war is also something that is used to consolidate public opinion. And it's something that is used to make us think, think, think.

Carpani: You mentioned that war is about who profits the most. I appreciate your statement that war is about who reaps the most benefits. However, if we examine the ongoing struggle between Israel and Palestine, it never ends – and no one benefits. Natu-rally, Israel is expanding and seizing more land, while the Palestinians are defend-ing themselves. But this has been the case since the 1940s. The Israeli military forces include both males and females. During my trip to Jerusalem in the spring of 2006, I saw young soldiers carrying large rifles. I also recall them visiting Yad Vashem, wearing civilian clothing, with these big guns. They entered the museum and were told about our history, the presence of an enemy, and their obligation to fight. And it was quite surprising to see those youngsters carrying a firearm. It felt like something out of a dystopian novel. However, the concept of a dystopian society is likely to emerge later. According to Freud's theory, aggression is considered one of our natural instincts. Do you believe it is more associated with the masculine or feminine?

Lakh: I was thinking about this question since we began corresponding, and when we think about war as masculine or feminine, it doesn't have to do only with men and women. Because the women in the Israeli army can engage in masculine activities. We can think about masculine forms of war, which are the majority – like what we see in Russia and Ukraine, and what we see throughout history: aggressive, violent attacks. But there is an interesting myth about Krishna, an Indian Hindu myth. The myth is about a demon that was killing many people and they asked for the help of the gods. But this demon was promised by the gods that no man can kill him, only a woman. Krishna asked his wife to interfere. And she went into a mania of killing and she couldn't stop. So, would we say that killing is masculine? I don't know.

Carpani: *We could also examine Medea or Athena, as you rightly suggested. This is why I pose the question. Although some argue that warfare is masculine due to males possessing a particular body and aggressiveness compared to females. During the police's attack on protesters in Maidan Square in 2014, men automatically circled around the women. There is also the occurrence of Kurdish women safeguarding their nation against the Islamic State, and the Colombian FARC containing female fighters. These fighters are women, too. But if only women go to war, the nation will perish, as there will be no women left to reproduce. This is a crucially important point that is not merely symbolic.*

Lakh: I want to tell you something about masculinity and femininity. Because this idea of women conducting things differently, we find it in *Lysistrata* and we find it in the Liberian resistance. Women have a different way of doing things. I'm always participating in demonstrations against the occupation. But in 2002 or 2003, I attended a demonstration against the separation wall in Jerusalem. General thinking was that Israeli women would not be beaten up by the police. So, they asked us to go ahead in the first rows of the demonstration. It was a frightening demonstration with tear gas and – and the clubs of the police. The idea that white women are protected from violence was not true. It wasn't true back then. Now, of course, we are less prone to being beaten than the Palestinians, of course, and less prone to being beaten than the men in the demonstrations. But . . .

Carpani: *This brings me to the question why wars turn normal people into beasts. Why does it escalate?*

Lakh: It doesn't come out of nowhere. It's a gradual act of seeing people as non-human beings. In order to beat a woman, you need to see this woman as non-human. You need to separate humanity and attribute humanity to one group; the others are non-human. And the practice of acting like that in demonstrations has been around for years in the demonstrations with the Palestinians. Palestinians [have been] beaten and shot in demonstrations ever since. In the past few years, this practice also entered demonstrations inside Israel: People who were demonstrating against Netanyahu in the past few years were beaten, not in the same way, but also; the ultra-orthodox people who were demonstrating, as well. The police become more violent because of the dehumanization of what is called the enemy. And then you come to beating up a coffin carrier. I think that's how I see it. In order get out the savage in you, you need see the other as non-human. And there are various mechanisms to do that. We see it in Ukraine. We see it also in the way that the Ukrainians perceive the Russians. It's interesting, I just now heard that the Ukrainians, they called the Russian soldiers orcs. Did you hear that?

Carpani: *No, but what I understand is that it's challenging for our Ukrainian colleagues to communicate with Russians. One colleague talked to me, but the Ukrainians were not interested in conversing with the Russians. They stated: "Perhaps in a few years, but for now, permit us to battle this war to safeguard ourselves." I have a patient who escaped from Ukraine two months ago, and she mentioned: "Three months ago, killing was against the law and you would be sent to prison for it. However, now we have to take a life to protect ourselves." I would like to discuss this with you because it goes against our Judeo-Christian values. Nevertheless, in Ukraine, it has become a daily occurrence. How do you feel about this? The danger you mentioned is indeed present, not only for non-military individuals. However, in Israel, every civilian is considered a soldier.*

Lakh: Not every, but many.

Carpani: *Not every; this is important. Because if you say you don't want to be a soldier, you are considered a traitor.*

Lakh: Yeah, that's true, in the Israeli public opinion. But you know – the discourse of refusing to join the army is a small one, but it exists since the past 20 years. When I was a teenager growing up, it wasn't even a possibility. I didn't think about it at all. Refusing to join the army was not thinkable. Everyone did as everybody else, but I knew very little back then when I went to the army. Growing up in Israel, I didn't know anything about the disaster of the Palestinians. I didn't hear the word Nakba. I didn't know about the remains of the villages around. I grew up in Be'er Sheva, a southern city that was a Palestinian city up until 1948. And there are Palestinian homes that were demolished. I never asked the question. I didn't know that these houses were inhabited by other people. I didn't know that these people were refugees. And that's the way we all grew up. Looking for this information is an act of wishing to know.

Carpani: *When did you discover the potential to see things from a new perspective? When did you ask yourself these questions? When did you begin your activism?*

Lakh: In high school, I was in a youth movement of one of the political parties, the leftist ones. It doesn't exist anymore. Then I began asking questions, but mostly [it was] when I was a student, a freshman student at university. I went working for one of the human rights organizations called Center for Defense of the Individual. It's an Israel-based organization that was receiving complaints of Palestinians, of things that the government, the police, the army did to them. And my job back then was to follow these complaints and to help these people try to get justice, which they never got. But when the army forces would arrest somebody in the middle of the night and take them somewhere, they would never tell the parents where they would take the arrested person. The parents would call the office in the morning, and I was the one to find out where they were and then we sent a lawyer to help. Or when back then there was still a practice of torture, which was, in a way, legal – it was before the Supreme Court banned torture. We would document torture incidents and submit them to [the] Supreme Court. I began understanding things that I didn't see, and I was shocked. I was trying to tell people about what I learned, but people thought that I'm lying: It's impossible; it can't be; we don't do that! People cannot stand the idea of us being perpetrators. There is a phrase that I heard from a friend of mine, an elderly woman who survived Auschwitz. She says that the Israeli people do not want to know – we actively want not to know.

Carpani: *David and Goliath come to mind, and I often think that it is not Israel itself, but rather the state and government that have become Goliath, no longer David. It is difficult to understand this due to the thousands of years of being under threat, and the Holocaust. Interestingly, Southwest Ukraine, Poland, Galicia, the Czech Republic, and Austria were home to the largest population of Jewish intellectuals in Europe. Removing Jews from these areas would leave them empty, as has happened before.*

Lakh: I was born in Kyiv, but I was 3 years old when we came here, and I don't remember anything. My father was 4 years old when the Nazi people came – only now he begins talking about it. He's 85 years old now, and he never said anything. Only now he is beginning to talk about it. I know that his grandfather was murdered by Ukrainians. And now around this actual conflict, in our institute, there are many connections with Ukrainians and Russians. And I began supporting some Ukrainian analysts. And I participate in a project of training people to work with victims of sexual violence, because that's what I do here in Israel. Usually, I don't perceive myself as of Ukrainian origin, and I'm not connected to this notion of the victim that is a cultural complex here. But I was listening to stories that Ukrainian people were telling me, and I was thinking about that my father's grandfather was killed by Ukrainians. In a way, I connected it in my mind in a very strange way, which is totally alien to what I usually think.

Carpani: *You mean killed by the Ukrainians that joined sides with the Nazis?*

Lakh: No, before, in 1903.

Carpani: *Anti-Semitism has existed for centuries, and it became more prevalent in central Europe at the end of the 19th century. Although many people believe that the Nazis were the only anti-Semites, this is not the case. It is crucial to note that Putin's claim that Ukrainians are Nazis is false. While there are groups such as the Azov Battalion and neo-Nazis in Europe, these movements are currently under control. I recall a time from my childhood when someone from my neighborhood, with whom I had played football, turned into a neo-Nazi, got arrested for several years, and hopefully underwent a change of heart. I believe that those that join neo-Nazi or super left movements are really people that need psychological help, because they hold a vision that is 100 years old, but perhaps we need a psychosocial effort and I take this opportunity to ask you: Did the Ukrainians, who experienced suffering under the Russians due to communist-style oppression, turn to the Germans and embrace the Nazis because they offered assistance and protection against communism? During the Austrian-Hungarian Empire, Ukrainians and Poles received better treatment compared to under the Russians. Putin's narrative that Ukrainians are all Nazis is false, as their president is Jewish. But I wonder if Ukrainians who leaned towards right-wing and Nazi ideologies did so consciously or unconsciously to oppose their enemy – in this case, Stalin. We should recall the famine of the 1930s. Although it's a complex psychosocial and political discussion, it's worth having to avoid superficiality.*

Lakh: I don't know about this movement in Ukraine in the 1930s, but Ukraine – like every other part of the Soviet Union – suffered from severe poverty and starvation and also suspicion. The policy of telling the government about which of your neighbors is withholding money or food was rewarded. It was a terrible thing. I don't know what drove Ukrainians to the Germans, if it did. I know about terrible suffering of Ukrainians and Jewish Ukrainians back then. But today saying that somebody is a Nazi is kind of marking him like the ultimate evil. And if somebody is the ultimate evil, you can do anything to him. That's a justification. That's dangerous in the same way as to call people not human.

Carpani: *During the Nazi era, Jewish people were labeled as the epitome of evil. Whenever I converse with colleagues, I wonder about the true purpose behind wars, including the Russo-Ukrainian and Israeli-Palestinian conflicts. There's a notion of reparation or reset to fix the political and psychosocial chaos of the early 20th century. Joseph Roth, a Jewish writer born in the Austrian Empire's Lviv, is someone I deeply admire, and he said that the longer the Austrian army wasn't at war – from the mid-19th century onwards – they were waiting for it. They trained, gambled, and there was prostitution, corruption, and a loss of honor. Of course, the Austria Felix[1] we all have in mind under emperor Franz-Joseph, is very much a time of peace, development and culture. However, as society became more like the Roman Empire towards its end, we lost our core values. According to Joachim Fest, a German historian and journalist, both Nazi and communist regimes emerged at the beginning of the 20th century to address the negative effects of industrialization and the loss of values caused by the bourgeoisie. Putin is also attempting to do the same, by rectifying Ukraine's deviation from European values. I want to read something from an essay that is very dear to Jungians: Wotan. Jung writes[2]:*

When we look back to the time before 1914, we find ourselves living in a world of events, which would have been inconceivable before the war. We were even beginning to regard war between civilized nation as a fable, thinking that such an absurdity would become less and less possible in our rational, internationally organized world.

Isn't this what we European thought until months ago? Then Jung continues[3]:

But in the sphere of religion, we can see that once some very significant things have been happening, we need feel no surprise that in Russia, the colorful splendor of the Eastern Orthodox Church has been superseded by the Movement of the Godless [the Red Revolution].

Doesn't Putin want to revive previous values through religion and spirituality? To fail to foresee a world of peaceful coexistence, he resembles the Nazis in his desire to return to the Middle Ages, just like Vladimir. It is simple to find commendable past values, but hard to envision the future. Is this also an issue in Israel and Palestine?

Lakh: That's the big disappointment, because many people in the peace movements thought that people are rational and want to live in peace – want to flourish. So, if people would understand that the war and occupation is bad for the economy and their life, then people would do the thing that is rational and we could develop. It's not only about religion. Nationalism is very strong and mostly it overcomes the forces of rationality. But you are quoting Jung and it gives me the shivers, because this war is in the West. But we are not feeling the same when it happens in Africa, when Sudan is being attacked, when Eritrea is fighting. Terrible incidents of rape by the Russian army have been taking place in Chechnya before and we weren't so shocked.

Carpani: *Why?*

Lakh: Why? Because we look away.

Carpani: *I have another suggestion. Tom Wolfe once commented on the left-wing New York-ers, saying they spent weekends in the Hamptons drinking champagne. We, too, are radical chic. Our activism is what the French call gauche caviar. Yes, we wash with caviar, and many people hope the war will end soon, not because we want peace, but because we want to keep watching Netflix.*

Lakh: And Zelenskyy doesn't allow us to do that. He pushes the reality in the West.

Carpani: *He is a skilled communicator who uses relatable language to connect with people. However, I believe there is another reason why we overlook certain issues. It is possible that as white Europeans with a Judeo-Christian background, we may perceive ourselves as morally superior to others.*

Lakh: True.

Carpani: *When there was a war in Yugoslavia, we had the moral high ground: These people are not like us, they are inferior, they are black – I use this term because it is etymologically problematic. And we Jungians, when we look at the concept of the primitive, it is problematic. We can no longer use it. And rightly so. But I think that Ukraine has been telling us since the early 2000s, and especially in 2014, that they do not want to be – directly or indirectly – under a state that is an authority. They want to have the right to self-determination. Ukraine wants to belong to certain values. There are the values that – according to Jürgen Habermas and Jacques Derrida – bind Europe together. But we did not see them as Western Europeans. We saw them as Russians: You don't belong to our club. But for Ukraine, it's not about BMWs or McDonald's, but is according to Habermas and Derrida: a political mentality, different from the American one, and based on some common traits. An aversion to the use of force, first of all, and therefore an insistence on law and respect for international legality; support for a global system based on "liberal" multilateral institutions and on human rights.*

   Fruit of a past inspired by opposite principles and characterized by centuries of bloody carnage, precisely, this mentality had to push Europe to build a foreign and security policy anchored to the EU, overcoming the "stupid" and simplistic opposition between war and peace.

   What do you think?

Lakh: That white people wrote this. There is a kind of a blindness in this idea because Europe as well as America and Israel have other ways to kill. Not violently with knives but by economic and other means. This war reminds us that we cannot dissociate. It reminds us what we don't want to see and we don't want to hear, and it puts the savageness of men, which Europeans and Israelis don't want to see, into our faces: They are the savages, not us. We are, because killing with bombs from the plane is not different than killing by hands.

Carpani: *What is the difference between the Russian invasion of Ukraine and the crap perpetrated by the U.S. since World War II in Vietnam, Iraq, Afghanistan? What about the U.S. invasion, occupation, or political destabilization of many countries under the pretext of liberation, under the protection of spreading liberal democracy? Addressing the nation on September 11, 1989, George Bush Senior [President George H.W. Bush] talks about a new order of peace among nations. And then came Iraq and Afghanistan. This really psychotic view is important. Maurizio Ferrara, an Italian journalist, wrote [in]* Corriere della Sera *on March 7th that the U.S. invasion of Iraq underlined that it was a predetermined carnage just like that of the last 90 days in Ukraine, which took place after months of preparation. Ferrara reminds us that Nobel Prize for literature winner Svetlana Alexievich has drawn a very clear picture of the imperialist impulses, not only of Putin but also of the large part of the Russian population – the long wake of the Homo Sovieticus and before that of the tsarist tradition. What about the imperialist impulses of the United States? Every time the British invaded and occupied a country, they created problems. Your region for example, Kenya, Cyprus, you could say they are still occupying Ireland. Is the death they cause any different from that of Stalin then or Putin now? The capitalist USA invaded countries for geopolitical reasons. I think it's very important to wash away this righteousness and moral superiority.*

Lakh: I think it's not totally the same because Stalinist practices and also Soviet practices later are not the same compared to that of the USA. The kind of distrust and paranoia they installed in the people of the Soviet Union is different than the way that the West does it. Still, we have this separating thinking of the West as good and Soviets or Muslims or every other opposing the West as bad. We don't see the evil we are causing. And if you ask me about my activism – that's my purpose. My purpose also as an analyst is to bring into consciousness the things that the consciousness doesn't want to see. I think it's also the purpose of analysis to let the ego encounter the things that are shadowy, that the ego resists. But it's difficult. In analysis, we need to do that in a very careful way, because if we confront somebody with his shadow, it can overwhelm the person and create resistance. I'm looking for ways to bring into consciousness this hidden feeling of superiority of white Westerners.

Carpani: *How do you do that? Because you are a white Westerner. Would you like to tell me what you are doing right now? Because I know that you are very active.*

Lakh: I was not so active in the past few years because I got tired of talking to people about what they don't want to hear. It is really difficult. I try to tell people around me about the life of Palestinian people, about what they live through every day, about how it is when soldiers enter your house in the middle of the night and take your 13-year-old son to interrogation and talk to you in a language you don't know and you are dead terrified. I'm trying to tell what's going on. I have a very good friend in Gaza who I talk to on a regular basis, and when he tells me about recent events, I try to tell my friends. Most of what I do is I trying to be a witness of what's going on and to participate in peaceful demonstrations.

Carpani: *Before, I suggested that David became Goliath. We know David succeeded through an act of violence. If we say the Palestinians are the new David, we also know violence will not change the situation. One might ask: Why the bombings, why the terrorist attacks, and why the rockets? I believe it's due to desperation, but as a white European, I recognize my lack of understanding. That's why I'd like to chat with Palestinians residing in Palestine to hear their perspective. I'm also curious about your life experiences, our present actions, and war's concept as a form of reset – I believe that culture solely follows as a reset. Culture provides insight into understanding our actions. Connecting the conscious to the subconscious is significant to know what we might not want to acknowledge. However, there are times when war is inevitable. Sometimes, in certain regions, it is impossible for culture to prevail, leading to war as the only option.*

Lakh: What do you mean by culture in that way?
Carpani: *Everything that brings people together.*
Lakh: May I propose another aspect of that? Not only culture, but symbolization. Symbolization is very important and the way to represent emotions, and history can be a way out, but it's very difficult around here because representation is only allowed on the Israeli side and Palestinians are fighting for representation of their narrative and for the survival of their human spirit, as well as Ukrainians.
Carpani: *Thank you. It was a wonderful conversation. I thank you very much.*
Lakh: Thank you, Stefano. It's been a pleasure talking to you.

Notes

1 The term Austria Felix (literally "Happy Austria") refers to a historical period of the Austrian Empire, particularly during the Habsburg rule in the 19th century, before the outbreak of the First World War. This era is often idealized as a time of relative political stability, economic prosperity, cultural and artistic tolerance, and a sense of social security within the vast and multiethnic Austro-Hungarian Empire.
2 Jung's CW 10, para 371.
3 Jung's CW 10, para 371.

Part II

Chapter 1

Analysis in the Shadow of Terror
Clinical Aspects[1]

Henry Abramovitch

Abstract

Drawing on the works of Carl Jung, clinical aspects of doing Jungian analysis in the shadow of war and terror are discussed. Terror places special demands on the analyst. These demands include encountering one's death, regressive restoration of persona and the reality of evil. Special emphasis is placed on difficulties in maintaining the *temenos*. Clinical vignettes are presented, along with patients' dreams. The threat of what Jung called "recollectization" is described in the treatment of Palestinian patients. Some patients became frozen looking back at the traumatic situation, like Lot's wife, while others are able to form a transformative, survivor mission that gives meaning and transcendence. As collective violence, in war and terror, becomes more common, we need to learn how to do analysis in their shadow.

At Jerusalem's gate, a black sun has risen

– Osip Mandelstam[2]

To do analysis, we need safe container and magical enclosure of the *temenos* (Abramovitch 1997, 2002) that says: "Here, we are safe; Here, we can explore, even the most violent fantasy" – so long as it remains fantasy. Once the boundaries between fantasy and reality are gone, analysis becomes impossible. Violence and analysis do not mix. Yet in Jerusalem, Kyiv, and elsewhere, we are forced to do analysis in the shadow of terror.

My Death

My first reaction to the recurring violence and suicide bombings is a relentless confrontation with "my death." Every day, I wake with a lingering sense that this day will be my last. This death anxiety is no paranoid fantasy but very much based on synchronous events and many near misses: A horrid devastation occurred in a popular café, meters away from where I regularly attend demonstrations against the Occupation of West Bank and Gaza. A second bomber entered another popular café near my office. He asked only for a glass of water, which aroused the waiter's suspicion, and in the ensuing struggle, the bomb failed to detonate. Another bomb, which with a certain poetic justice killed only its perpetrator, a young Palestinian woman seeking revenge for the death of a family member, was targeted at a bus which runs right past my house. One of the most distressing occurred when I was showing an analysand out of my office; a voice distorted by a loudspeaker called out to us, "Please remain indoors!" Stepping back, we watched together as the bomb squad robot examined and disarmed a suspicious object lying across the street. At that moment, we were no longer analyst and analysand, but victims of a *participation mystique* of helplessness that seemed to break down the boundaries between us.

DOI: 10.4324/9781003390039-17

Most spiritual traditions place a great value on the ongoing confrontation with mortality. Imagining the reality of my death keeps my priorities exquisitely clear. The Hebrew phrase "Repent on the day before your death" indicates that since we never know the time of our death, "returning" – as repentance (*tsuva*) is called in Hebrew – should be ever-present and ongoing. Confronting violent demise may be good for the soul, but it does take a toll. I feel like I am living on borrowed time, waiting until the next bomb. This chronic death anxiety may even give daily life an "as-if" quality of a provisional life, something that normally typifies *puer aeternis*. I dream of escaping to a world like the one Abraham envisioned in which the innocent never die and I know I am trapped in an naïve escape fantasy. "My death" seemed to be getting closer all the time, pursuing me, waiting for me, taunting me that I would never arrive to stand before you today!

I am possessed by another more terrifying and perverse fantasy. In this fantasy, I am speaking to a large audience at a conference when door of the hall opens. A member of the conference staff enters. She looks around and sees me at the microphone. Making eye contact, she indicates that she has an urgent note for me that cannot wait. She walks up to the podium and hands me a note. The note screams out at me that my dear wife and lovely children, my entire family have been destroyed in a terrorist attack in Jerusalem. I break down before you . . . . If the next time the door opens, you feel a twinge of anxiety, then I will have succeeded in infecting you and your psyche with the pervasive anxiety which living under the shadow of terror entails.

Terror attacks not only the body but also the psyche, attacking our primordial sense of security, that the very ground of our life can be so brutally and unexpected fragmented. At the height of the daily explosions, I would jump at any loud noise; I would become uncharacteristically fearful whenever an analysand was late, especially if I knew they were coming by public transportation. I would dread the phone call telling me that one of my patients was blown apart or dismembered. Temenos is a place where people are free to create and learn spontaneously, like children, without criticism or judgment. Jung, in Dream Symbolism in Relation to Alchemy (CW 12) discussed the temenos via a patient's dream series. Dreamer was a young man with an excellent scientific education who produced more than 400 dreams over a period of ten months. The first, initial dream was: *The dreamer is at a social gathering. On leaving, he puts on a stranger's hat instead of his own.* The dream clearly indicates a disturbance of identity and persona. The next dreams involve blocking view of fellow passengers on a train, and sitting on a lonely island. The fourth dream immediately preceding the dream in question was: *The dreamer is surrounded by a throng of vague female forms. A voice within him says, "First I must get away from Father."* This dream reveals an archetypal tension between the feminine and the Father, perhaps with a need to first separate from Father before being able to deal with an overwhelming feminine presence.

The fifth dream that dealt with persona was as follows: *A snake describes a circle round the dreamer, who stands rooted to the ground like a tree.* Here is Jung's comment:

The spellbinding circle is an ancient magical device used by everyone who has a special or secret purpose in mind. He thereby protects himself from the "perils of the soul" that threaten him from without and attack anyone who is isolated by a secret. The same procedure has also been used since olden times to set apart a place as holy and inviolable; in founding a city, for instance, they first drew the *sulcus primigenius* or original furrow. The fact that the dreamer stands rooted to the center is a compensation of his almost insuperable desire to run away from the unconscious. He experienced an agreeable feeling of relief after this vision – and rightly, since he has succeeded in establishing a protected *temenos*, a taboo area where he will be able to meet the unconscious. His isolation, so uncanny before, is now endowed with meaning and purpose and this robbed of its terrors.

(Portable Jung pp. 333–4)

Historically, the temenos was the inner sanctuary of the temples of the ancient Greeks, a place of shelter and eternal truth, walled off from the temptations and shifting winds of the temporal world: the most sacred part of the temple, where the presence of the gods can be experienced. Temenos is one of the most important ways in which Jungians conceptualize therapeutic containment. Like the holy of holies, it is a place both sacred and inviolate. The etymology of the Hebrew word for holy and sanctification – "kadosh," meaning "set apart" – conveys the sense that something holy is set apart from everyday life; so, too, Jung felt therapeutic temenos was set apart for a higher purpose. Just as the temple is a place of meeting between believer and the divine, so too, analytic temenos was a place where an analysand might encounter the self. The numinosity and danger of the encounter require maximum containment and enclosure. The temenos of the Hebrew temple was empty all year except for one day. Only on the Day of Atonement did the high priest enter to ask for forgiveness for sins. As analysts, we know there are occasions when the sacred containment is violated and even lost. Scottish-born, Rome-based Jungian analyst Angela Connolly, newly arrived in Moscow, described some of the difficulties she encountered in trying to maintain the temenos of her therapeutic space (Connolly 2006). Guards at her gate would refuse to allow analysands to enter, or demand identification; her maid would disturb her during sessions. Worse, the terror which Moscow experienced at that time carried a terrifying threat, that Angela would be forced to leave and to abandon not only the temenos, but all those who had entered it. My article "Temenos Lost" described what may happen to the temenos when you move, or in some way change your clinical space so that the temenos is lost; a second article, "Temenos Regained," dealt with what happens to temenos when analyst is absent and describes a highly unusual set of circumstances in which I gave the keys to my office to an analysand, who functioned like a priestess guarding the temenos during my absence.

Confronting Evil

The bombings also forced me to confront evil – both the evil done to us and the evil that necessarily arises in occupying another people. Jungian psychology has much to say about evil, both personal and archetypal, and I feel we in Israel are having an intensive daily semina in which attendance is compulsory.

My colleague, Avi Baumann, draws distinction between archetypal and personal evil. Archetypal evil is an unconquerable, supra-human force that functions in a way similar to the monsters of Greek mythology, or the Devil in Christianity. Personal evil such as cruelty can and should be resisted. The ever-present danger is to be drawn into the clutches of archetypal evil under the control of the victim–victimizer archetype. In this pattern, an endless cycle of violence is created in which victims of violence victimize others out of a consuming sense of their own victimhood. The dehumanizing acts of one side draw out an inhuman response by the other. As Jung wrote:

When evil breaks out at any point in the order of things, our whole circle of psychic protection is disrupted. Action inevitably calls up reaction and in the matter of destructiveness, this turns out to be just as bad as the cause and possibly even worse, because the evil must be exterminated root and branch.

(CW 10, para. 411)

It is important that most Israeli analysts have a direct connection with Holocaust, either as child survivors, Kindertransport, children of survivors, Auschwitz survivor, or indirectly living in a society permeated by Holocaust imagery so that we have all lived and worked in the shadow of that archetypal evil.

When a suicide bomber explodes, their body is literally mixed up with those of her victims. This horrendous loss of boundaries between the bodies of victims and victimizers has, I believe, a psychic equivalent, in which the perspectives of victim and perpetrator – or even observer and victim – become fused. This psychic merging can be seen in dreams in which the dreamer is drawn against their will into the vortex of the violence, as both perpetrator and victim. A student of my analysand, an educational specialist, was blown apart in the horrendous attack at the Mt. Scopus campus of the Hebrew University. Soon after, my analysand had two dreams. In the first, she dreamed that the walls of her house were suddenly destroyed and all that was left was a hidden pile of old photographs. In normal times, such a dream might be understood to reflect personal issues connected with the breakdown of her marriage, or inner feelings of being exposed. But following the funeral of her murdered student, the dream clearly had a collective significance, indicating the abrupt loss of security which a home is supposed to provide. As analysts, parents, and teachers, we often feel complicit in the deaths of our analysands, children, and disciples – shamed that we somehow did not protect them from their fate. This negative participation-mystique in the terror itself was reflected in the analysand's second dream:

In my dream, I had bombed an educational institution in the Old City of Jerusalem, at which I had studied. In the next scene, I was helping children escape from the building that was in danger of collapse, until someone came to stabilize it.

The name Jerusalem itself means "peace" and it is known as the place where heaven and earth meet. It includes the holiest places for Judaism, Christianity, and Islam. It is place of pilgrimage even if many like Jung's strange visitors at the beginning of the *Septem Sermones* said, "We have come back from Jerusalem where we found not what we sought" (*Memories, Dreams, Reflections*, p. 190–1). Jerusalem over generations has been a natural symbol for the self – at its best, holding together the opposites, pointing to something higher; at its worst, gripped by violence committed in the name of a jealous God. I must confess that I have my own love affair with this beautiful and terrible city.

Returning to the dream. The dreamer, although not based in the Old City, would have spent considerable time there. In this dream, my analysand is both attacker and rescuer, seemingly caught in a Sysphisian endeavor. In normal times, one might see a cycle of destructive rage and the compensatory, reparative rescue fantasy, with the analyst saving the psychic structure from collapse under the weight of her own destructiveness. But in the shadow of the bombing, the dream also reflected her sense of being complicit in the events – that somehow, irrationally, she occupies both sides of the victim-victimizer archetype. A similar sense of psychic confusion is reflected in the dream of another analysand:

There is terrible fighting between Israelis and Palestinians and I do not know which side I am on. In a large field, a plane strafes the center again and again and again . . . . The Queen's guards prepare to storm a school.

Not knowing which side one is on points to psychic confusion – even a refusal to choose sides and so implicitly, the urgent need for a third perspective. The ambiguous image of the Queen's guard about to attack a school highlights this ambiguity: Are they aggressors or rescuers, or both? The image of "a plane strafes the center again and again and again" – which in a personal sense might be understood as the paternal sky aggression turned inwards – now seems as the archetypal damage done to the self by the repeating cycle of violence and counter-violence. Under the control of the victim–victimizer archetype, there is an actual and imminent danger of being possessed by the shadow. This danger was poignantly illustrated in the dream of another Jewish patient:

> In this dream, a friend asks me to pick him up and take him by car to an undisclosed destination. Suddenly, we are in the middle of an Arab village. My friend reveals to me that he is a suicide bomber who will blow us up along with the villagers in an act of revenge. I awake in horror.

This dream points out clearly how a weak ego can be taken over and commandeered by the shadow into committing immoral acts. In a different sense, it can be seen as a "big dream" in that it predicted the coming of a secret Jewish underground which sought to kill innocent Palestinians in retaliatory vengeance. During times of terror, there is an ever-recurrent threat of "re-collectivization," in which an individual is swallowed up in collective identity. Jung introduced this term in the context of the dissolution of the persona, writing:

> For the development of personality, then, strict differentiation from the collective psyche is absolutely necessary, since partial or blurred differentiation leads to an immediate melting away of the individual in the collective . . . through his identification with the collective psyche, he will inevitably try to force the demands of his unconscious upon others for identity with the collective psyche always brings with it a feeling of universal validity "godlikeness" – which completely ignores all differences in the personal psyche of his fellows . . . the suffocation of the single individual, as a consequence of which the element of differentiation is obliterated from the community.
>
> (CW 7, para. 240)

Re-collectivization occurs when a person – due to the stress of the individuation process – is re-absorbed back into the collective identity of his group. It can provide a wonderful and profound sense of belonging and togetherness that the abdication of individual ego may bring about. But the process "can be a numbing, soul destroying experience – a robot-like fate imposed by a society in which individual capacities are numbed, destroyed or turned away from the task of creation" (Friedlander 1987, p. 138).

Jamilla

I felt the pull of re-collectivization most strongly with Palestinian clients (Gorkin 1986; Gorkin et al. 1985; Pérez Foster et al. 1996).

I want to illustrate this dynamic first through an incident in my work with a Palestinian woman who I will call Jamilla, which means "beautiful" in Arabic. She was attractive, professional, in her late 20s, married with one young child. She elected to come to a Jewish Israeli analyst since she felt she could not trust the confidentiality of Arab therapists, some who she knew personally. As a result, the decision to enter analysis was situated on contrast between how the familiar was unsafe, while the stranger was safe. The analysis, from the outset, lay in the shadow of terror since Jamilla's father had spent her entire childhood and early adolescence in an Israeli prison as a suspected member of a terrorist organization. He had not engaged in acts of violence but was a member of a banned organization. Jamilla, his first and then only child, would visit her father in prison and grew to dread these visits. She became fearful of this stranger called "father" who treated as an idealized self-object. During her visits, he would say things like "I am surviving only because of you!" Only she, his innocent anima child, gave him the strength to persevere during his long incarceration. Her unseen inner distress was expressed in her recurring dream:

> There is a baby in my parents' arms at the edge of the sea. My parents do not notice as it slips from their arms and fall to the sea bottom, where it seems in suspended animation. The parents notice that the baby is missing but say that it is not a problem and when they see the baby desperately swimming on the surface of the water, they say, "See, we told you there was no problem!"

The dream gives a moving depiction of Jamilla's psyche. Her baby self loses its maternal protection and falls into the unconscious, leading to a splitting of her consciousness. Part of her was "drowned," a dissociated "baby" autonomous complex in suspended animation, and part of her was swimming desperately, over coping, just to stay on top of the water. Her parents never saw the desperate intensity of her situation. She lived without any confirmation of her inner life and whole self, with a sharp split between a compliant persona and a hidden self.

At home, she grew up alone with her mother in a symbiotic style "great round" (to use Neumann's term for the maternal uroborus). Without the intrusive influence of father or paternal uroborus, she took on the role as always coping, never making trouble child. Growing up without a male presence made her culturally, as well as psychologically, anomalous, since she was accustomed to being an independent woman, free to make her own decisions, without the relentless concern with the family honor.

Immediately following each suicide bombing, I felt the intense force of re-collectivization, when Jamilla saw me as the "enemy." In the countertransference, I struggled against seeing Jamilla, not as a fellow human struggling to individuate, but as one of "them." The analysis really began to take off only when Jamilla felt safe enough to blurt out "I hate all the Jews." Later, there was another dramatic incident. Jews place a "mezzuza" on the lintel at the entrance to their home (following the biblical commandment in Deuteronomy 6:9; 11:20). Typically, it is a small rectangular container, containing key passages from the Bible written on parchment that is ritually hammered into the lintel. It is a visible sign of a Jewish home. Observant Jews touch the "mezzuza" and then kiss their fingers as they enter and leave, as my observant Jewish analysands do at my office. On one occasion, on her way out, Jamilla reached over and "kissed" the mezzuza with great seriousness. Then, in the next moment, she turned and spat on the ground in disgust. This mimetic device of imitating a "Jew" was, I believe, a symbolic act, showing the psychic pressure on every ethnic minority to become like the dominant majority. Yet, her seeming act of religious devotion was revealed as a "trickster-like" act, which revealed her disgust toward to her own tendency to imitate the Jews, their use of their religious practices, and their dominant/oppressive position in her homeland. Because I understood how this act was her way of working out her identity conflict, I could accept it without evoking a collective, negative response.

Imagine, for yourselves, how you would feel should an analysand violate some aspect of your home culture or collective identity. A few weeks ago, I did have an experience with another Palestinian analysand, who asked to be called Mohammed. It was on Israeli Memorial Day – a day taken very seriously, as almost everyone knows someone who died. The power of the collective is strongly felt. The collective marker is a two-minute siren when citizens stand in silent remembrance. I realized that our next session would occur during the memorial siren. I wondered what to do. I am ashamed to confess that I tried to reschedule the session to avoid facing the dilemma. However, we were unable to find an alternative time and it did not seem right to cancel. So what should I do?

At the start of our session, I said we would have the siren in the middle of the session and that I would stand while he should do what he felt most comfortable doing. When the siren came, I stood and he sat. After I sat down, I asked how he had felt with me standing. He said he preferred to continue discussing the personal issues for which he had come. But weeks later, we were able to have an open and frank discussion of his feelings, which brought us closer.

Similarly, because I was accepting of Jamilla's symbolic violence, Jamilla also began to feel closer to me. Jamilla began to fantasize that I was not really Jewish, but perhaps Christian. I understood this as an attempt to contexualize our analysis outside the Arab–Israeli conflict, as a defense against re-collectivization. It was safe to "love" a Christian, whereas it was treason to "love" a Jew! Gradually, she came to see me more as an individual, less as a Jew – and this, in turn, helped her to explore her own path within her own conformist culture. Her marital and sexual dissatisfaction led her to have affairs. She had an orgasm for the first time. For any Arab woman, sexual freedom comes up against the cultural complex in which woman symbolizes family honor. Muslim patriarchy views female sexuality as extremely powerful but at the same time subversive to the social order (Mernissi 1987). Women are taught from childhood that their sexuality is an inalienable and permanent property of the *hamula,* or extended family, and a woman's sexual identity is not her private concern but of concern to all (al-Krenawi & Graham 2000). The concept of individual and individuation so strongly emphasized in the West, and Jungian psychology does not easily apply. Instead in many Arab or highly collective societies have what has been called "collective ego identity." Women are expected to act according to the principle of "*mastoura*" – a kind of tight lipped, learned helplessness, as illustrated in the proverb, "A mouth is for eating but not for talking" (Abu-Baker 2005). Women who violate the family honor codes are a dangerous threat to the whole system of ascribed identity and entitlement, based on to whom you are born. Russian use patronym; Arabs called themselves after their children. Her father might say Abu-Jamilla, "Father of Jamilla" except her father would never use a daughter's name, only firstborn son's name.

Sexual liberties endangered Jamilla's life, since dishonor must be wiped out by purifying acts of violence by her closest male relative. In a sense, women have most to fear from those they love most: their father, brother, husband, or even sons. Honor killings only occur when affairs become public knowledge. There were times when I genuinely feared for her life. Ultimately, Jamilla considered breaking another social taboo and asking for a divorce. Divorce is not forbidden in Islam, but is considered an extreme act. Her husband, who she saw as sensitive if weak, did not oppose her. She told her father that she wanted a divorce and received an initial supportive reaction. Driving on the way to speak with her father, her car was stoned by 20 rock-throwing Palestinian youngsters, who mistook her for a Jewish Israeli. One rock smashed through the windscreen and narrowly missed her and her young son. When she arrived at her father's house, he now suddenly turned against her and said: "It would be better if you had died than if you had divorced." At this point, she went into her room; locked the door; put on her favorite music, a romantic Lebanese singer of Arabic music; got into bed; and made an unexpected, impulsive suicide attempt. It had the quality of uroboric suicide, desiring to merge with Great Mother. This close encounter with intrusive aggression, I believe, destabilized her and made her feel unsafe from their rage, as she felt unsafe from her father's homicidal threat. One of the worst fates is to be violently mistaken by one's own people as one of the archetypal "Other." One can try to protect oneself from violence; it is much harder to protect oneself from being misperceived. It also had stark overtimes of internalized Laius complex, of a father killing his daughter.

She survived, returned to Jerusalem and continued her analysis, inventing what it might mean to be a Palestinian feminist, a woman not under masculine control and yet living within a patriarchal society. Other Palestinian patients of mine, however, were forced to break off their analysis because they literally could not come, held back by curfews, check posts, and perhaps the unresolved sense of betraying secrets to one of the enemy. I suspect the issues of doing analysis with "enemies" – whether current, historical, class, or otherwise – is worthy of a conference of its own.

It is important to note positive responses to terror. I recall one patient who was present at a bombing at a café, synchronistically sitting at the same table I myself had been sitting just hours before. She was hardly hurt physically, saying, "only a few scratches, my purse was covered with blood." She fled and her first reaction was what literary theorists call inability to narrate. She could not put together the pieces. "If I wasn't really hurt physically, then nothing really happened. Shouldn't I have seen more, heard more?" she would say. She went to great lengths to investigate what exactly happened, to construct a narrative but for many months she was unable. At the same time, she began to be afraid of going to cafes, going downtown – "how *can* you know anything is safe?" After another bomb, she entered my consulting room and asked half joking, half seriously, "Are there any terrorists here?"

Ultimately, like Abraham in the Bible, she was able to construct a survivor mission, a reason why she was saved. She began telling her story as part of campaign to raise funds for a children's hospital that treats victims, Jews as well as Arabs. Similarly, the common task of treating actual victims of trauma has led to unprecedented atmosphere of cooperation between Israeli and Palestinian mental health professionals, e.g., with Palestinian Counseling Center or Mobile Clinic for Mental Health. My Institute were able to organize the first joint workshop on Jungian psychology with Palestinian colleagues in the heart of the Old City. I am co-facilitator of an interfaith group with a Palestinian from the Mount of Olives.

When I gave an earlier version of this talk, I made slip saying "death" instead of "thanatos." A member of the audience commented and asked about the difference. Death implies grief, but also mourning and some possibility of rebirth. Thanatos does not. It is death-like, through and through.

At times, I feel overwhelmed by the hopeless of thanatos, that there will never be an end to the suffering but doomed to pass on this terrible heritage to the next generations.

This mix of terror, happiness, and sadness reminds me of a poem by Osip Mandelstam,[3] with which I would like to close:

Mounds of human heads are wandering into the distance.
I dwindle among them. Nobody sees me. But in books
Much loved, and in children's games I shall rise
From the dead to say the sun is shining.

I believe that we are only beginning to understand the collective impact of terror and understand the depth of its shadow.

Notes

1 From: Abramovitch, H. (2002) *Analysis in the Shadow of Terror: Living and Working in Israel in Terrible Times*, Copyright (© 2002) by the San Francisco Jung Institute Library Journal. Reproduced by permission of Taylor & Francis Group.
2 Osip Mandelstam, *Sobranie sochinenii*, Vol. 1, 2nd revised and expanded edition, Inter-Language Literary Associates, 1967, Poem 91.
3 Osip Mandelstam, *Sobranie sochinenii*, Vol. 1, 2nd revised and expanded edition, Inter-Language Literary Associates, 1967, Poem 341.

Bibliography

Abramovitch, Henry (1997). 'Temenos Lost: Reflections on Moving' *Journal of Analytical Psychology* 42:569–584.
Abramovitch, Henry (2002). 'Temenos Regained: Reflections on the absence of the Analyst' *Journal of Analytical Psychology* 47:583–597.
Abu-Baker, Khawla (2005). 'The Impact of Social Values on the Psychology of Gender among Arab Couples: A View from Psychotherapy' *Israel Journal of Psychiatry and Related Sciences* 42(2):106–115.
al-Krenawi, A. & Graham, J. R. (2000). 'Culturally Sensitive Social Work Practice with Arab Clients in Mental Health Settings' *Health Social Work* 25(1):9–22.
Connolly, Angela (2006). 'Psychoanalytic Theory in Times of Terror' *Journal of Analytical Psychology* 48:407–431.
Friedlander, Albert (1987). 'Destiny & Fate' in *Contemporary Jewish Thought* (Eds. Arthur A. Cohen & Paul Mendes-Flohr). New York: Scribners Sons.
Gorkin, Michael (1986). 'Countertransference in Cross-Cultural Psychotherapy: The Example of Jewish Therapist and Arab Patient' *Psychiatry* 49(1):69–79.
Gorkin, Michael, Masalha, Shafiq & Yatziv, Gabi (1985). 'Psychotherapy of Israeli-Arab Patients: Some Cultural Considerations' *Journal of Psychoanalytic Anthropology* 8:215–230.
Mandelstam, Nadezha (1999). *Hope Against Hope*. New York: The Modern Library.
Mandelstam, Osip. "341" from The Selected Poems of Osip Mandelstam p. 84, translated by Clarence Brown and W.S. Merwin. Translation copyright @ 1973 by Clarence Brown and W.S. Merwin, used by permission of The Wylie Agency LLC.
Mernissi, Fatima. (1987). *Beyond the Veil: Male-Female Dynamics in Modern Muslim Society*. Bloomington, Indiana: Indiana University Press.
Pérez Foster, Rose Marie, Moskowitz, Michael & Javier, Rafael Art (Eds.) (1996). *Reaching Across Boundaries of Culture and Class: Widening the Scope of Psychotherapy*. Lanham, Maryland: Jason Aronson.

Chapter 2

Donbas in the Battle for Cultural Identity, or Cultural Identity in the Battle for Donbas

Natalia Bolycheva

Translation from Russian into English by Elena Grishina

Abstract

The chapter looks into the process of surviving and reflecting the Russian-Ukrainian conflict in the symbolic space of the Russian culture. The key image is the collective image of Donbas defending its cultural identity. Its relationship with Russia is seen as the "Center/Edge" liminal dynamics. In the traditional culture, these dynamics shapes the worldview and cultural identity. It sets the dichotomy Friend/Foe, Ours/Theirs, Cosmos/Chaos between the Self and the Shadow. Rite, myth, and fairy tale reveal their identity and conflict between them, the paradox of reciprocity between the Edge and the Center. The necessary journey from the Center to the Edge constellates and regenerates the image of the sacred Center – *Axis mundi*. In conscious (values) and unconscious (cultural complexes) aspects of cultural identity, this is manifested as the restoration of kinship in the current war poetry (Melnikov, Dolgareva, Rubanov, Starushko, etc.) and its call to the unity of the living and the dead in the Soviet and Christian cosmos.

This article was written in 2022, the year of the outbreak of the military conflict between Russia and Ukraine.

Introduction

Today's tragic events in Donbas and Ukraine are the boundary of the split that has yielded two completely opposite worldviews. The image of the boundary/edge, which I discuss in what follows, creates tension of the traditional polarities and shadow projections in the collective unconscious. Hence, it is inherently difficult to discuss this topic without losing academic or analytical neutrality, since we are all immersed in the collective processes. It is also impossible to claim any comprehensive coverage of the conflict with such a complex cultural and historical background, so the necessary historical details can be only outlined and sketched. Besides, whatever discourse we use, it will inevitably overlap with the realm of ideology, news, and political vocabulary. With this in mind, I do not expect neutrality of myself but rather prefer not to stick to my personal involvement but to retain research attitude towards the symbolizing process with regards to the current Russian–Ukrainian conflict and Donbas integration in the Russian culture space and in the dynamics of the images in the Russian cultural unconscious.

The lyric of the song by Donetsk authors Vladimir Skobtsov and Mikhail Khokhlov says that no one can force people in Donbas to bow down, even if they must die on the cross, and Donbas is grasping the skies and standing in the bloody field, one for all.[1] This song was first performed in 2020 to celebrate the 77th anniversary of the Donbas liberation from the Nazi occupation and is very popular locally, and it evokes the image of a lone hero holding an overwhelming load on his shoulders. Donbas is seen as a mythical Atlas holding the sky and thus the whole true order of things. Not only this is an image of a great burden, but also of a blood-drenched battlefield

DOI: 10.4324/9781003390039-18

where, as a Russian proverb says, "one man is no warrior". Still, Donbas, like an epic Russian warrior, goes on fighting with the Beast, who "woke up in total darkness", as the song has it. The image of the Beast here is obviously a reference to the Apocalypse symbolism, and we can see both a ubiquitous image of a solar serpent fighter and a Russian fairy tale narrative about the battle between the Hero and the Serpent at the edge of the world, on the Kalinov Bridge, on the river Smorodina.[2] Exhausted in the battle, he calls his brothers for help, but they sleep and cannot hear him. The fairy tale is quite expressive here – while the hero sheds his blood in the battle, blood also pours (in a symbolic, magical manner) from the walls of the house where his lazy brothers sleep, to finally awaken them.

The Donbas narrative during many years of its resistance to the Ukrainian authorities has conveyed a very clear motif of a lonely hero's fight and call for help, of bitterness about his brothers (i.e. about Russia), of hope they will awaken, which finally happened when the Russian Federation recognized the independence of the Donbas people's republics. However, this has been (and still is) an uneasy process – not only in the political space, but also in the cultural unconscious.

One cannot really say that the cultural unconscious in this case is clearly defined by the national boundaries. Boundaries of the independent states have run through the body of one big country with shared culture, shared history, families whose members now live in different countries, and no wonder that the image of the boundary/edge itself is "floating", whatever our conscious attitude to it can be.

Despite the international status of the current conflict, both hypothetical experts and men in the street often say it is largely a civil war, since the split may run between family members or old friends, and there are people with both Russian and Ukrainian names in trenches on both sides of the frontline. The Russian Orthodox Church calls civil war fratricidal, but these very words contain the split line, since it is the solidarity and brotherhood between the Ukrainian and Russian peoples, or even the unity we had in mind when we grew in the Union of Soviet Socialist Republics (USSR) which Ukrainian identity results from. But Donbas is standing for this unity by defending its civil, cultural identity. I believe that in Jungian terms, we should rather talk about cultural identity which includes both the feeling of belonging to the Russian culture, continuity of its history, values and meanings on the conscious level, and meaningful identifications functioning as unconscious defensive and adaptive mechanisms manifested on the level of the cultural unconscious (Henderson, 2007) and cultural complexes (Singer & Kimble, 2004). I also believe it is hard to differentiate between these two aspects within the identity. It would be more accurate to say that the unconscious and conscious enantiodromiacly flow into each other, and that the philosophical and ontological dimension of the culture is rooted in the individual and family experience, and vice versa, just as the affective content of a complex reinforces rational self-awareness.

Thus, I would like to explore the ways the collective image of Donbas defending its cultural identity and the dynamics of its relationship with Russia unfolds in the Russian cultural field. To do this, I suggest we add culture science's perspective to a Jungian one and view the collective and cultural unconscious processes through the prism of the worldview of the traditional culture. Starting with Mircea Eliade, it has convincingly been reconstructed to be a helpful model for the analysis of in-depth sources of both individual and collective problems. We can say the traditional worldview makes a historical foundation for the modern cultures, but it also functions as a more profound structure that preserves invariables within the changing realia and connects today's events with the archetypal core at the level of the cultural unconscious and cultural complexes. This is very evident in times of an internal or intercultural crisis. Nowadays, when the idea of otherness (the image of the Other) has been easily trashed in the international and intercultural communication by mutual Shadow (Foe) projections, it has been especially obvious.

The problem of the identity boundaries (frontier identity) and cultural field in the discourse of culture science and social studies is discussed in terms of the frontier thesis (Turner, 1962) and liminality theory (Turner, 1983). Since the concepts of frontier and liminality describe fundamental invariables of the same traditional worldview, this is what I will largely discuss, as well as the ways this structure appears today in the collective unconscious.

I refer to the materials available on the Russian, Donbas, and Ukrainian Internet and media. These are largely not official sources, but social media and Telegram channels' texts about Donbas and SVO,[3] poems, songs, fragments from philosophical debates and papers, etc., by Donbas and Russian authors. I view these materials as the manifestation of the cultural identity at work to integrate Donbas experience and Donbas as experience: both its unconscious (cultural unconscious and cultural complexes) and conscious (values and meanings) aspects. I think it is particularly important to focus on the poetic breakthrough, the sudden popularity of Donbas poetry in Russia, since it is the poetic language by which both the affective/unconscious level and the level of philosophy and values are fully manifested.

Center/Edge Dynamics in the Traditional Culture as a Factor of Forming Worldview and Cultural Identity

From the Jungian perspective, the traditional culture worldview is a kind of projection of the psyche on the external (space-and-time–bound) world. It is built on the fundamental binary oppositions (Cosmos/Chaos, Friendly/Alien, Sacred/Profane) and based on the Center principle. This means that the traditional culture space is built around the sacred Center, which sets the image of cosmos as "friendly" cultural space. The time of the traditional culture is cyclical; it revolves again and again around the static axis of the Center, so the end becomes the beginning and the world collapses into chaos, from which cosmos is created again following the sacred pattern, in the center of the world. This Center – the "*Axis mundi*" (Eliade, 2000) – connects with the myth's space and time, brings together the human and the divine, the profane and the sacred, eschatology and cosmogony.

This relation regulates significant events in the individual and collective life and is a symbolic representation of the transition from one state into another through a culturally legitimized transformation, death, and rebirth in a ritual or festive act. In other words, the *Axis mundi* center imagery works as a symbolic container (rite, festivity) for going through a crisis as a way of entering controlled liminality where something old dies as Sacrifice and something new is reborn as a Hero. All these qualities may be viewed as the manifestation of the symbolism of the Self, which presents itself both as a kind of centered wholeness, as the regulating Center itself, as a reference point and the goal, and as the dynamic transformation process. I would say this sacred *Axis mundi* image that connects the human and the divine is the projection of the ego–Self axis. Besides, as we move down the axis, like down the shamanic tree or the tree of the Cross following Jesus, who descends into Hell, we enter the Shadow space – but since the Center image is the heart of our familiar world, the Shadow is adapted as part of the known, restricted by the ritual framework, and the symbolic descent to the underworld of chaos only precedes the ascent to the sacred one. Thus, the Center principle that unites the culture space is tension between the opposites, where the archetypal energy is symbolized and contained following the cultural pattern. This is the way the image of "Friendly" is set and the individual (age, gender in transition rituals) and general cultural identity is formed.

Symbolism of the Center, however, corresponds to the symbolism of the Edge, the boundary of the inhabited space as the boundary of the human, beyond which there is a different world, one no longer blessed by the image of the Center and the ritual, but dangerous and frightening realm of the Shadow. This sets the dichotomy of Friendly/Alien, Cosmos/Chaos between the Self and the Shadow. Rites, myths, and fairy tales reveal their identity and conflict between them, the paradox of interchangeability between the Edge and the Center. The fairy tale Hero travels from the Center of the inhabited world to its boundary, crosses it to fulfill his task, to confront the forces of chaos – fight the Beast and receive the magical gifts of the Self (in a rite this journey to the Edge is enacted symbolically as the rite subject entering the liminal space of death and rebirth). This constellates the eschatological myth, and the Hero's battle with the Beast regenerates and improves the entire cosmos. It is the Hero's journey to the edge of the world that re-builds this sacred *Axis mundi*, since his journey in space represents both the connection between the Center and the boundary and the down-and-up movement along the world axis, or catabasis and anabasis. Thus, on the border between the Cosmos and Chaos, Friendly and Alien, the eschatological and cosmogonic imagery constellates, the new world emerges and the Edge becomes the heart of the Center, the sacred, Friendly.

The Hero's transition in space follows a certain logical sequence: from a sort of center of "this" world to "that" world and back. The Hero's actions themselves have structuring function and world-creating force. He participates in creating "this" world within "that" world, brings there his values and ideas, and destroys the other world's foundations and principles. Thus the Hero, armed with a certain scale of moral and ethical values, builds his own world, or substantially adds to it. The fairy tale Hero's journey becomes a world-creation force.

(Suslov, 2013)

At the Edge

When we speak about the liminal Center/Edge dynamics in a fairy tale space of the unconscious with regards to this chapter's topic, we need to mention that the name "Ukraine" means in Russian (and Polish) "periphery", place "at the edge" ("owkraina" in Old Russian). According to the Old Russian chronicles and later historical references, the word "Ukrainian" universally meant for many centuries all marginal lands. During the 17th and 18th centuries, the words "Ukraine" and "Ukrainians" were used both as synonyms for the entire Little Russia (Malorossiya)[4] and Little Russians (Malorosses), and as synonyms only for Cossacks,[5] Slobozhans,[6] peripheral people. The framework of this chapter leaves no opportunity to reveal details known to historians and linguists, let alone that their opinions vary greatly. However, it is important that this word – which has become a country name – contains all the liminality problem areas with regards to the image of the Edge. To be near the Edge means to be in a space where the polarities "Friendly/Alien" manifest, and the worldview of the marginal lands population is of liminal quality, including alterability, or shapeshifting. They are seen as both "friendly" and "alien", and can lean to one side or the other. "And it was no longer a symptom, but diagnosis: within a nice, kind, wonderful brotherly nation – who are my blood – there have settled demon-possessed people" (Prilepin, 2021).

This phrase by Z. Prilepin, Russian writer and political activist, expresses the traditional ambivalence of "The Edge" ("wonderful" – "demon-possessed"), fascination, and fear. In his

"Evenings on a Farm near Dikanka," Nicolay Gogol fascinates his readers with vivid characters, but at the same time he creates local, peripheral, Ukrainian images of Trickster and Shadow (Pannochka, the Witch, the Devil in a red scroll, and other characters who are ordinary people during the day and "evil spirits" at night). In the 20th century, this image of "Alien/Shadow" pretending to be "Friendly" resurfaced in the story of Bandera supporters who would go on reprisal raids at night and hide among civil people even after the war was over. This is also true of the image of a civil man with a machine gun buried near his house – this image was present in the cultural unconscious since the October Revolution and Civil War in many trickster-like jokes. The Trickster image manifests itself during the Donbas uprising, when for some reason no one is surprised by the weapons of the militia who "excavated" their supplies.

And at the same time, identity labeled as "peripheral" is in the center of the culture, among us and in ourselves: Ukrainians and Russians live together, in mixed families, and have mixed origins in both Ukraine and Russia. For example, the author of this text is half-Russian and half-Ukrainian, both on her father's and mother's sides.

Gogol wrote:

> I do not know what kind of soul I have, Uke[7] or Russian. I only know that I would never prefer a Little Russian to a Russian or a Russian to a Little Russian. Both natures are too generously endowed by God and, as chance would have it, each of them individually has something the other one lacks – a clear clue they should complement one another.
>
> (Karpov et al., 1988)

Gogol has become a paradoxical phenomenon for modern Ukraine – a national writer whose heritage bears something "strange" in it. His great works written in Russian need not only to be translated, but also censored; where Gogol wrote "Russia", "Russian", "Russian land", "Russian soul," etc., the translation is "Ukraine", "Ukrainian", etc.

Donbas Liminality Symbols in the Space of Ukraine

Neither Ukraine nor Russia – I'm scared, Donbas, scared of you.[8]

Rebellious Donbas, in its turn, is symbolically located between Ukraine and Russia. It is a Russian-speaking region (Russian was assumed to be native for 74.9% of the population according to the 2001 census – the only one during the post-Soviet period), multi-national, with complex history of settlement and part of the territory where its borders had not been stabilized for a long time. But it was quickly developing and shaping as part of Novorossia (New Russia) – lands were recognized as part of the Russian Empire after the Russian–Turkish wars in the 18th century. In 1827, it was named Donbas ("Donetsk coal basin"), a geological term that determined the lifestyle and specifics of this industrial region. Already in the 19th century, it became kind of a melting pot for peoples (even referred to as the Russian "Wild West" at that time), where the identity shaped rather based on one's activity and job than the origin. This process became even more intense in the Soviet period, and workers from all over the Soviet Union arrived in Donbas. At the same time, the region was actively Ukrainized during the Soviet era, because imposing the Ukrainian language was seen as a necessary thing to integrate Donbas into the Ukrainian SSR. The Donetsk/Krivoy Rog Republic was established in Donbas in 1918 as an autonomy within the Russian Republic, but by the decision of the Soviet government, it was included into the USSR as part of Ukraine. Still, one of the first 1920s Soviet posters read: "Donbas is the heart of Russia" – a slogan that has regained its relevance today. However, due to its late assimilation, Donbas is also associated with such characteristics as "wild" and "marginal" space. For Ukraine, they are aggravated with the Soviet, international, and Russian cultural markers and lack of roots in the Ukrainian culture. After the collapse of the Soviet Union, Donbas remained part of independent Ukraine, but almost immediately becomes symbolic "borderland". In the 1990s, threats came from Donbas miners' strikes and Donetsk and Luhansk regions' referenda in 1994 on Ukraine's federal structure and the status of the Russian language as a second national language. In the 2000s, there was dangerous electoral influence of this region (and of south-east of Ukraine in general, including the Crimea with its strong pro-Russian sentiments and Russian majority) on the political process. Even the Donbas population's worldview seemed increasingly threatening since it conflicted with the promoted narrative built around the confrontation with everything Russian as imperial, Soviet, and necessary to be separated from. Eastern Ukrainians (in particular, the industrial and multi-national Donbas) were named "sovoks"[9] whose mentality had to be eliminated. As for Eastern Ukrainians, they saw threat in "zapadnenska", "bandera"[10] culture, with its glorification of the Nazi collaborators and slogans like "Moskalaku na guilyaku".[11]

This growing tension between the conflicting identities within the same country led to a situation whereby Donbas was symbolically separated, liminalized, and labeled as "Alien" in the mainstream media and political space long before it expressed its own will in the 2014 referenda. The difference in positions on the issues important for the country was attributed to the lack of "svidomost'" ("awareness") of the industrial "sausage" region population and their low cultural and intellectual level. A popular slogan in the days of the Orange Revolution said: "Don't piss in the doorway; you're not from Donetsk". Donbas was described as a land of bandits deported "to the mines" since the time of Catherine the Great.

Thus, the traditional edge/borderline qualities constelled in the collective image of Donbas, and those who used to represent "Friendly" space before started to represent "Alien". The Donbas population marginalization process means their symbolical displacement to the Edge, to the borderline position, and bears the symbolism of the ritual separation of the rite subjects from the community. Similarly, Donbas population have been placed in the liminal space; they lost their previous features and identity and found themselves between the two positions, between "Friendly" and "Alien".

> The attributes of liminality or of liminal personae ("threshold people") are necessarily ambiguous since this condition and these persons elude or slip through the network of classifications that normally locate states and positions in cultural space. Liminal entities are neither here nor there; they are betwixt and between the positions assigned and arrayed by law, custom, convention, and ceremony.
>
> (Turner, 1983, p. 168, 1969/1991)

With a rite or a fairy tale, this betwixt-and-between position is often represented as a denigrated (ridiculed or berated), savage (loss of cultural persona), poverty-struck external condition of those who undergo the rite and are stripped of their previous status.

Liminal entities, such as neophytes in initiation or puberty rites, may be disguised as monsters, wear only a strip of clothing, or even go naked, to demonstrate that as liminal beings they have no status, property, insignia, [no marks] that may distinguish them from their fellow neophytes or initiands.

(Turner, 1983, p. 168, 1969/1991)

The monstrous, frightening, low, ridiculed image of a liminal figure in the traditional culture is both a fairy tale image of the younger son, fool, and a jester image, image of a beast emerging from darkness – a combination of Trickster and Shadow.

Donbas is generally seen as "local Ukrainian Mordor",[12] with most of its population clearly divided into bandits and "herd of cattle". Many Ukrainian citizens are convinced that it is Donetsk where all the evil and negativity affecting the rest of the country comes from. In the view of an average Kyiv resident, people in Donetsk and Luhansk regions are stupid, undeveloped, uneducated and can only swill vodka, have drunken brawls, and, armed with a piece of water pipe, rob random passers-by.

(Tkachyov, 2014)

R. Minin, a Ukrainian artist born and raised in Donbas, said in an interview:

In fact, Donetsk region is something that arouses fear and disgust, hatred and contempt in people. I am disgusted when people say there are no gifted people in Donbas, only nitwits. I am hurt because it is not true, and an image of Donbas is deliberately created as a horde of cattle only to be fenced with barbed wire. It is a hard task to change this attitude.

(Minin, 2014)

And then the interviewer said something surprising: "I talked to some culture scientists, and they think that Donbas population has no identity" (Minin, 2014). The implication must have been lack of Ukrainian identity, but it sounds as no status at all – ritual liminality.

When the open confrontation after the Maidan protests in 2014, anti-Maidan protests in southeast Ukraine, unrecognized referendum in Donbas, and a counter-terrorism operation announced by Ukrainian authorities began – the pejorative language of hate which placed "Luganda" and "Donbabwe" behind the cultural fence intensified. These names are allusions to the African countries Uganda and Zimbabwe, emphasizing the symbolic remoteness and edge characteristics (Africa here is an image of a distant, wild space, dangerous and full of crime). Donbas becomes "Downbas" with focus on inferiority, underdevelopment, though this name was mirrored by "Maidown" – the word used by Maidan opponents about its supporters. "Vatnik" and "Kolorada" are pejorative, depreciating language. "Vatnik" is warm working clothes in the Soviet period, "Kolorada" (Colorado bugs) are pest bugs whose color is orange and black like St. George Ribbon, worn as a memorial sign and symbol of the Russian military glory, victory in the Great Patriotic War. The old war of slogans and memes becomes more violent, from exhibitions of posters[13] inviting to destroy "slaves" and "koloradas" up to sadly famous memes like "the Kolorada barbeque" with regards to people (pro-Russian activists) burned in Odessa Trade Union House.

Pejorative, depreciating, and dehumanizing rhetoric on the part of political and media figures increasingly constellate the image of liminal separation of Donbas residents from the human world, and words are used known as Nazi vocabulary like "subhumans" about the militia (A.P. Yatsenyuk, prime minister of Ukraine in 2014–2016), "zooids" about people subject to economic sanctions (Volodymyr Zelensky), etc. Neither word meant Donbas population in general, but to the collective memory, they still sound as generalization.

In 2016, E.N. Nyshchuk, Ukrainian Minister of Culture, stated on the Ukrainian television program "Freedom of Speech" that the situation in the east and south is an abyss for consciousness, and added that "there is no genetics there". This statement contains both archetypal symbolism of the Edge – an abyss of the unconscious between the country and its rebellious region – and a cultural complex connected to the Nazi ideology of genetic adequacy/inadequacy. This inferiority and underdevelopment may also be linked to the "evolutionary immaturity".

A popular Ukrainian blogger wrote in 2014 that the Donbas populace was frequently criticized for being slaves, "sovoks", weak-willed, accepting of violence, etc.; that many of them were unable to pursue careers in small business, creativity, or the arts because of their relatively low IQs; and with regards to their political views, they were still essentially children, that they regardless of their age needed guidance and care throughout life. He attributes this to the fact that the Donbas people had evolved considerably more recently compared to Ukrainians.[14]

For many years, Donbas has been given little focus anywhere but Russia. The position and arguments of the unrecognized republics cannot be heard against the background of what is seen as confrontation between Ukraine and Russia. Donbas is "turned off" and "silenced" like a subject of the rite, staying "in liminality", devoid of human traits.

Igor Gomolsky, a Donetsk journalist, wrote in his Telegram channel that Donbas had not been able to speak up for itself for years and that its image had been made up by Ukrainians. He asked: Who cares about Putin's orcs in filthy vatniks who were drunks and drug addicts?[15]

Two poems discussed in what follows by Donbas authors illustrate how this displacement beyond the boundaries of the human world feels from the inside.

Alexandra Khairullina writes bitterly about armistice that marks the oecumene boundary and represents peace on the outskirts of the world: beyond that boundary there are creatures with hideous heads, non-humans with the voices of birds, birds' rights, or just with no rights, no identification, no driver's license to prove they are human. They are on the edge of the world. She calls this an unbelievable article in the newspaper about people who don't exist for the world.[16]

Dmitry Tribushny, an Orthodox priest and poet, refers to Shakespeare and his characters. There would be stories of more woe, he says, alluding to a famous line in *Romeo and Juliet*, if Shakespeare came back to life in Donetsk Republic and wrote in Twitter about the ghetto and the struggles of the miners. Then Kansas and Arizona would come to know about the area where trains are prohibited and where people have to follow stalkers who guide them to foreign cities to obtain their pension allowances. Or maybe even Shakespeare would not be able to fix the time that is out of joint. Despite what Hamlet may say, the GRADs[17] are the ones who ultimately determine whether "to be or not to be".[18]

Both authors describe this exclusion from the human world, invisibility, voicelessness, and powerlessness that people in Donbas feel. The image of Shakespeare, a great master of tragedy, is there as a symbol of an absent witness who could reveal the truth about the Donbas tragedy to the world. The word "ghetto" adds tragic connotation and takes me as a reader somewhere beyond just reflecting about the liminality images and the ways for the collective unconscious to deal with the traditional polarities and the images of "transition" in conflict. Here we enter the level of the cultural complexes, the recent past of World War II, when the liminal space did not suggest any ritual symbolism of transformation. The image of ghetto takes us to the concentration camp, "subhumans" remain "subhumans", and liminality can no more be contained in the symbolic space of culture, and it lives in the memory of many generations as trauma.

Donbas Liminality Symbols in the Space of Russia

The image of Donbas in the Russian cultural unconscious has long been polarized, too, with all sympathy toward Donetsk and Luhansk victims of cannonades and persecution on the part of Ukraine, with all respect to the heroism of ordinary people and the militia. "Peripheral" people, with their liminal shapeshifter dynamics, betwixt and between, are inherently suspicious: Are they friends or foes? Can they be trusted? "The Edge" is associated both with Trickster's volatility and Shadow's danger (the motif of secret "double loyalty" – refugees, media figures, spies).

Donbas is often a collective image of a hero, militia man, volunteer, and passionary. It gains solidarity, sympathy, and admiration. But it is also the image of a marginal person, a border violator associated with the threat of chaos invasion – violation of the existing order within Russia itself. These threats are associated with historical mythologized images of disorder and autocracy in the region (in the history of Cossacks, Slobozhans, during the Civil War). If we go to the archetypal level represented by a fairy tale, then both the one who resides on the boundary of cultural space and the one who crosses it as a fairy tale hero and engages in battle on the fairy tale bridge carry the projection of the Shadow, by law of similarity between the hero and the antagonist. He is ambivalent in his qualities for those who live in a safe cultural space, because crossing the border between "friendly" and "alien" implies the threat of the chaos invasion, manifestation of the Shadow outside the ritual container. It is literally movement to the edge of the known, conscious, familiar, and ambivalent public attitude toward Russian volunteers who go to Donbas and media celebrities who help people in Donetsk and Luhansk People's Republics has many years been oscillating between the two poles.

On the one hand, it is the fear of the Edge as an abyss. On the other, like in the fairy tale, where the hero fights the dragon on his own and tries to wake up his brothers, but they would not, "Donbas is waking up Russia".

A Donetsk author shares in his Telegram channel that he believes that Moscow is scared of Donbas because it could set up the system that works for a lot of people, that Donbas is Russia's conscience. Like a loud alarm clock to sleeping Russia, it annoys, torments, and reminds of itself.[19]

The Miner's World

It is important to say that Donbas is largely a miners' region, its foundation is the mining business, and one of its main holidays is the Miner's Day. The Soviet image of a miner is idealized; he is a hero of labor, proletarian elite. With post-Soviet loss of this position in the community, there surfaces the reverse side of the ideal miner's image – "dark": unthinking, uneducated, lacking culture. But the image of the miner is inherently pregnant with ambivalent symbolism of liminality: miners are associated with the underground world, with the dynamics of death and rebirth. When underground, they belong to the realm of the Shadow, but extract fire (coal) from its darkness to warm and illuminate the world. Thus, in the symbolic space of this region the main metaphor of work and lifestyle itself is catabasis, the descent into the underworld, and anabasis, ascent to the divine light.

This symbolism has shown itself in all its ambivalence: from a dangerous image of "mining cattle" pulling "down" the whole country (in Ukraine); to ambiguity, uncertainty, and fear of being "infected" by disasters of Donbas (in Russia); to idealization in philosophical and poetic images of workers of spiritual mines. A Donbas philosopher and defender writes in his Telegram channel about Donbas miners who stack up dirt-piled pyramids and refers to the Russian metaphysicists and German philosopher Martin Heidegger, according to whom heaven is the highest depth and Mother Earth is the deepest height. Thus, the earth that the miners extract from the depths is reaching on high.[20]

Another Telegram author, a Russian historian of art, asks: Why did the Donbas region develop into a potent hub for profound and potent processes that rewrote history and delineated boundaries that were not just physical but also moral, ethical, and spiritual? He wonders why it was Donbas and not Moscow, the nation's capital, which united people and bound together Crimea, Moscow, and all of Russia. He attributes this to brave people of the Donbas, who boldly descend into the earth's depths every day, who are individuals with a seasoned spirit, more resilient than anthracite and steel, who made 40,000 dirt mounds in the steppe bigger than the Giza pyramids. They raised their own children to love the Motherland with selfless love, planted beautiful gardens, and grew millions of roses.[21]

Of course, people in Donbas are not only militia and miners. Thus, as early as in the 1970s, the Donetsk population counted more than one million, and it was an industrial center, a city of colleges, research institutes, factories, theaters, libraries, etc. – a city of people with very different jobs. The edge/boundary polarity manifests in the fact that they are seen as non-differentiated, lacking status inhabitants of the "liminality desert", the "gray zone". At the same time, there emerges an image of a heroic city, as enduring as besieged Leningrad was. For example, in years of cannonades by Ukraine, Donetsk National University annually held conferences called "Donetsk Readings". Dmitriy Tribushnyi reveals the image of Donetsk through biblical images, in which the Liminality desert – where Moses led the people of Israel – and the Promised Land merge together. Ring, his poems says, the Promised Land of Donbas, all the way to the top. Manna was promised by prophets, but snow was given instead. Rejoice, blameless city, with every building being put to the test by GRADs and then burned to ashes. For a brief hour, cherubim have unlocked the gates to paradise. Burn, Donetsk, but never be burned.[22]

Donbas in the Rite of Passage

The core of the ritual liminality is both sacrifice and heroism. Something should be sacrificed to be reborn as a hero within the rite. Years of being in legal lacuna, in the "gray zone", are symbolically identical to the "passage" rite in the traditional culture, for people in Donetsk People's Republic (DNR) and Luhansk People's Republic (LNR) live beyond time and normal society with its usual order and status, between life and death.

> Van Gennep has shown that all rites of transition are marked by three phases: separation, margin (or limen), and aggregation. The first phase of separation comprises symbolic behavior signifying the detachment of the individual or group either from an earlier fixed point in the social structure or a set of cultural conditions (a "state"); during the intervening liminal period, the state of the individual subject (the "passenger") is ambiguous; he passes through a realm that has few or. none of the attributes of the past or coming state; in the third phrase the passage is consummated.
>
> (Turner, 1977, p. 69, 1983, p. 168)

Symbolic (but unfortunately, not only symbolic) disparagement, deprival of status, of language, of the right to vote, and excluding Donbas from the cultural field of Ukraine corresponds to the first phase. The second one is liminal position – between and betwixt. The DNR and LNR are no longer Ukraine in terms of self-identification, but still no one recognizes these republics (which means they do not exist), as well as "nonexistent" people. Finally, the third phase is the drama of restoration and reuniting we are all going through today. Donbas is moving out from liminality, receiving a new status (independence), new identity (the names DNR and LNR have been recognized by Russia), and a new group where it belongs (by joining Russia through referendum). At the same time, the image of Sacrifice devoid of human features (together with Trickster and Shadow figures associated with liminality) has transformed into the Hero, who is welcome and waited for.

At the same time, the image of the borderland is changing; what used to be the Edge now becomes the frontline, a symbolic vanguard, and the space of Friendly, built around the Center, becomes the rear. By law of the fairy tale and myth, the principle of their identity and inter-changeability has constelled here; it is the boundary between "friendly" and "alien" – the Edge, which has now become the forefront, where the sacred content of culture is shaped and defended. This is where that the symbolic Center, the *Axis mundi*, is now.

Anna Dolgareva's poem about young soldiers killed in Luhansk brings together the image of the local River Donetz, the fairy tale image of the River Smorodina, Russian Styx as the boundary between life and death, and opening of the boundary between heaven and earth: They are marching off beyond the Donets River, the Smorodina River, the dead waters, the deceased Soviet homeland, and everyone's freedom. They bump and wave from armored cars on dry land, in the clouds, and on the water. Still, life is going on, and Ksyusha is pushing her stroller.[23]

"We are All Donbas"

This is where values and meanings that can be described as culture-shaping are crystallized. Everything that has been attacked – the language, memory, history, family, and cultural continuity – is what people in Donbas defend and feel as nuclear, irreducible values. What is taken for granted and not seen in the Russian cultural space becomes value not to be given up in a situation when it can be taken away. Donbas becomes a symbol of resilience of the Russian people, the center of identity.

A humanitarian volunteer states in his Telegram channel:

> Win or lose, for the people of Donetsk and Luhansk, means life or death. Standing through to the very end. Because they are in charge, so they risk everything. It is not because Moscow is behind them that they can back down. They are the Moscow of today. And because we are all from Donbas, society and the nation will quickly come to understand that there is no other option – nor a step back – the sooner victory is in the rear and in our thoughts.[24]

What is experienced in Ukraine as a separation process from history shared with Russia and liberation from the Russian influence on the Ukrainian cultural identity, for the Russian identity in Donbas and in Russia itself is an attack on the values and memory as sacred cultural foundations and the image of a familiar existential threat. Unfortunately, building of its own identity by Ukraine after the collapse of the Soviet Union occurred, among other means, through the integration of previously unacceptable nationalist and Nazi contents into its cultural field. Anti-heroes of the past, once associated with the Shadow, were now honored, while the image of liberating soldiers was devalued, monuments were desecrated, war veterans insulted, etc. Torchlight nationalist marches, Bandera slogans, the use of the Nazi symbols by the militia and the Ukrainian army, burning people in the Trade Union House with no punishment, arrests, and, finally, years of shelling civilians in Donbas – all this evokes the images of the World War II and cultural complexes associated with it.

As one of Olga Starushko's poems say, World War II will never end as long as there are torches in the dark, executioners eagerly await executions, and people are burned alive.[25]

Protection of memory leaves the symbolic space and becomes an urgent need, individual and collective.

Another Donbas author writes in his Telegram channel that when he sees videos of Pushkin monuments demolished in Ukraine, it makes him feel resentful and hurt. But if the Zoya Kosmodemyanskaya monument is destroyed, with a noose around the figure's neck and disgusting yelling, he is terribly ashamed of the disrespect for the memory of people who protected and saved his family. The Nazi soldier who was burned in the house Zoya set on fire could have been the one who had killed his grandfather, and the Soviet soldier who was a model for Valentin Znoba's warrior/liberator could have saved his wounded great-grandfather. He wonders what he can do to honor and remember them.[26]

The Holy War

For Russian cultural identity which survived in the 20th century, the biggest inner split between the two images of its history – the Russian Empire and the Soviet Union, the Great Patriotic War and the victory that cost 27 million lives – has been an unambiguous integrating symbol and a historical moment. In the cultural space, it is felt both as an ideological and moral victory over fascism (Nazism) and an existential victory over death, salvation of people. It is the point where different views on national history are reconciled and where the vast majority of family chronicles come together ("There is no family in Russia without a hero to remember"[27]). This is also a point of synchronization with the world on the other side of the "border" because the victory shared with the allies created a new world – divided by the Cold War, polarized, but nevertheless with a shared frame of reference. Nuremberg as the legal and symbolic end to the Nazi crimes, the establishment of the UN, the inadmissibility of institutionalization and adoption of the Nazi ideology for countries and societies – all these things are manifestations of the images of the Center and the Edge in the symbolic space. They also marked a security frontier – rigid and backed on both sides by nuclear projects. "NATO's movement to the East", as well as constellation of nationalist, Nazi cultural complexes (now filled with the pathos of resistance to the Soviet, Russian authorities) in the countries bordering Russia (Ukraine, the Baltics), shifts this security boundary, as Russia sees it. In the context of the traditional structure, it is the balance between the cosmos and the chaos in the collective psyche that is shifted, and eschatological and cosmogonic logic is unfolding in the liminal space.

The Victory Day

Holidays in their traditional meaning – as a sacred Center and ritual container – are the key to understanding any culture. Inexhaustible significance of the Victory Day in Russia shows that this is the point where the *Axis mundi* principle largely implements today. The vertical of this axis reads as the connection between ancestors and descendants, representing both a purely individual and collective experience of the cultural identity – through kinship, cultural continuity, and ideological and moral solidarity. "The Immortal Regiment" movement makes it very obvious. A memorial movement started by ordinary people in several years became an annual manifestation of many millions of people, with streets throughout the country crowded with people carrying portraits of their grandfathers and great-grandfathers, relatives, close friends, and winner/warriors. Victory Day comes on the same time as Orthodox Easter, and in religious and poetic language is often referred to as "second Easter" – victory over death. This Easter character expands the symbolic range of the holiday, associating it not only with historical and ancestral experience, but also with the Christian depth of culture. Sacrifice and self-sacrifice of the Soviet people, who paid 27 million lives, is associated at this depth with Christ's sacrifice, and Christ's "shattering of iron shutters" of hell is an Orthodox image of concentration camps gates opened by the Soviet soldiers.

The same encounter is expressed in the Donbas poetry; Donbas people, in their position of Sacrifice, integrate with the image of Christ. He descends to them "to the very bottom" even after His Resurrection. He is one of them.

In Dmitriy Tribushny's poem, doctors, cleaners, and drivers won life by having gone down lower than miners, and the Creator of all life left his home in heaven to be here.[28]

The miracle of Christmas is also there, in this liminal space, in a Donbas basement. In Boris Bergin's poem, the Son of God was born to help people, and now he is in a cold basement in the village of miners, be it Shakhtersk, Pervomaisk, or Octyabr', where children are happy when there is no gunfire.[29]

And, of course, Christian symbolism unfolds in the images of the battle for salvation. Elena Zaslavskaya's poem "Donbas Easter" says:

GRADs are firing in Frunze village. Sakuras in the Donbas are withering away. Grandma Nata is standing by her dilapidated hut and glancing up at the light. Petals are falling like angel feathers, swirling around in the gunfire. Grandpa Ivan shields her with an invisible shield made of his immortality. Amidst the sounds of war, the message of Christ rising is heard. Russian soldiers, save Grandma Nata![30]

The image of Christmas and Easter as a battle, the descent into "Gehenna of fire" come together to explain precisely what the battle is for – for the salvation of the "divine child", not only for the memory that defines us, but also for the future. In Dmitry Melnikov's poem,[31] in the midst of an unending battle, either in Gorlovka or in Popasna or Rubezhnoe, there is a baby wearing a pink coat sleeping and shining in the dark, while soldiers burn in the fire, shielding him with their bodies.

Sobornost: The Unity of "The Living and the Dead" in the Soviet and Christian Russian Cosmos and in the SVO Poetry

Our dead will not leave us in trouble,
Our fallen are our guards.
(Vladimir Vysotsky's song)

The pathos of "SVO poetry" (and humanitarian reflection of what is going on in general) goes in with the theme of the holy war space. What the official language calls "the special military operation" is constellated in the cultural unconscious as an image of an existential battle, where the dead and the living come together, and where the connection between the earth and the sky, past and present, ancestors and descendants, is restored. This is the restoration of the vertical *Axis mundi* – as a historical and kinship continuity, loyalty to the values of ancestors who defeated the fascism, and as the sobornost[32] (collegiality, unity) – a religious and philosophical concept of the Russian culture. "Friendly" world appears as an omniscience, bound by shared meaning, and poetically seen as the Divine "sobor" (assembly, gathering) where both the heavenly host and the earthly host belong. The living and the dead pray about each other.

Another Telegram subscriber from Donetsk wonders she has never seen as many candles in the church in front of St. George and Alexander Nevsky[33] icons as she has in the past few months. The only thing she can do is pray for the Donbas men, as well as for Zoya Kosmodemyanskaya and liberator soldiers, so that they are their protectors in heaven.[34]

Not only soldiers of the Great Patriotic War, but its poets are also there in the poetic space as images of the link of times witnessing the past and the present. In another poem by Olga Starushko, Simonov and Selvinsky stand, hugging and looking at the snow and tank tracks. "Kostya, who is firing shots at our guys?" / "Those who escaped death back then are firing at them".[35]

This image of a common deed, of common war against "the same evil", of "standing in prayer" for one another is represented as "heavenly infantry" that consists both of soldiers of the Great Patriotic War, buried under obelisks and "heights", and soldiers dying today. All unity also gives the strength to accept and comprehend death that opens the door to true immortality and reconnects us with "friendly" in the military poetics. "You will fall asleep and God will take you home to our heavenly boys", promises Roman Rubanov,[36] and Dmitry Melnikov writes: "Right on the battlefield, he was drafted in the heavenly infantry, in the troops of angels".[37]

All the preceding quotations, as well as poems beyond the scope of this chapter, share the same motive – transformation of the terrible, morbid war experience into the "opening of heavenly gates" as profound meanings that manifest the sacred. Thus, the image of a volunteer, a defender with ambivalent qualities, with trickster and shadowy liminality components, gives place to the lofty image of a Russian soldier, an immortal cultural hero. Like epic warrior Donbas who carries the sky in the song, he is ready for his burden and archetypal dimension, like in Dmitry Melnikov's poem where he has reached cosmic heights and have been given the eternal home, so now he is "as sky-high as the Russian heaven".[38]

A Russian journalist comments in his Telegram channel that once again we see a superb Russian soldier who was given a new life. He always comes back when he is needed, no matter how easy or difficult the times are, and this is what makes Russia strong. In times of peace, there may be no soldiers around, but in times of trouble and danger, the warrior is always there to protect. He must have lived among common, mortal men.[39]

His mission is not only military, but a spiritual one, too; the image of the soldier is associated with the spiritual host and goes back to the images of warrior-monks Peresvet and Oslyabya.[40] The same journalist continues that today's Russian soldier is accomplishing a moral as well as a military feat. Mean and untrue things are said about him. He is the target of all the devil's schemes, with a web of lies dragging him down and millions of mean people surrounding him. The peacemaker and the noble hero is labeled an aggressor, a rapist, a criminal.[41]

This Russian soldier, slandered and thrown off his pedestal, wakes up to do "God's work" again, because the first battle is fought in heaven: "And there was a great battle in heaven, Michael and his angels fought with the dragon, and the dragon fought and his angels" (Apocalypse 12:7). Thus, in one of the poems by a Donetsk author, God asks Michael the Archangel himself to release from duty a tired soldier. The image of a soldier becomes huge, ideal, heavenly, but at the same time is balanced and filled with personal, concrete content, which resonates with the Soviet system of images and familiar features. War correspondents' telegram channels remind readers of the novel by K. Simonov "The Living and the Dead", and the SVO participants' poetry brings in mind the stylistics of "Vasily Tiorkin",[42] whose protagonist image is far from the image of a "monument", but rather represents the "Russian character": kindness, sense of justice, worldly wisdom, humor, the ability to overcome difficulties, etc. Famous Tiorkin travels from one new poetic text to another devoted to "Vasily Tiorkin in Donbas". He comes back as a new, but recognizable image. In Priest Vladimir Rusin's poem, instead of a worn-out shirt, he is donning a bulletproof vest. Nobody still questions that it is Tiorkin and no one else.[43]

But an ordinary Donbas man, a peaceful worker, also becomes God's "co-worker", the one who fixes the world and time "out of joint" both on earth and in heaven.

In Anna Dolgareva's poem, there's war in the city now, and mines are all around. With the water supply ruined, there are floods of blood mingling with the water. Seryoga, who is no hero or warrior, fixes water pipes. He is doing his job beneath the fire, beneath the stuffy, hot steam, with blood-tinged water flowing through the streets. Naturally, he ends up on one of the mines. He gets up, shakes off the blood, and walks, with shining behind him and a hole from shrapnel in his eyebrow. Seryoga heads for heaven – where else would he go? His silhouette is made darker by the earthly shadow. He sees it rains with blood from heaven and asks God to let him fix it.[44]

This is an archetypal, fairy tale transformation of the hero, invisible to the world behind the dirt and ashes covering him, into a "wonder hero" who saves the world. And, of course, in the Russian cultural unconscious this image of "a man of Donbas" resonates with the role of the Soviet man in World War II, with a huge cultural trauma and dealing with it.

The traumatic experience, constellated in the image of the struggling Donbas as a kind of split complex engaging in a dialogue with ego's mundane world that holds homeostasis, is gradually shifting in the symbolic space from the Edge to the Center and becoming the focus of general attention. The experience of Donbas becomes also a bottom-upwards and part-to-whole movement:

> From the crucible of the Donbas war, a new idea of life is sure to emerge – not from states and empires, but from ordinary people fighting for the right to choose their destiny. This is the answer to the need for the "ideology from below," which this state so often talks about – and which can greatly transform this state.
>
> ("Man of Donbas", 2022)

This constellates the entire corpus of culture, the *Axis mundi* principle, which connects the individual and collective psyche with multi-level but unity-bound cultural meanings. The vertical link between ancestors and descendants, embodying both the *senex* principle and the image of inheriting to "fathers", "parents", symbolic and akin, affects horizontal, "brotherly" ties in the multi-national and multi-confessional space of culture, and the activation of the civil society, which previously seemed inert. Here we can speak of the Russian image of sobornost as a consolidating and meaning-generating image of the Self in the cultural space. It is both immanent and transcendent; it links with the divine (ideal) and contains the cultural unity. This is the liturgical unity of the Church for believers, and the integrity of historical experience, cultural and family connection of generations, common heritage and individual search, etc. The collective image of Donbas worker, warrior, patriot who starts this consolidation process becomes the cultural Hero, in whom and toward whom the Russian cultural identity is now accumulated.

Notes

1 The original text in Russian in the public domain at: https://stihi.ru/2020/09/08/8426. Accessed November 30, 2023.
2 In Slavic mythology, Kalinov Bridge and the Smorodina River are the image of the boundary between the world of the living and the dead.

3 SVO (in Russian, *CBO*) is an acronym for the "Special Military Operation", the official name for the Russian Army operation in Ukraine.

4 "Little Russia" ("Malorossia") is a Greek term dating to the 14th century; since 1654, it was officially used as the name of the territory of modern Ukraine. "Kiev Synopsis" ("A Brief Description of the Beginning of the Russian People") in 1674 describes Great, Little, and White Russia as three branches of the same ethnic group.

5 Cossacks – a sociocultural group of Russians and Ukrainians; an independent military community that became a military social stratum in the Russian Empire.

6 "Slobozhans" – residents of a "sloboda" ("free settlement") who provided the defense on the state boundaries; "Slobozhanshchina", "Slobodska ukraina" (owkraina, periphery) – area in the territory of modern Russia and Ukraine.

7 Uke ("hohlatsky") – Urkainian.

8 These lines are attributed to Nikolay Domovitov (1918–1996), a Soviet war veteran poet.

9 "Sovok" – short form "soviet", "sovetsky"; disrespectful slang of the 1990s.

10 "Zapadnenska" – western Ukrainian; "bandera" – associated with the nationalist community led by Stepan Bandera known for its collaboration with Hitler's Third Reich and punitive expeditions against the Polish and Russians.

11 "Hang Russians" in Ukrainian slang.

12 Interestingly both in Ukraine and in Europe, Tolkien's images from *The Lord of the Rings* are used to talk about the confrontation between Ukraine and Donbas, and especially now about the Russian–Ukrainian conflict. "Putin's orcs", "Russian orcs", and "Mordor" are images, the fear of which is balanced by the sense of superiority over an underdeveloped antagonist. In turn, Russian social media react with sarcasm to "fair-faced elves from the West", with their sense of superiority over the under-developed east. "Elves" has become an inverted, abusive image that has lost its positive mythological meaning the author had in mind, and "orcs" has become an image of workers.

13 Exhibition "100 Best Patriotic Posters" devoted to the conflict in eastern Ukraine in December 2014.

14 The original text in Russian in the public domain at: https://pikabu.ru/story/donbass_voyna_za_poteryannyiy_ray_2397806. Accessed December 4, 2023.

15 The original text in Russian in the public domain at: https://t.me/harry_homolsky/2482. Accessed December 4, 2023.

16 The original text in Russian in the public domain at: https://soldatskyprival.ru/news/stixi-o-donbasse-chas-muzhestva-sbornik-stixov.html. Accessed November 30, 2023.

17 "GRAD" is a Soviet and Russian 122mm caliber multiple rocket launcher system.

18 The original text in Russian in the public domain at: https://stihi.ru/2014/12/29/4831. Accessed November 30, 2023.

19 The original text in Russian in the public domain at: https://t.me/hrisma/108. Accessed November 30, 2023.

20 The original text in Russian in the public domain at: https://t.me/korobov_latyncev/2302. Accessed November 30, 2023.

21 The original text in Russian in the public domain at: https://t.me/ossinoe/5510. Accessed November 30, 2023.

22 The original text in Russian in the public domain at: https://stihi.ru/2022/04/15/6442. Accessed December 1, 2023.

23 The original text in Russian in the public domain at: https://annadolgareva.ru/poetry/10142/. Accessed December 1, 2023.

24 The original text in Russian in the public domain at: https://t.me/diakon_Maslov/152. Accessed December 1, 2023.

25 The original text in Russian in the public domain at: https://stihi.ru/2015/05/02/5767. Accessed December 1, 2023.

26 The original text in Russian in the public domain at: https://t.me/zakharprilepin/16294. Accessed December 1, 2023.

27 Lyrics from a famous Soviet song.

28 The original text in Russian in the public domain at: https://stihi.ru/2014/12/29/4852. Accessed December 1, 2023.

29 The original text in Russian in the public domain at: https://stihi.ru/2015/01/08/1159. Accessed December 1, 2023.

30 The original text in Russian in the public domain at: https://stihi.ru/2022/09/27/7121. Accessed December 1, 2023.
31 https://moskvam.ru/publications/publication_2939.html (It is in the public domain but you need to click «Читать дальше» [Further reading] after the third poem – then more poems open, and the cited one is the sixth from the top).
32 From "sobor": (1) cathedral; (2) meeting, assembly.
33 Saint George the Victorious, a martyr; in iconography, however, he is depicted as dragon-slayer, an image of a holy warrior, the patron saint of Moscow. Saint Blessed Prince Alexander Nevsky (1221–1263), known for his victories over the Germanic and Swedish invaders, protector of Russia.
34 The original text in Russian in the public domain at: https://t.me/zakharprilepin/16294. Accessed December 2, 2023.
35 The original text in Russian in the public domain at: https://stihi.ru/2022/03/11/6876. Accessed December 2, 2023.
36 The original text in Russian in the public domain at: https://t.me/poemsnewday/2782. Accessed December 2, 2023.
37 The original text in Russian in the public domain at: https://t.me/Dmitry_Melnikoff_Poetry/928. Accessed December 2, 2023.
38 The original text in Russian in the public domain at: https://t.me/margaritasimonyan/11830. Accessed December 2, 2023.
39 The original text in Russian in the public domain at: https://t.me/ossinoe/8537/. Accessed December 2, 2023.
40 Peresvet and Oslyabya – monks, participants of the Battle of Kulikovo (1380), disciples of Sergius of Radonezh (one of the most famous Russian saints).
41 The original text in Russian in the public domain at: https://t.me/ossinoe/8537/. Accessed December 2, 2023.
42 "Vasily Tyorkin" is a poem by Alexander Tvardovsky (1910–1971), written in 1942–1945, about the front-line life of a soldier whose name has become a generic name.
43 The original text in Russian in the public domain at: https://stihi.ru/2023/02/10/457. Accessed December 2, 2023.
44 The original text in Russian in the public domain at: https://stihi.ru/diary/marukava/2016-02-05. Accessed December 2, 2023.

References

Eliade, M. (2000). *Mif o vechnom vozvrashchenii [The Myth of the Eternal Return]*. Moscow: Ladomir. [Eliade, M. (1971). *The Myth of the Eternal Return: Cosmos and History*. Princeton, NJ: Princeton University Press].

Henderson, J. L. (2007). *Psihologicheskij analiz kul'turnyh ustanovok [Psychological Analysis of Cultural Attitudes]*. Moscow: Dobrosvet. [Henderson, J. L. (1984). *Cultural Attitudes in Psychological Perspective (Studies in Jungian Psychology by Jungian Analysts)*. Toronto: Inner City Books].

Karpov, A. A. et al. (eds. 1988). *Perepiska N. V. Gogolya. V dvuh tomah [N. V. Gogol's Letters. In two volumes]*. Moscow: Hudozhestvennaya Literatura. https://ru.wikisource.org/wiki/Переписка_с_А._О._Смирновой_(Гоголь)

"Man of Donbas" (2022). Expert, Dec. 26. Человек Донбасса (26 декабря 2022) | Monocle.ru

Minin, R. (2014). *The Image of Donbas as a Herd of Cattle is Artificially Created*. https://topnews.kr.ua/society/2014/04/15/21748.html

Prilepin, Z. (2021). *Vsyo, chto dolzhno razreshit'sya . . . Hronika idushchej vojny [Everything That Has to be Resolved. . . The Chronicle of the Current War]*. Moscow: AST, p. 8. https://www.rulit.me/books/vsyo-chto-dolzhno-razreshitsya-hronika-idushchej-vojny-read-438367-8.html

Singer, T. & Kimbles, S. L. (2004). *The Cultural Complex. Contemporary Jungian Perspectives on Psyche and Society*. Hove & New York: Brunner & Routledge.

Suslov, A. A. (2013). Tempos of the Russian Fairy Tale. *Teoriya i praktika obshchestvennogo razvitiya*, no. 1, pp. 257–261.

Tkachyov, Y. (2014). *Those Horrible 'Donetsk People,' or Are Ukrainians Told the Truth about the Donbas Populaion*. https://mifov.net/?p=2925

Turner, F. J. (1962). *The Frontier in American History*. New York: Holt, Rinehart and Winston.

Turner, V. (1969/1991) The Ritual Process: Structure and Anti-Structure. In Turner, V. (ed.) *Liminality and Communitas (Foundations of Human Behavior)*. Cornell: Cornell University Press.

Turner, V. (1977). Process, System, and Symbol: A New Anthrolological Synthesis. *Daedalus*, vol. 106, no. 3, p. 69.

Turner, V. (1983). *Simvol i ritual [Symbol and Rite]*. Moscow: Nauka.

Chapter 3

Sicily's Infinite War

A Neo-Jungian Point of View

Chiara Capri

Abstract

This chapter explores the mafia's enduring presence in Italy, particularly in Sicily, from a neo-Jungian perspective. It sheds light on the complex interplay of family, society, and criminality within mafia culture. Examining its historical origins and growth alongside changing power dynamics, the chapter uncovers critical psychological factors contributing to the perpetuation of mafia involvement. Focusing on the psychology of mafia individuals, it questions whether the mafia can be viewed as a psychopathological phenomenon. Notably, the absence of fathers and the influence of mothers within mafia families are discussed as key factors hindering personal growth and development. Additionally, the chapter analyses the symbolic and archetypal aspects of mafia culture, emphasizing the role of the "Great Mother" archetype in shaping mafia members' psychological landscapes. In conclusion, it highlights the importance of ongoing research and understanding of the mafia phenomenon. It underscores the need for addressing cultural and psychological factors within the mafia as a means to drive societal transformation, ultimately offering insights into the roots of these criminal organisations.

Introduction: Between History and Legend

Why discuss the mafia in a book that deals with war?

Because there are some loud wars, with explosions, bombs, and migration of civilians, and others that are silent and subterranean. Both are dangerous and terrifying.

For centuries in Sicily, and now also in the rest of the world, a war has been fought against the mafia. To those who have not experienced it at first hand, the mafia may seem something very abstract. A cultural heritage, a strange creature, a legend, even, at times – an invention or something from the past that one finds hard to believe is still relevant and present in our century.

An astonished colleague of mine told me one day: "you Sicilians are obsessed with the mafia! You can't get it out of your heads".

He is right. But how can we think of "getting it out of our heads" when from childhood we come into contact with it? The city of Palermo, for example, is an open-air cemetery. So many tombstones, so many cement wreaths like *memento mori*, are scattered around the city to commemorate the various mafia murders. They are so commonplace that by now (almost) no one pays attention to them. Nevertheless, if one stops for a moment, they stand out in their power of remembrance. The chasm on the Capaci motorway, in which Judge Falcone, his wife, and their bodyguards lost their lives; the explosion in Via D'Amelio, in which Judge Borsellino and his bodyguards were killed; the murderous shooting in which Dalla Chiesa, his wife, and his bodyguards lost their lives, are only the best known. However, between 1861 and 2002, 1,069 people were killed, 133 of whom were women and 115 minors; today 50, of these victims are

DOI: 10.4324/9781003390039-19

still without justice. In total, the victims killed by Cosa Nostra are 519; if, instead, we also count among the dead the mafiosi themselves who were killed in the various mafia wars, we arrive at 5,000 people.[1] Missing from this number are all the other silent victims: those who were wounded, those who fell ill with depression or complicated grief at the loss of loved ones. Finally, those who could not have a destiny outside crime because they were forced into the mafia yoke.

Many more lives have been ruined by the mafia. And it is not possible to enclose all this in a precise number. When the great and terrible mafia attacks took place, everyone remembered where they were. Time stopped, lives changed. Just as when bombs fall or wars break out: there is a before and an after.

I personally remember very well the afternoon of 23 May 1992.[2] I was just 5 years old and I was doing my first ballet performance. I remember the excitement as a child, the stage lights and the loving looks of the audience. Soon after, the music was disturbed by police and ambulance sirens. I was too young to understand what was happening, but the atmosphere in the theatre changed: I remember the murmur, the fear, that turned into terror. Those were very, very difficult years, when the city of Palermo was besieged by the violence of the murdered dead.

These were the years that will go to be known in history as "la mattanza", a term usually used in Sicily to indicate the manner in which tuna were killed.[3] Between 1981 and 1984, around 1,000 people were killed; the era of Riina's rule had begun: The Corleonesi clan, of which he was head, swept the previous ruling families out of Palermo, murder after murder.

This was also the beginning of the era of collaborators of justice such as Tommaso Buscetta and Salvatore Contorno, who helped the magistrates of the anti-mafia pool to define, from the inside, the structure of the Cosa Nostra criminal organisation and to trace those responsible for 120 murders.

The following year, on 9 November 1985, the preliminary investigation of the maxi-trial was concluded. The order/sentence for committal to trial would be signed by the head of the investigation office in Palermo, Judge Antonino Caponnetto; it had been written by Giovanni Falcone and his colleague Paolo Borsellino on the island of Asinara, a former maximum security prison, where they had both been transferred with their families for security reasons. The order is called "Abbate Giovanni + 706".

For the first time, the mafia was brought into the courtrooms of a courthouse, built especially for the occasion: the Aula Bunker. Here, 460 men of honour are tried and 346 are sentenced. The judges impose 19 life sentences and 2,265 years in prison on mafia bosses.

For the first time, the Italian state responded unitedly to criminal violence.

The mafia's response would not be long in coming. After a few years, the magistrates of the anti-mafia pool will be mown down by bombs or left alone in a welter of suspicions and unanswered questions.

So many years have passed since that period of bloodshed and war; today, everything seems to have gone back under the radar, but mafias continue to live and proliferate.

In Italy we have four local mafia: Cosa Nostra, Camorra, 'Ndrangheta, and Sacra Corona Unita. In the last year, the National Anti-Mafia Division (Direzione Nazionale Antimafia, in Italian) discovered other foreign mafia in Italy: Chinese, Nigerian, Romanian, African and ex-USSR mafia.[4]

Our considerations can be valid to the Italian mafia, which is the subject of our study. Italian mafias are spread across different territories; we find Cosa Nostra in Sicily, Camorra in Campania, 'Ndrangheta in Calabria and Sacra Corona Unita in Apulia.

There are some legends related to the origin of the Italian mafias.

The legend of Osso, Mastrosso, and Carcagnosso began in the 15th century in Toledo, Spain, home to an association called the Garduña, founded in 1412. The members of this society acted according to a precise behavioural model referring to the chivalric epic and its main peculiarities: honour, a certain code of ethics, and the strong link to religiosity. The members were outstanding personalities of the society of the time: judges, governors, and churchmen. Among them were three brother knights: Osso, Mastrosso, and Carcagnosso.

For years, the three lived by the rules of Garduña; one day, the honour of our protagonist's sister was outraged by a man protected by the King of Spain. Nevertheless, the brothers decided to redeem the offence suffered by their kinswoman. After committing the crime, Osso, Mastrosso and Carcagnosso were found guilty of murder and sentenced to serve almost 30 years in the prison on the island of Favignana.

They were locked up in a cell in the bowels of the earth. They lived in total darkness for 30 years.

In that gloomy place, the three knights cultivated a deep hatred for the unjust society that had forced them into captivity. In the darkness of their cell, they decided to write the commandments of a new society, in which every man could obtain the justice he desired as long as he was willing to perform esoteric rituals. These included sacrifices and blood tributes.

After that forced sojourn on Sicilian soil, the brothers had devised laws of blood and war that would allow all future followers to grow and multiply, and not to be stopped by anyone. Finally freed from the Spanish irons, the three knights embraced each other and decided to separate, swearing on the sacred bond of family and faith, mutual respect, and loyalty. The brothers wrote entire manuals outlining the rules and disciplines of their new society, arcane rituals, secret formulae, and occult symbols for joining it. They separated: Osso decided to remain in Sicily to become the founder of Cosa Nostra; Mastrosso crossed the Straits of Messina and settled in Calabria, where he laid the foundations for the creation of the 'Ndragheta; Carcagnosso continued his ascent up the boot stopping in Campania, where he gave life to the primordial underworld structures of the Camorra.

Of course, this legend attempts to give a mythological origin to criminal organisations.

From a symbolic point of view, this legend can be interpreted as a fairy tale of evil.

From Jungian studies on fairy tales,[5] we know how usually the protagonist, if he is stuck for a long time such as 30 years, will use that time as part of his own evolutionary process of individuation.

In the mafia fairy tale, on the other hand, confinement in prisons only leads to a structuring and amplification of the shadow aspects: violence, oppression, arrogance, and the pursuit of power at any cost become the guidelines of the mafia criminal's personality.

These characteristics can always be found in the history of the mafia.

We know from historians that mafia societies always grew in the folds of power. The Sicilian mafia, for example, was structured from the power vacuum between the aristocratic landowners who lived in the cities and the peasants who worked the countryside. These large plots of land, which represented the economic power of the Sicilian landowners, were far removed from the cities. In order to control the territory from cattle theft and fires, and to manage economic administrations, the noblemen of Palermo preferred to leave these tasks to trusted persons. These "capibastone" were none other than the very perpetrators of the thefts and damage; a vicious circle was thus created that fed the idea of insecurity and the need to receive protection, which was, however, given precisely by those who were responsible for the crimes.

This capillary control of the territory, typical of every mafia, is one of the most important features; this power was also exploited both during the turbulent years of the unification of Italy and later in the liberation from fascism.[6]

As we have seen, the mafia is a patchwork of behaviour, rituals, rules, and ways of being. It therefore seems interesting to us to also take up the etymology of the term mafia.

Its official registration in lexicography is due to the *Nuovo vocabolario siciliano-italiano* (*New Sicilian-Italian Vocabulary*) by Antonino Traina (Palermo, 1868–1873) with the meanings of "bravery, ruthless, tracotance, smugness", and finally "collective name of all mafiosi".

Since the origin of the term, therefore, the personal characteristics of the members and their way of doing and showing themselves have always determined recognition within society.

But do you really know the personal features of a mafia man?

The Personality of a Mafioso

We ask ourselves a question: is mafia a psychopathology? But what is definitely meant by psychopathology?

As can be seen, it is not easy and straightforward to answer this question, since this construct takes on different resonances. For some, it is synonymous with psychic suffering; for others, it is the result of a behavioural disorder. In the *Statistical and Diagnostic Manual for Mental Disorders* (the DSM-5), psychopathology is "a syndrome characterised by a clinically significant alteration in an individual's cognitive sphere, emotion regulation or behaviour, reflecting a dysfunction in psychological, biological or developmental processes".[7] As further confirmation of the significant relationship between psychopathology and culture, a "cultural framing" of mental disorders is proposed as a necessary prerequisite for effective diagnostic assessment and clinical management.

One cannot therefore ignore the study of the culture one belongs to and its resonances on the subject. Indeed, we Jungians know how crucial the ancestral context is in the construction of the Ego.

Many authors have researched a "specific" psychopathology of criminals belonging to Cosa Nostra. In our studies,[8] however, no features have emerged to date that would suggest a purely psychiatric symptomatology specific to organised crime.

Let us now analyse the results of our first study.

We started with an interesting and descriptive definition of a psychopath by Robert Hare, which in many ways reminded us of what a shopkeeper might make of a mafioso who demands protection money from him.

> He will choose you, disarm you with his words, and control you with this presence. He will delight you with his wit and his plans. He will show you a good time, but you will always get the bill. He will smile and deceive you, and he will scare you with his eyes. And when he is through with you, and he will be through with you, he will desert you and take with him your innocence and your pride. You will be left much sadder but not a lot wiser, and for a long time you will wonder what happened and what you did wrong. And if another of his kind comes knocking at your door, will you open it?[9]
>
> (From an essay signed, "A Psychopath in Prison", Robert Hare).

In our study, we decided to use a validated test to score the personality aspects.

We employed Robert Hare's Psychopathy Checklist-Revised (PCL-R); this test enables quantitative exploration of a wide range of features that all concur to diagnoses of psychopathy. The use of PCL-R test, together with the investigation of additional sociological features, will allow us to characterise these subjects in terms of some psychiatric or sociological features. During the interviews, we also recorded body language, reactions to the questions, and how crimes were described; we also tried to get them to talk as much as possible to see if they contradicted what we had previously read in their dossiers.

We can consider as psychopathic any subject with a total score larger than or equal to 30.

The total score obtained by the 30 subjects condemned for mafia shows that no one can be classified in the psychopathic category. Only two of them obtain a borderline point. Both of them spent their childhood in a reformatory and committed robberies well before the age of 18. A traumatic childhood could explain how they have a higher score.

Another interesting features is an overall low level of education and many of them started to work very early.

In another study entitled "Free to Choose"[10] carried out on a sample of teenage offenders within criminal organisations, we had similar results.[11]

We find that difficult and traumatic childhoods, school drop-outs, and increased substance use (this figure seems to be in line with the rest of the non-criminal adolescents where substance use seems to be increasing) seem to point to more sociopathic than purely psychopathic aspects of mafia subjects. For sociopaths, we mean individuals who assume a deviant cultural pattern primarily as a result of socio-economic conditions of deprivation (Birnbaum, 1909).[12]

Probably more psychopathic characteristics can be found in the leaders than in the operational base we interviewed.

In both studies, what was surprising was the availability to be interviewed and to talk about their lives; the interviews also revealed the presence of solid moral values within the family of origin and the internal rules of the criminal organisation: the founding pillars of the mafia had been incorporated and made their own; they had not been altered by time or the changes in contemporary society.

Indeed, it is astonishing that men who have committed extremely heinous and bloody crimes – who have threatened other men with death, beaten, and stolen – can then hold their own family in such high regard and sacred respect. They seem to have compartmentalised affections, and such compartmentalisation could be considered as an important psychological dimension in mafia culture. For example, it could explain why many mafiosi show lower rather than higher psychopathic traits, demonstrating positive care toward the family while they are able to commit ignominious crimes.[13]

Group strength has always been one of the pillars of organised crime. Perhaps also because, as Luigi Zoja recalls: "evil has an advantage, an asymmetrical force compared to good. Impulses for peace are not accompanied by strong emotions. Instead, destructive impulses are intoxicating, especially within a crowd that dilutes responsibility and reinforces emotions".[14]

The mafia has always had a power and strength fortified by the number of its members. Its domination and control are founded on the group. The individual loses his characteristics, becomes a member of the "family", a superior body that dictates the law, which is unassailable, unshakeable, imperishable. Individuals pass away, families and gangs remain solid.

But in this group dynamic, what can remain of the path of individuation?

The aim of individuation is nothing less than to divest the self of the false wrappings of the Persona on the one hand and the suggestive power of primordial images on the other, according to Jung.

By Persona, Jung means "a kind of mask, designed on the one hand to make a definite impression upon others, and on the other to conceal the true nature of the individual".[15]

The development of a viable social persona is a vital part of adapting to, and preparing for, adult life in the external social world. "A strong Ego relates to the outside world through a flexible Persona; identifications with a specific Persona (doctor, scholar, artist, etc.) inhibits psychological development".[16] For Jung, "the danger is that [people] become identical with their Personas – the professor with his textbook, the tenor with his voice".[17] The result could be to structure oneself and identify with the Persona, i.e., having a fragile and conformist personality; a false self that would result in the inability to self-determine and distinguish between who one is and the world in which one lives. This sets the stage for what Jung called enantiodromia:[18] "the individual will be completely suffocated under an empty Persona".[19]

This could be an interpretation of the failed path of individuation of the mafia subject.

Growing up in a context in which the individuality, the particularity of each subject, and the aspirations are stifled for the greater good such as "the stability of the family", there are few possibilities of undertaking the path of individuation.

Absence of Fathers, Absence of Separation from the Great Mother

In many of the stories collected in our studies, we noted the absence of fathers in the raising of children.

Mafia fathers are often distant, either physically because they are in prison, or if present, they are unapproachable with their indestructible and unquestionable laws.

"In the cruel code of antiquity, the fatherless child also remains without identity and without honour. Since father and society are one, without a father one falls out of society, out of respect into nothingness".[20] Jung emphasised the importance of the father figure for the individual's destiny: It blocks the child's identification with the mother and allows him to acquire the Logos – rational and epistemological thinking, which enables him not to lose himself or regress into maternal aspects. If the father, however, represents a figure that is too strong and terrible for the son, he will be distressed and frightened and can only take refuge in the maternal. For Lacan, too, the father has a regulatory function: by imposing the law of the symbolic, he leads the son to separate and turn towards the outside.

Paternal absence and neglect at this stage may lead to lifelong "father thirst" (Abelin, 1971) or "father hunger" (Herzog, 1980; Schore, 2003), and may lead to experiences whereby it becomes difficult for a child to separate from the mother, as there is a lack of Oedipal experience that will permit the symbolic understanding necessary for this next developmental step (Woodhead, 2005).

Mafia fathers seem reminiscent of the archetypal Zeus man, according to the reflections of author Jean S. Bolen.[21] This type of man would put power – command – first; he demands authority and does not care if in order to obtain it he has to run risks or take someone out (symbolically or physically). There is no space for feelings of weakness or remorse. The world of affection remains external and unknown. Even the choice of wife will often be dictated by power dynamics and political calculation. "For the man Zeus, finding a suitable wife is not a matter of the heart or of affinity, but a matter of state, an alliance that serves the cause of the foundation and consolidation of the kingdom".[22] Wives must therefore be able to stay in their place, to manage and raise children according to the rules of the mafia family. There can be no other way.

The mother educates her children giving them the values of the mafia culture, in a condition of acceptance of the family hierarchy, preventing them from developing autonomy and keeping them forever tied to herself; the father, mostly idealised, is the model to imitate and provides rules and values; the child is educated to the code of silence, to show virility, strength, and opposition to the other power that of the legal state.

Often, therefore, we can assume that children find themselves bound to a motherhood that leaves no escape, from which it is difficult to distance themselves. They are unwieldy mothers because they are over-presented in the mind of the child who cannot question them. An interesting evaluation of the mafia from a Jungian perspective was precisely that of evaluating it as an expression of the Great Mother archetype.

In this archetype, we find at the positive pole the magical authority of the feminine, the wisdom and spiritual elevation that transcends the limits of the intellect, that which is benevolent, fecund, protective, tolerant, and that which favours growth, fertility, and nurturing – all qualities we have in the good mother.

At the negative pole we have the bad mother: that which devours, seduces, intoxicates, and generates anguish and inescapable destruction – mother of death, in short.

The mafia symbolically could embody these two polarities: while on the one hand it is experienced as generating protection, salvation, and even work, in reality it only entails blockade and destruction. The children of the mafia would therefore be conditioned by this contradictory and shattered collective maternal imago, with no clear separation between what is good and what is bad.

But, as long as one is a prisoner of the Great Mother, the negative experience of the feminine world explains the sentimental aridity and the inability to grow manifested by an individual who, incapable of a spiritual ascent and a critique of consciousness, experiences the influences of the unconscious in an extremely permeable way, in a fantastic limbo where the object and the subject come to interpenetrate. But if this differentiation is not implemented, there can be no true conscious life.[23]

Mafia children therefore without a father, without the Logos principle, could be stuck in a suffocating motherhood that does not allow them to grow, symbolically forced to remain in the ranks of the criminal organisation.

"Everything is a message, everything is pregnant with meaning in the world of Cosa Nostra", said Giovanni Falcone.[24]

It is therefore interesting to recall that there is another way of calling each other in the ranks of the criminal organisation: the "Holy mothers", a curious appellation considering the predominance of men within the mafia. We could read it as another symbolic reference to the Great Goddess: soldiers at the service of the Great Mother Mafia.

Soldiers who, however, stop living fully, who become dissociated from the fullness of life, who stop feeling their affections, who become insensitive machines capable of killing their relatives and friends if the Great Goddess of Death commands them to do so.

Soldiers who seem to have drunk the opium potion prepared by Helen, described thus by Homer in the Odyssey:

She suddenly threw into the wine, of which they drank, a medicine
which anger and pain calmed, oblivion of all sorrows.
Whoever swallowed it, once mixed with wine,
down from his eyelids wept not that day,
not even if his father or mother died,
nor if in front of him with bronze they would slaughter
a brother or a son, and he saw with his eyes.[25]

Conclusions

The mafia phenomenon in Italy and the world is complex and multifaceted.

Jungian, post- and neo-Jungian authors have always tried to analyse the social phenomena of the eras in which they lived and the effects on individuals, starting from Jung's assumption that if the subject is blocked in his own individuation process, society will also suffer. "The change must begin in individuals, and must consist in a transformation of their personal tendencies and aversions, their way of seeing life, their values; only the sum of these individual transformations will produce a collective solution".[26]

The fight against the mafia, against the human and social misery that leads to swelling the ranks of criminal organisations, cannot be separated from the need to continue to study it, to get to know it. This is the only way to really get to the end as Falcone hoped, because as Jung says:

> The idea that one can simply turn one's back on evil and in this way avoid it, belongs to the long list of the most obsolete naivety.[27]

Notes

1 *VITTIME DI COSA NOSTRA*, (PDF), su *osserbari.files.wordpress.com*, Centro Studi dell'Osservatorio per la Legalità e la Sicurezza, accessed on 9.5.2023, https://osserbari.files.wordpress.com/2013/11/vittime-di-cosa-nostra.pdf.

2 Date of the Capaci massacre, in which judge Giovanni Falcone, his wife Francesca Morvillo, and escort men Rocco Dicillo, Antonio Montinaro, and Vito Schifani died.

3 The work of the tuna fishermen began in April when a series of nets that could be as long as 5 km are placed in the sea to form the various "chambers" and, given their arrangement, induce the tuna to go deeper and deeper into the inner meshes until they reach the so-called death chamber. In May, boats departed from the tuna fleet and, under the orders of the rais, the chief, took part in the slaughter. This was carried out by encircling the nets of the last chamber, the death chamber, and gradually pulling the outer flaps over the boats until the tuna surfaced and were then pierced with harpoons that caused the loss of the fish's blood and their death.

4 https://direzioneinvestigativaantimafia.interno.gov.it/relazioni-semestrali.

5 Cfr: Von Franz, M. L., *Shadow and Evil in Fairy Tales*, Shambhala, 1995, Von Franz, M. L., *The Interpretation of Fairy Tales*, Revised Edition, Shambhala, 1996; Von Franz, M. L., *The Feminine in Fairy Tales*, Shambhala, 2001; Von Franz, M. L., *Individuation in Fairy Tales*, Shambhala, 2001; Von Franz, M. L., *Psychological Meaning of Redemption Motifs in Fairytales*, Inner City Books, 1980; Von Franz, M. L., *Animus and Anima in Fairy Tales*, Inner City Books, 2002; Kast, V., *Through Emotions to Maturity: Psychological Readings of Fairy Tales*, Fromm International, 1993.

6 Please refer to the work of John Dickie for a more comprehensive discussion.

7 American Psychiatric Association, *Diagnostic and Statistical Manual of Mental Disorders: DSM-5*, Amer Psychiatric Pub Inc, 2013, p. 22.

8 Cfr: Capri, C., Micciche', S., Sideli, L. & La Barbera, D., Using network community detection to investigate psychological and social features of individuals condemned for mafia crimes. *Clinical Neuropsychiatry*, 15(2), pp. 111–122, 2018; Schimmenti A., Caprì, C., La Barbera, D. & Caretti V., Mafia and psychopathy. *Criminal Behavior and Mental Health*, 24(5), pp. 321–331, Dicembre 2014; Caprì, C., Schimmenti, A., Caretti, V. & La Barbera D., Il mafioso come sociopatico. Considerazioni relative alla personalità dei mafiosi sulla base di una ricerca empirica. In G. Craparo, A. M. Ferraro & G. Lo Verso (Eds.) *Mafia e Psicopatologia. Crimini, vittime e storie di straordinaria follia*. Francoangeli, Milano, 2017.

9 Hare, R., *Without Conscience: The Disturbing World of the Psychopaths Among Us*, Guilford Press, New York, 1999.

10 Freedom to Choose project was a National Operational Programme (PON) legality 2014–2020, carried out in the regions of Calabria, Campania and Sicily; Cfr: https://www.giustizia.it/giustizia/it/mg_1_11_1.page?contentId=SPR130775&previsiousPage=mg_1_11

11 Caretti, V., Carpentieri, R., Caprì, C., Fuso, C. R., Bracalenti, R, & Sideli, L. Recovery e disturbi di personalità antisociali e psicopatici negli adolescenti e nei giovani adulti. In L. Grassi & P. Carozza (Eds.) *Psichiatria di comunità basata sulle evidenze e orientata al recovery*. Edizioni Minerva Medica, Torino, 2023. pp. 243–255.

12 Christopher, J. P., *Handbook of Psychopathy*, Second Edition, Guilford Press, 2019; Millon, T., Simonsen, E. & Birket-Smith, M., Historical conceptions of psychopathy in the United States and Europe. In T. Millon, et al. (Eds.) *Psychopathy: Antisocial, Criminal, and Violent Behavior*, Guilford Press, New York, 2003, p. 11.

13 Cfr: Caprì, C., Schimmenti, A., Caretti, V. & La Barbera D., Il mafioso come sociopatico. Considerazioni relative alla personalità dei mafiosi sulla base di una ricerca empirica. In G. Craparo, A. M. Ferraro & G. Lo Verso (Eds.) *Mafia e Psicopatologia. Crimini, vittime e storie di straordinaria follia*. Francoangeli, Milano, 2017; and Schimmenti, A., Caprì, C., La Barbera, D. & Caretti, V., Mafia and psychopathy. *Criminal Behavior and Mental Health*, 24(5), pp. 321–331, 2014.

14 Zoja, L., *Violence in History, Culture, and the Psyche: Essays*, Paperback Spring Journal Books Analytical Psychology & Contemporary Culture English, 2009, p. 140.

15 Jung, C. G., *Collected Works of C. G. Jung, Volume 7: Two Essays on Analytical Psychology*, edited and translated by Gerhard Adler R. and F. C. Hull. Princeton University Press, New Jersey, NJ, 1972, p. 195.

16 Jung, C. G., *Collected Works of C. G. Jung, Volume 7: Two Essays on Analytical Psychology*, edited and translated by Gerhard Adler R. and F. C. Hull. Princeton University Press, New Jersey, NJ, 1972, p. 200.

17 Jung, C. G., *Memories, Dreams, Reflections*, Fontana Library Theology & Philosophy, 1983, p. 416.

18 Jung defines enantiodromia as "the emergence of the unconscious opposite in the course of time". Jung adds that "this characteristic phenomenon practically always occurs when an extreme, one-sided tendency dominates conscious life; in time an equally powerful counterposition is built up which first inhibits the conscious performance and subsequently breaks through the conscious control". In Jung, C. G., *Psychological Types*, Princeton University Press, New Jersey, NJ, 1976, p. 426.

19 Hannah, B., *Striving Toward Wholeness*, Chiron Publication, Asheville, NC, 2001, p. 263.

20 Zoja, L., *The Father: Historical, Psychological and Cultural Perspectives*, London, Routledge, 2000, p. 100.

21 Bolen, J. S., *Gods in Everyman: Archetypes That Shape Men's Lives*, New York, Harper Paperbacks, 2014.

22 Bolen, J. S., *op.cit.*, p. 52.

23 Baratta, S. & Ermini F., *La grande madre. Convergenze*, Vol. 5, Moretti &Vitali, 2009, p. 15.

24 Falcone, G. & Padovani M., *Cose di Cosa Nostra*, Milano, Rizzoli, 1991.

25 Homer, *Odyssey*, Canto IV, vv.220–229.

26 Jung, C. G., Psychology and religion. In *Collected Works*, Vol. 11, 1938–1940.

27 Jung C. G., Conscious, unconscious, and individuation. In *Collected Works*, Vol. 9, part 1.

Bibliography

Abelin, E., The role of the father in the separation-individuation process. In J. B. McDevitt & C. F. Settlage (Eds.) *Separation-Individuation*. International Universities Press, New York, 1971.

American Psychiatric Association, *Diagnostic and Statistical Manual of Mental Disorders: Dsm-5*. Psychiatric Association Publishing, Washington, 2013.

Baratta, S. & Ermini, F., *La grande madre. Vol. 5 Convergenze*. Moretti & Vitali, Bergamo, 2009.

Bolen, J. S., *Gods in Everyman: Archetypes That Shape Men's Lives*, Harper Paperbacks, New York, 2014.

Capri, C., Micciche, S., Sideli, L. & La Barbera, D., Using network community detection to investigate psychological and social features of individuals condemned for mafia crimes. *Clinical Neuropsychiatry*, 15(2), pp. 111–122, 2018.

Capri, C., Schimmenti, A., Caretti, V. & La Barbera, D., Il mafioso come sociopatico. Considerazioni relative alla personalità dei mafiosi sulla base di una ricerca empirica. In G. Craparo, A. M. Ferraro & G. Lo Verso (Eds.) *Mafia e Psicopatologia. Crimini, vittime e storie di straordinaria follia*. Francoangeli, Milan, 2017.

Caretti, V., Caprì, C., Carpentieri, R. & Schimmenti, A., Attendibilità, validità e proprietà psicometriche della versione italiana della HCR-20 v3. In V. Caretti, S. Ciappi, A. Schimmenti, L. Castelletti, R. Catanesi, F. Francesco Carabellese, S. Ferracuti, F. Alfonso Nava, G. Nicolò, R. Paterniti, G. Rivellini & F.

Scarpa (Eds.) *Italian Version: HCR-20 v3 Assessing Risk for Violence. Checklist per la valutazione del rischio di recidiva di un crimine violento*. Hogrefe, Florence, 2019, pp. 92–126.

Caretti, V., Carpentieri, R., Caprì, C., Fuso, C. R., Bracalenti, R., & Sideli, L., Recovery e disturbi di personalità antisociali e psicopatici negli adolescenti e nei giovani adulti. In L. Grassi & P. Carozza (Eds.) *Psichiatria di comunità basata sulle evidenze e orientata al recovery*. Edizioni Minerva Medica, Torino, 2023.

Christopher, J. P., *Handbook of Psychopathy*, Second Edition. Guilford Press, New York, 2019.

Di Maria, F. & Falgares, G., Mafia organization: Governance processes and leadership function. *Integral Leadership Review*, 9, 2009. http://integralleadershipreview.com/8287-mafia-organization-governance-processes-and leadershipfunction

Di Maria, F., Menarini, R. & Lavanco, G., Sindromi depressive etniche e Sentire Mafioso. Un modello esplicativo gruppoanalitica. *Archivio di Psicologia, Neurologia e Psichiatria*, LVI(5–6), 1995.

Dickie, J., *Cosa Nostra. A History of the Sicilian Mafia: The Definitive History of the Sicilian Mafia*. Hodder & Stoughton, London, 2007.

Falcone, G. & Padovani, M., *Cose di Cosa Nostra*, Rizzoli, Milan, 1991.

Hannah, B., *Striving toward Wholeness*, Chiron Publication, Asheville, NC, 2001.

Hare, R. D., *The Hare Psychopathy Checklist-Revised (PCL-R)*. Multi-Health Systems, Toronto, ON, 1991.

Hare, R. D., *Without Conscience: The Disturbing World of the Psychopaths Among Us*. New York, NY: Simon & Schuster (Pocket Books). Paperback Published in 1995. Reissued in 1998 by Guilford Press, 1993.

Herzog, J. M., Sleep disturbance and father hunger in 18-to-28-month-old boys. In A. Solnit et al. (Eds.). *Psychoanalytic Study of the Child*, 35, pp. 223–230, 1980.

Jung, C. G., *Collected Works of C. G. Jung, Volume 11: Psychology and Religion: West and East*, edited and translated by Gerhard Adler R. & F. C. Hull, Princeton University Press, Princeton, NJ, 1970.

Jung, C. G., *Collected Works of C. G. Jung, Volume 7: Two Essays on Analytical Psychology*, edited and translated Gerhard Adler R. & F. C. Hull. Princeton University Press, Princeton, NJ, 1972.

Jung, C. G., *Psychological Types*. Princeton University Press, Bollingen, Princeton, NJ, 1976.

Jung, C. G., *Memories, Dreams, Reflections*. Fontana Library Theology & Philosophy, London, 1983.

Jung, C. G., *Collected Works of C. G. Jung, Volume 9, Part 1: Archetypes and the Collective Unconscious*, edited and translated Gerhard Adler R. & F. C. Hull. Princeton University Press, Princeton, NJ, 1969.

Kalsched, D., *The Trauma and the Soul: A Psycho-Spiritual Approach to Human Development and Its Interruption*. Routledge, Taylor & Francis Group, London, 2013.

Kast, V., *Through Emotions to Maturity: Psychological Readings of Fairy Tales*. Fromm International, 1993.

Millon, T., Simonsen, E. & Birket-Smith, M., Historical conceptions of psychopathy in the United States and Europe. In T. Millon et al. (Eds.) *Psychopathy: Antisocial, Criminal, and Violent Behavior*. Guilford Press, New York, 2003.

Neumann, E., *The Great Mother*. Princeton University Press, Bollingen, Princeton, NJ, 1951.

Schimmenti, A., Caprì, C., La Barbera, D. & Caretti, V., Mafia and psychopathy. *Criminal Behavior and Mental Health*, 24(5), pp. 321–331, 2014.

Schore, A. N., *Affect Regulation and the Repair of the Self*. W. W. Norton & Company, New York, 2003.

Von Franz, M. L., *Psychological Meaning of Redemption Motifs in Fairytales*. Inner City Books, Toronto, 1980.

Von Franz, M. L., *Shadow and Evil in Fairy Tales*. Shambhala, London & Boston, 1995.

Von Franz, M. L., *The Interpretation of Fairy Tales*, Revised Edition. Shambhala, London & Boston, 1996.

Von Franz, M. L., *The Feminine in Fairy Tales*. Shambhala, London & Boston, 2001a.

Von Franz, M. L., *Individuation in Fairy Tales*. Shambhala, London & Boston, 2001b.

Von Franz, M. L., *Animus and Anima in Fairy Tales*. Inner City Books, Toronto, 2002.

Wilkinson, M., *Coming into Mind: The Mind-Brain Relationship: A Jungian Clinical Perspective*. Routledge, London, 2006.

Woodhead, M., WooEarly childhood development: A question of rights. *International Journal of Early Childhood*, December 2005. https://doi.org/10.1007/BF03168347

Zoja, L., *The Father: Historical, Psychological and Cultural Perspectives*. Routledge, London, 2000.

Zoja, L., *Violence in History, Culture, and the Psyche: Essays. Analytical Psychology & Contemporary Culture English*. Paperback Spring Journal Books, Thompson, CT, 2009.

Chapter 4

The Preserved Moment Through Art

Looking at Jungian Arts-Based Research and the Articulation of Inherited War Traumas

Roula-Maria Dib

Abstract

Not all wars are experienced directly, as the reverberations of traumatic histories may be passed down for generations. Much history is taught and learned through art, which reflects the mental states of victims, warriors, and traumatized individuals and peoples. As Jung says, "Just as psychological knowledge furthers our understanding of historical material, so, conversely, historical material can throw new light on individual psychological problems" (*The Collected Works*, Vol. 5). Looking at the making of art from a Jungian perspective, we can see how arts-based research can be a method that heals as it teaches and creates. Whether it is through producing non-fiction, fiction, photography, painting, music, or basically any other form of creativity, a significant road towards healing the traumatic wounds of war (both experienced or inherited) lies in the engagement of the creative self in the process of restoration and reintegration. Using the creative voice through various mediums articulates unspoken impressions, and rebelliously refuses to carry on the silence of generations. It is a path towards understanding the self and the collective, which helps move the artist out of the place of victimhood.

Not all wars are experienced firsthand, as the echoes of traumatic histories often reverberate across generations. Contemporary artists from diverse disciplines frequently delve into these traumatic histories, whether they stem from their native lands, the personal complexities observed in their parents or distant ancestors, or even responses to historical events that transcend the personal sphere. Their aim is to forge connections with different eras and people, driven by empathy and the art of expressing the unsaid. This endeavour enables the archetypes of the collective unconscious to voice a shared experience, even if the sharing occurs on an impersonal level. Each moment possesses its unique reverberations, perpetually echoing through the medium of creativity. This phenomenon is notably evident in photography – not exclusively of the professional variety – but rather in the act of people capturing, preserving, and engaging with past moments through the lens.

Art has long been a powerful medium for conveying the human experience, and during times of war and conflict, it takes on a profound role as a conduit for understanding, healing, and transcending trauma. In the context of Jungian arts-based research (ABR), art becomes a vehicle for exploring the depths of the human psyche, connecting with ancestral histories, and articulating the unspoken. This chapter delves into the significance of art created during or about war, emphasizing its ability to bridge the gap between knowing and being, to communicate profound emotions, and to preserve the essence of moments that might otherwise be lost to history's conventional narratives. Through various artistic forms, from poetry to visual art, artists embark on a transformative journey, using their creations as both a means of personal healing and a way to contribute to a collective understanding of the human condition in times of crisis. Ultimately, art

DOI: 10.4324/9781003390039-20

serves as a testament to the resilience of the human spirit and a timeless reflection of our shared history, identity, and capacity for transcendence.

Much history is taught and learned through art, which reflects the mental states of victims, warriors, and traumatized individuals and peoples. As Carl Jung says, "Just as psychological knowledge furthers our understanding of historical material, so, conversely, historical material can throw new light on individual psychological problems" (Jung, 1956, p. 5). Looking at the making of art from a Jungian perspective, we can see how ABR can be a method that heals as it also teaches and creates. Whether it is through producing non-fiction, fiction, photography, painting, music, or basically any other form of art, a significant road towards healing the traumatic wounds of war (both experienced or inherited) lies in the engagement of the creative self in the process of restoration and reintegration. Using the creative voice through various mediums articulates unspoken impressions, and becomes a path towards understanding the self and the collective.

According to Jung, "It seems to me perfectly possible to teach history in the widest sense not as dry-as-dust, lifeless book-knowledge but to understand it in terms of the fully alive present" (Jung, 2015, p. 37). While history itself is inherently rooted in the past, art, on the other hand, serves as a preserved fragment of the past within the current moment. It encapsulates the "alive present" that Jung deems essential for comprehending history. As Susan Rowland aptly posits, "Fortunately, history can be recollected through various languages" (Rowland and Weishaus, 2020, p. 118). This is where ABR comes into play.

ABR, for the artist, is the "practice" of history – the *making of* the moment in a forever present tense. For the observer, an artefact serves as more than a reminder: It becomes an informer, a teacher, and a window into the past that is not really portrayed or is that effective in traditional "dry-as-dust" methods of carrying or educating history. Traumas, losses, psychic changes, and collective damages are not "news" by nature – they are states of mind, affects, and emotions that can only be articulated and shared collectively by proper storytelling methods, i.e., through creative artefacts. In other words, wars in history books serve as states of *knowing*, rather than *being*. Creative responses to wars, however, serve as meaning-making, which combines both aspects of knowledge and being.

The creation of art has often been perceived as largely disconnected from other forms of meaning-making within academic disciplines, including history. History is typically regarded as the sanctioned avenue for understanding the past, delving into details about battles, wars, and their socioeconomic and demographic repercussions on societies. However, it frequently falls short in addressing, approaching, resolving, or imparting knowledge about the adverse psychological impacts that may have endured through generations. In contrast, art, including the realm of ABR, steps in to bridge this gap. Susan Rowland observes that despite its historical role as the essence of any era and epoch, "Art in Western modernity has been stripped of many of its therapeutic, sacred, and social functions . . . . Indeed, today, Western art often succumbs to conventional perceptions that view it as mere ornamentation, a financial asset, or mere entertainment" (Rowland and Weishaus, 2020, p. 1). Returning to Jung's perspective, it becomes evident that this standpoint grossly dismisses the inherent nature and purpose of creative endeavours. Jung contends that "The creative process, insofar as we can discern it, consists of the unconscious activation of an archetypal image, which is then molded and shaped into the final work" (Jung, 1978, p. 82).

So, we must consider: How many of these archetypal movements that occur within an artist during times of war can also resonate with the observer, reader, or listener? To what extent can we perceive, learn from, and utilize the emotional releases contained within these archetypal movements for our personal and collective healing? All this hinges on the amalgamation of *knowing* and *being*, which art provides, particularly within the realm of Jungian ABR. In his essay, "Psychology and Literature," Jung posits that the collective unconscious in art can facilitate the individuation of both the observer and the artist: "To grasp its meaning, we must allow it to shape us as it shaped him. Then we understand the nature of this primordial experience" (Jung, 1978, p. 105). While artists create during times of war, they are not only reaching inward but also outward to the world. They narrate stories, not necessarily through words alone but via various forms of expression that convey messages in ways uncharted by conventional history. This transformation affects the creator and simultaneously engages and instructs the observer by evoking these archetypal movements, transitioning from the personal to the collective. The artefact created becomes a transformed entity in itself – it transcends mere objecthood, embodying a fragment of the artist's essence while manifesting the collective unconscious:

> Art of the collective unconscious is transformative of both creator and consumer. With art, knowing is connected to being as a mutually informing process. Ultimately, visionary art provides an epistemological process of social, historical and cultural concerns by and through transforming being into knowing. . . . Jung's visionary art is made of the archetypal stuff that weaves us into the fabric of the universe.
>
> (Rowland and Weishaus, 2020, p. 72)

So art is a live way of knowing, discovering, and telling. The war artist creating is actually on a journey towards individuation through ABR. Rowland also suggests that:

> Jung and arts-based research already share ideas and strategies about knowing and being that can be furthered by forging a relationship between them . . . and Jungian studies expands by seeing how the creativity innate to Jung emerges into knowing through arts-based research.
>
> (Rowland and Weishaus, 2020, p. 1)

If Jung believes that "the human psyche is the womb of all the arts and sciences" (Jung, 1978, p. 86), then we can also say that he believes in it as the source for both *knowing* and *being*, both which come through artmaking. ABR uses various forms of art (such as poetry, fiction, creative non-fiction, filmmaking, dance, sculpture, painting, and music) integrating it into the research process while also making it a part of the research work's data representation. ABR methods provide a powerful way to mine historic moments and present deep, nuanced understandings, allowing space for play and ambiguity, revealing fresh and surprising ways of thinking about phenomena. It crystallizes history into a different form and remodels it not only through conscious means, but also through voices in the collective unconscious. After all, it is also Jung who said:

> It is evident, of course, that history takes on a new aspect when considered not only from the standpoint of our conscious reason, but also from that of the phenomena due to unconscious processes which never fail to accompany the peripeteia of consciousness.
>
> (Jung, 2015, p. 510)

Regrettably, it is a prevailing misconception, particularly during times of crisis, that art is a superfluous indulgence rather than an essential, despite the stark reality that art – indeed, a significant portion of it – is conceived precisely in response to these tumultuous periods, if not explicitly *for* them. Another crucial facet to consider is that many wars can metamorphose into crises centred around identity and self-determination. These conflicts and wars have, in many instances, wrought traumas that echo across generations, for crises pertaining to identity and self-determination are not ephemeral or fleeting by nature. Regardless of whether one actively participated in, was victimized by, or was even born during a time of war, the profound and enduring identity crisis left in the wake of certain historic junctures remains relatively unchanged.

For artists navigating through the desolation of such times, art serves as a pivotal conduit for expression and the preservation of these pivotal moments. Moreover, art embodies a universal function, as articulated by Rev. Mitri Raheb, in enabling:

> genuine self-expression and authentic communication of one's story. Art is the means through which we establish our own identities, defining ourselves independently of external influences. It provides a medium to convey our genuine desires, distinct from the lexicon of political rhetoric and religious dogma.
>
> (Chrabieh and Salibi, 2021, p. 14)

This universality of archetypes is indispensable for comprehending both the individual and collective aspects of identity, facilitating the emergence of inherited voices that have traversed generations and coalesced within the tapestry of the artwork.

In the realm of poetry, for instance, this process is realized through poetic inquiry, which serves as an instructive tool shedding light on the events of history, whether we lived through them firsthand or not. Beyond that, poetic inquiry and other expressions of ABR assume a vital role in narrating stories of identity. In the words of Jung, "Just as an individual's conscious perspective is balanced by responses from the unconscious, art represents a mechanism of self-regulation in the context of nations and epochs" (Jung, 1978, p. 83).

Times of war can prove profoundly disruptive to one's sense of identity and self-regulation across various dimensions of existence and understanding. In this context, art assumes the role of a transformative process, capturing moments and subsequently transmuting them into a means for realization, processing, and representation within the broader journey towards meaning. This pursuit encompasses self-knowledge, historical awareness, and the comprehension of the other through the lens of self-realization. It stands as a restorative journey, seeking to regain the equilibrium that often falters amidst the turmoil of chaos. As articulated by Rowland:

> Art that draws upon the collective unconscious possesses the capacity to unveil new insights precisely because it fills the voids of the current era. Jung contended that the psyche, whether collective or individual, perpetually strives for equilibrium. Thus, visionary art plays a pivotal role in fostering the equilibrium necessary for collective psychic stability. Visionary art has the power to reveal hitherto undiscovered facets of our society. Jung described visionary art rooted in the collective unconscious as "a creative act of epochal significance".
>
> (Rowland and Weishaus, 2020, p. 71; Jung, 1978, p. 98)

While our knowledge of societies, history, and the wars that have punctuated the centuries is sourced from history books, our creative artefacts manifest in order to illuminate and convey aspects we are yet to fully comprehend, sense, and encounter. In one of his correspondences with Helton Godwin, Jung passionately declares, "I loathe the new style, the new Art, the new Music, Literature, Politics, and above all the new Man" (Jung, 1973, p. 286). Jung harboured a vehement aversion to modern art, particularly between 1938 and 1955 (Van den Berk, 2012, p. 105), during a period when he associated modern art with the aftermath of war, perceiving it as a reflection of the interwar collective psyche. He posed a profound question: "How are we to explain the blatantly pathological element in modern painting . . . [t]he far-reaching influence of Joyce's unfathomable *Ulysses*?" (Jung, 1964, p. 210). It can be posited that, through grappling with the enigmatic nature of modern art, Jung himself gleaned valuable insights through his visceral aversion, allowing him to directly confront the post-war "new Man" of modernism and its concomitant manifestation of fragmented psychic states.

According to Jung, the creative process inherent to an artist is driven by the unconscious. However, this creative impetus is especially potent in individuals of genius who possess the capacity to gather scattered fragments of ideas and latent memories, weaving them into novel and meaningful structures. As Jung aptly expressed, "The work of genius is distinguished by its ability to retrieve these remote fragments and assemble them into fresh and significant con-figurations" (Jung, 1957, p. 105). Some poets and artists navigate creative chaos by channelling it into more structured forms of expression, perhaps as an instinctual response to the need for order within a disordered world. Notably, Irish poet Christine Murray asserts,

> While history books are indispensable, human witnesses and artistic expression remain equally vital during turbulent periods and global catastrophes. Poetry reminds us that even in moments of vulnerability and amidst trauma, the human voice possesses the capacity to craft beauty.
>
> (Dib et al., 2021)

The experiences of trauma and global cataclysms resonate with a significant portion of the diaspora population, particularly those forced to leave their homelands. For numerous poets, writers, visual artists, filmmakers, and creatives, both with and without direct wartime exposure, the concepts of diaspora, home, history, and citizenship have become compelling subjects of exploration. They reflect these themes through their artistic endeavours, shedding light on the inherited quest for home and identity. As Jung observes, "Psychology and aesthetics are destined to forever complement each other, with neither diminishing the significance of the other" (Jung, 1978, p. 87).

For many within the diaspora, identity and home evolve into a mental image – a yearning for a specific place at a particular moment. Home, in this context, transcends the confines of a mere physical space one can step into; instead, it assumes the ethereal essence of a cherished memory. As Gaston Bachelard suggests:

> Memories of the outside world will never have the same tonality as those of home and, by recalling these memories, we add to our store of dreams; we are never real historians, but always near poets, and our emotion is perhaps nothing but an expression of a poetry that was lost.
>
> (Bachelard, 2014, p. 28)

To artistically express the clashes of culture and the anxieties of separation using the unify-ing language of symbols and metaphors demonstrates the therapeutic potential of poetry. This phenomenon is vividly evident in the works of poets, writers, and artists featured in the London Arts-Based Research Centre's literary and arts journal, *Indelible*. Many of these individuals, particularly those living in the diaspora or with a history of war and migration, grapple with fragments of moments, pondering and occasionally wrestling with matters of identity, belonging, and citizenship. Through their ABR, they delve into themes such as history, war, ancestry, inheritance, and home, echoing poet Victoria Chang that "Poems function as time machines, traversing individual memories while also engaging with historical timelines and collective cultural memory" (Chang, 2018).

Before briefly zooming in on contemporary poetic works that explore the dynamics of war and migration, it is worth emphasizing Sandra Faulkner's perspective: "I firmly believe that poetry can serve as a form of social science and that its true value lies in its capacity to conduct research on relationships" (Faulkner, 2017, p. 148). Notably, the acclaimed poet Ruth Padel's poetic inquiry, titled *We Are All from Somewhere Else*, comprises a collection of poetry and prose seeking to establish a biological and historical framework for understanding contemporary large-scale human migrations. It underscores the religious, spiritual, and psychological ramifications of migration for all individuals involved. A central theme in this work is migration resulting from war, illustrating how "every life undergoes transitions that offer a new sense of self" (Padel, 2020, p. xx). These newfound senses of self are achieved through the creative process, where war, migration, and art intertwine as crucial elements in comprehending histories. As Padel aptly observes, "While physical displacement primarily propels migration, psychological displacement fuels imagination and creativity" (Padel, 2020, p. 251). Consequently, the act of creation can serve as a defence mechanism against the traumas of displacement for the artist, serving as a means to comprehend and "rescue" the self. Padel expounds on this idea by stating,

> Displacement, a term that echoes in discussions of migration, serves as one such psychological defense. We "displace" our unsettling emotions into areas where they feel less menacing. By distancing ourselves from trauma, our minds redirect these emotions towards something else, thus transforming them.
>
> (Padel, 2020, p. 251)

Art, not only as a vessel for the artist's emotions and a compelling medium to convey stories and experiences to observers, possesses the power to effect transformation. Whether the artist has directly encountered displacement or has no recollection of its occurrence, the trauma can be intergenerational and still evoke emotions that find expression in their artistic works:

> We all displace unpleasant emotions into areas where they don't feel so destructive and may even be creative . . . if you displace pain you escape it, not by denying it or distorting reality by migrating to a different (and with any luck creative) perspective. You're not misunderstanding or denying the world, you know the massacre has happened, but you're making something of it. Goya paints *The Disasters of War*.
>
> (Padel, 2020, p. 252)

Padel's *We Are All from Somewhere Else* meticulously depicts and encapsulate the intricate histories of global migration. This work can be readily appreciated as a blend of historical research, performed and conveyed through the medium of poetry. Within these verses, the poet illuminates the narratives and histories of migration and survival, rendering them poetically in both verse and prose.

Returning to our initial assertion, capturing moments through art liberates the artist from victimhood by preserving the vision through the aesthetic portrayal of moments. This process unfolds past histories and shapes future narratives, echoing Jung's wisdom that "In our most intimate and subjective lives, we are not just passive witnesses of our era, enduring its trials, but also active contributors. We shape our own epoch" (Jung et al., 1934/1992, p. 149).

Wars, undoubtedly, constitute some of the most devastating chapters in artists' lives – periods that compel them towards metamorphosis and reconstruction. Through their artistic creations, they pose inquiries, safeguard memories, seek resolutions, and strive to forge new realms. They acquire knowledge and bequeath their own creations as instructors to others, whether from their contemporaries or posterity. This can be perceived as a method of clutching onto life, of safeguarding and conserving a fragment of the self, encompassing not solely the individual but also the collective. Ultimately, art becomes the lasting testament. As Jung expresses:

> The great events of world history are, at bottom, profoundly unimportant. In the last analysis, the essential thing is the life of the individual. This alone makes history, here alone do the great transformations first take place, and the whole future, the whole history of the world, ultimately spring as a gigantic summation from these hidden sources in individuals. In our most private and most subjective lives we are not only the passive witnesses of our age, and its sufferers, but also its makers. We make our own epoch.
>
> (Jung et al., 1934/1992, p. 149)

And art is again a testimony to this "epoch-making". An artefact is a living piece of the times from which it was made, as this is also reflected in the different styles, depending on the historical epochs they depict. Wilhelm Worringer, modernist art theorist particularly influential to Jung, wrote *Abstraction and Empathy: A Contribution to the Psychology of Style* in 1908 (reprinted in 1953). Worringer's argument centred on the idea that the prevalence of abstraction in early 20th-century art and in ancient periods did not result from a straightforward, logical progression in the history of aesthetic representation. Instead, he suggested that art's nature should be viewed in the context of the society in which it was created. In times of confidence and security, art tended to lean towards realistic, empathetic styles, while periods marked by insecurity often saw a shift towards abstraction (Rowland and Weishaus, 2020, p. 88). Therefore, as "makers" of our own epochs, the things we "make" also become testimony to that, as our creations, while reflecting our personal psyche, usually transcend the individual with an archetypal language different than that of conventional history books:

> our personal psychology is just a thin skin, a ripple upon the ocean of collective psychology. The powerful factor, the factor which changes our whole life, which changes the surface of our known world, which makes history, is collective psychology, and collective psychology moves according to laws entirely different from those of our [individual] consciousness. The archetypes are the great decisive forces, they bring about the real events, and not our personal reasoning and practical intellect.
>
> (Jung, 1977, p. 163)

An artefact, during its making and after its completion, shows the artist's desire to go beyond the individual voice; ABR also serves a para-expressive cause, which is to teach via meaning-making, in a more holistic method rather than merely looking at facts and statistical idea – it speaks a different language, and it is more inclusive of all the knowledge disciplines (hence the term "transdisciplinarity", described by Nicolescu as a non-hierarchical access to knowledge and meaning in the academic disciplines. The sciences, for example, are not seen as more truth-yielding than the arts). As described by Cathy A. Malchiodi in *The Handbook of Arts-Based Research*, "arts expression is not only a method of knowing and making meaning but also a form of communication when words do not convey the totality of human experience" (Leavy, 2019, p. 69). This is how the artist's psyche expresses, asks, seeks answers, and makes meaning during times of war. This can be seen as a "consciously unconscious" process – a conscious act of creation while tapping into the unconscious world of archetypes. The war psyche is a shifting psyche in a changing world; it preserves "present moments" that remain to show and share stories, feelings, and certain truths that cannot be communicated otherwise. This is how it "writes" history through the unconscious, if we are to take Jung's words that "history takes on a new aspect when considered not only from the standpoint of our conscious reason, but also from that of the phenomena due to unconscious processes which never fail to accompany the peripeteia of consciousness" (Jung, 2015, p. 510). The war psyche, therefore, is one that can speak in this "archetypal language" that Jung mentions, which often leads to the creative, meaning-making journey of the arts-based researcher. As observed by Patricia Leavy (2015), "when subjective, inter-subjective, socioemotional, spiritual, interpersonal, and artistic experiences are difficult or impossible to quantify, arts-based inquiry provides a pathway to exploration, description, and discovery that other methods cannot" (Leavy, 2019, p. 68).

So, can one view war art (poetry, novels, short stories, music, sculpture, painting) as historic and psychological "data" that looks at the human condition from a more holistic angle? Can the creative psyche showing these histories cause change? Do the artists during war learn from their artefacts, thereby making the creative process an inquiry, or a research process in a non-conventional way? In considering all forms of art created during or about war as historical narrative, I would agree with Patricia Leavy:

> Arts-based practices can be employed as a means of creating critical awareness or raising consciousness. This is important in social justice-oriented research that seeks to reveal power relations (often invisible to those in privileged groups), raise critical race or gender consciousness, build coalitions across groups, and challenge dominant ideologies.
>
> (Leavy, 2009, p. 13)

In conclusion, the power of art created during or about war, particularly in the context of Jungian ABR, transcends time and place to become a vessel for the preservation of moments, emotions, and histories. Through various artistic forms, individuals connect with the past, whether it be their own personal experiences, ancestral traumas, or collective memories of wars and conflicts. Art serves as a bridge between knowing and being, offering a profound means of understanding and healing. It defies conventional historical narratives by delving into the realm of the unconscious and archetypal, providing a more holistic perspective on human experiences during times of crisis. As artists engage in the creative process, they embark on a journey of self-discovery and transformation, while their creations become a testament to the epoch in which they were made. Art, in all its forms, tells stories that historic records alone cannot convey, making it an invaluable tool for exploring, documenting, and transcending the complex interplay of history, identity, and trauma. Ultimately, art is a testament to the resilience of the human spirit, a means of finding beauty amidst chaos, and a conduit for understanding the past, present, and future.

References

Bachelard, G. (2014). *The Poetics of Space*. New York: Penguin Classics.

Chang, J. (2018, June 14). "Each time the light changed: A Micro(inter)view with Jennifer Chang, curated by Lisa Olstein." *Tupleo Quarterly*. [Online] Available at: http://www.tupeloquarterly.com/each-time-the-light-changed-a-microinterview-with-jennifer-chang-curated-by-lisa-olstein/.

Chrabieh, P. and Salibi, R. (eds.) (2021). *The Beirut Call: Harnessing Creativity for Change*. Bethlehem: Dar Al Kalima University College of Arts and Culture.

Dib, R. et al. (2021). "Poetry as food for the soul (various *indelible* poets)." *Indelible*. [Online] Available at: https://indeliblelit.com/2021/06/13/poetry-as-food-for-the-soul-various-indelible-poets/ (Accessed: June 27, 2023).

Faulkner, S. (2017). "Faulkner writes a middle-aged Ars Poetica." In *Poetic Inquiries of Reflection and Renewal*, edited by L. Butler-Kisber, J.J. Guiney Yallop, M. Stewart, and S. Wiebe. Nova Scotia: MacIntyre Purcell Publishing Inc.

Jung, C.G. (1956). *The Collected Works. Volume 5: Symbols of Transformation* (R.F.C. Hull, Trans.). London: Routledge.

Jung, C.G. (1957). "Cryptomnesia." In *The Collected Works of C.G. Jung. Volume 1: Psychiatric Studies.* New York: Pantheon Books.

Jung, C.G. (1964). "After the catastrophe." In *The Collected Works of C.G. Jung. Volume 10: Civilization in Transition*. New York: Pantheon Books.

Jung, C.G. (1973). *Letters. Volume 1: 1906–1950* (G. Adler and A. Jaffe, Eds.). Princeton, NJ: Princeton University Press.

Jung, C.G. (1977). *The Collected Works of C.G. Jung. Volume 18: The Symbolic Life* (R.F.C. Hull, Trans.). London: Routledge.

Jung, C.G. (1978). *The Collected Works. Volume 15: The Spirit in Man, Art, and Literature* (R.F.C. Hull, Trans.). Princeton, NJ: Princeton University Press.

Jung, C.G. (2015). *Letters of C. G. Jung. Volume 2: 1951–1961* (G. Adler, Ed.). London: Routledge.

Jung, C.G., Adler, G. and Read, H.R. (1934/1992). *Collected Works of C.G. Jung*. Volume 10. Princeton, NJ: Princeton University Press.

Leavy, P. (2009). *Method Meets Art: Arts-Based Research Practice* (1st ed.). New York: The Guilford Press.

Leavy, P. (2015). *Method Meets Art: Arts-Based Research Practice* (2nd ed.). New York: The Guilford Press.

Leavy, P. (ed.) (2019). "Cathy A. Malchiodi, creative arts therapies and arts-based research." In *The Handbook of Arts-Based Research*. New York: The Guilford Press.

Padel, R. (2020). *We Are All from Somewhere Else*. London: Penguin.

Rowland, S. and Weishaus, J. (2020). *Jungian Arts-Based Research and "The Nuclear Enchantment of New Mexico"*. London: Routledge.

Van den Berk, T. (2012). *Jung on Art*. New York: Routledge.

Chapter 5

Embodied Analysis

The Recovery of Early Psychological Functions Interrupted by an Experience of Early Trauma Due to State Terrorism

Karin Fleischer

Abstract

This chapter aims to account for the possibility that the body represents a via regia to aspects of the self that never access a representational level due to a collective trauma lived at an early age. It offers an introduction to some of the psychic consequences and clinical difficulties that may emerge when the early personal wound is deeply merged with a family and collective trauma. The chapter particularly focuses on the enforced disappearance of people occurred during Argentina's state terrorism which has seen hundreds of children stripped of their roots and true identity. When, as in these cases, the process of translation and integration from implicit to explicit autobiographical memory becomes interrupted, there is a need in the clinical work for new approaches that may facilitate the access to this preverbal-somatic-affective narrative. The chapter reflects on how, by including a non–verbal-symbolic approach such as Embodied Active Imagination, also known as Authentic Movement, sensations experienced by the living body of the patient can function as nodal centres of implicit memories that – under certain conditions – can release and give rise to long-buried impressions. The patient's body gestures and physical sensations act as a point of departure, which bridges the preverbal implicit knowledge with the emergent emotions and images, contributing to a restoration of early psychological functions that were interrupted due to the trauma.

Introduction

The reality of war has existed since the dawn of humanity, but we usually did not think of its consequences, when the conflict happened in a remote place and time without affecting us personally. These days, however, it is difficult to think of patients, as well as analysts, who have not been affected by an experience of war, forced migration, social violence, or state terrorism, either in their own life history or in that of previous generations. These experiences have probably marked us more than we are able to realise consciously.

All forms of collective violence from the extermination of native peoples, to the Holocaust, to the various military dictatorships in Latin America, to the current invasion of Ukraine by Russia – to name just a few – have something in common: They are part of a traumatic history that is experienced both individually and collectively. However, each of these horrific events has particular elements that influence and lead to diverse psychic consequences.

In certain cases, it is not only the experience of extreme violence, but its subsequent denial or the insistence on trying to prevent its symbolic recognition in the collective consciousness, that determines that certain psychic sequelae continue to be transmitted from generation to generation. Such impediment translates into the need to "leave things where they are", thus accepting denial, silence, senselessness, and death.

DOI: 10.4324/9781003390039-21

However, as Davoine and Gaudilliere write in *History and Trauma*: "no matter what measures are taken to erase facts and people from memory, eradications, even perfectly timed ones do nothing more than set in motion a memory that does not forget and that wants to make itself known" (2011, p. 37).

When in the clinical work we are faced with places of non-existence, due to early collective trauma, we are also faced with the challenge of how to access them and how to elaborate that which could not have been symbolized. In these cases, many times, words and verbal language are not enough.

In this chapter, I will begin by describing the particular exercise of violence that is state terrorism, based on the events that took place during the military dictatorship in Argentina, as well as some of the consequences of that violence, from an analytical perspective. I recognize that in writing this, my own need to continue weaving fragments of a memory – both individual and collective – that for a long time lived in the shadows and did not have the words to be named. In the second part, I will dwell on the possibilities that open up when the embodied dimension is considered in the clinical work.

State Terrorism and Its Consequences

State terrorism is a particular form of collective violence, with a particular logic of "crimes without crime" (Volnovich 1992, p. 138), which is sustained on the basis of the denial of truth and the imposition of fear. In this state of denial, the false treatment of reality establishes a process of alienation that affects the population in general and leaves its imprint in the collective-social fabric (Chevance-Bertín 1987).

One of the most invidious forms of denial during the military dictatorship in Argentina was the enforced disappearance of people. The concealment of the origin of life and death imposed a new logic of non-existence. "They are neither alive nor dead; they are disappeared", repeated the former dictator Jorge Rafael Videla, with the impunity of a god. It is alarming and disturbing to see how today we are witnessing those who, from potential public and political roles, try to impose this same negation and denial once again.

From psychology and from our clinical work, we know the psychic consequences of the concealment of a family truth, of a secret, or of the denial of an event. Can we imagine then the disturbance that takes place when such concealment produces the destruction of social and family ties through several generations – even more so when such concealment entails the abolition of the right to know about the origin of life and death, preventing the possibility of there being rituals concerning both?

In the forced disappearance of people, life and death are annulled, bringing about a negation whereby there can be no symbolization. In the non-death, a new form of mourning emerges, different from the pathological mourning studied by Freud and psychoanalysis (Freud 1917). It is rather a suspended or frozen mourning, since there is no possibility of burial without a body, and therefore of mourning without an acknowledged death.

In the absence of an acknowledged death of a body-subject-mother-father, another death is produced, that of time, or that of the capacity to historicise one's own life. The death of time takes the place of that which could not be assimilated, turning history into an atemporal and ahistorical narrative. I suggest that this suspension of the individual's capacity to envisage his or her own life, to imagine beyond the trauma – an expression of frozen mourning – corresponds to an unconscious way of sustaining the existence of those parents that are neither alive nor dead. It is the absence of the body of the missing person that becomes overly present in the internal experience of the descendants (Bianchi 2023). An "omnipresence" which – as an anchor to that which is impossible to process – prevents the development of life itself in those relatives who are still alive, particularly the children.

Clinical work with victims of early trauma due to state terrorism presents us with the challenge of how to understand and accompany this suspended mourning. Thus, analysis becomes a symbolic and relational space in which it is possible for the patient to recover the sequentiality of an interrupted life, or in Jungian terms, the potential for individuation.

In previous writings, I recognised and described, first, the need for differentiation that emerges in the elaboration of an early trauma when it is fused to a collective trauma (Fleischer 2022), and second, from the perspective of memory, the task of constructing one's own identity (Fleischer 2023). In this chapter, I would like to address this further by focusing on the role of early psychological functions: the translational function, and the transcendent function, in the development of the ego–Self axis.

Psychic Functions in Early Development, Their Interruption, and Their Possible Restoration

Analytical psychology understands the emergence of subjective life from the existence of an initial Self, which is of an affective nature, which provides the foundations for the infant's responses to the environment (Jung, CW 7, CW 9/1). Fordham (1976) introduces the idea of self-agency, or the intervention of the infant in his own development. From this perspective, the infant participates with his whole body-Self in the shaping of these contexts and is not considered as a passive being who only receives stimulation from the adult. Fordham (1976) described the mechanism of deintegration/reintegration through which – from this initial Self, and the bond with significant figures – the ego–Self axis will be shaped. Also from a post-Jungian perspective, Knox (2010) proposes that the infant, from its bodily action in space and with significant others, gives rise to the formation of image-schemas, which remain as a first foundation in later conceptual and symbolic development. These symbols may later become disconnected from motor action, "however, they remain embedded in it and are derived through it" (Knox 2010, p. 528).

From a neuroscientific perspective, and in parallel, Antonio Damasio (2000) described the nuclear self as the self of the beginning of life. This includes an internal, image-based sense that carries the experience of the relationship between the organism and the environment. He postulated that it has a biological underpinning in what he called the proto-self, which is a set of neural patterns responsible for mapping the state of the physical structure of the organism. Among the properties of the early self are sensory-affective states, which refer to rudimentary perceptual experiences that have an intrinsic affective value – for example, the state of being caressed, or lulled by a melody, or abruptly awakened by alarming sounds. These experiences are part of a somatic-affective core which, in an adequate environment, will develop first through a process of integration of subsequent image-based experiences and, later, through verbal-symbolic "translations", contributing to the gradual development of an autobiographical self (Damasio 2000, p. 206).

In addition to finding support and echo in neuroscientific findings, analytic theory and practice continues to develop. Recently, Cavalli (2020, 2023) seems to describe the same process as described by Damasio, speaking of a translational function that allows somatic events to be translated into feelings and thoughts. In her article, "Noah's Ark – Technical and Theoretical Implications Concerning the Use of Metaphor in the Treatment of Trauma" (2023), she distinguishes this function from the psychic capacity to sustain and integrate opposing tendencies, a capacity which Jung (1916) called the transcendent function. She adds that a well-organised psyche is a psyche capable of carrying out both functions (Cavalli 2023, p. 147).

So, we come into the world with the potential capacity to translate sensations into images and words – the translational function – as well as with the potential capacity to psychologically hold and bridge opposite aspects of the same experience – the transcendent function. However, in order for these functions to develop and contribute to the conformation of the ego–Self axis, an adequate environment is necessary in which this primary or nuclear self may have good enough experiences of reciprocity.

Several authors – from both psychoanalysis and analytical psychology – have given an account of this process, referring to it as the Alpha function and the capacity for reverie (Bion 1962, 2006), the process of deintegration/reintegration (Fordham 1985), the importance of affective attunement (Stern 1991), and the process of early intersubjectivity (Trevarthen 1998).

Thus, when the infant has enough good experiences to integrate into its primary self, they also develop the capacity to tolerate bad experiences. When both good and bad experiences can be integrated, and these integrations are repeated over time, the development of an ego in connection with the Self takes place (Fordham 1985). Later, in adult life, the individual will have the capacity to translate their affective experiences and bodily sensations into images and thoughts, and also to sustain, in the same symbolic representation, opposing affective aspects of the same experience.

Interruption

But what happens when, at an early age, due to traumatic experiences, this process is interrupted?

Current developments in psychology and analytical psychology have allowed us to more deeply and clearly recognise and understand the psychological consequences of early trauma. These studies have shown that no matter how much psychic organisation an individual has, they are shaken by traumatic experiences.

However, the impact of the dissociation and the possibility for recovery will depend essentially on the time in life and the psychological development of the individual when the trauma occurs (Cavalli 2020, 2023). Cavalli suggests that the process of analysis needs to explore the state of psychological organisation at the time of the trauma in order to understand the possible dissociative strategies and modes of unconscious adaptation undertaken by the patient (Cavalli 2023, p. 148).

In this chapter, I try to show what happens when a collective trauma occurs early in life, but after the child has had good enough experiences of reciprocity enabling the emergence of the ego–Self axis. I propose that it may cause an interruption both in the process of translation, which affects the capacity to convey bodily sensations into feelings and images, as well as in the operation of the transcendent function, which brings together opposing responses.

When the experience of psychic pain is so overwhelming, and the psyche cannot manage the intensity of such emotion, it activates defences tending to protect those good experiences already stored in the nuclear self, prior to the trauma, dissociating them from the subsequent traumatic experience. A displacement of the incipient ego–Self axis towards the ego-traumatic complex axis takes place, thus damaging the capacity for the transcendent function to operate.

As a consequence, a psychic state will be imposed that Luci (2023, p. 338) calls "monolithic", and which is characterised by: (1) the absence of spatiality and tension between oppositions; (2) identification with unilateral positions; and (3) an adhesive relational state. In these dissociative states, that which is impossible to symbolise is transmitted unconsciously to the analyst via states of participation mystique, or projective identification. The unconscious-to-unconscious communication plays a significant role, activating sensory perceptions that can become more intense. In other words, the body acquires a new relevance, either symptomatically and/or also offering the possibility of including its being registered and "heard" in the analytic relational encounter, challenging the analyst to a "beyond symbolic reflection". In the words of Luci (2023, p. 339), it is a matter of "two bodies together in a suspended time" in the therapeutic space.

Analysis: Recovery and Reinstatement

I agree with Cavalli when she states that "analysis becomes the place where life before and after the trauma can become a continuous sequence again" (2023, p. 150).

Being able to go beyond a suspended temporality confronts us with an unconscious that is impossible to be symbolised due to an early interruption in the process of translation. This relates to the early experiences that did not reach a representational level, as well as to the experiences that, due to their traumatic intensity, remain at the mercy of deep dissociative defences and consequently cannot be recovered symbolically. They will be part of the un-repressed unconscious, of traumatic complexes or areas of "un-formulated experiences" (Stern 2017). These primitives and/or excessive experiences, being unable to be expressed symbolically, may manifest themselves through psychosomatic symptoms, or be acted out in the transference: "If the patient cannot formulate the experience, he/she will be driven by it" (Lagutina 2021, p. 7).

Therefore, in clinical work, we are challenged to find new ways to facilitate access to these unconscious elements, when words and the symbolic-reflexive method is not enough. Jung had already emphasised the importance of unconscious-to-unconscious communication in the analytic relationship. In his introduction to *The Psychology of Transference*, he writes: "when two substances are combined, both are altered. This is precisely what happens in the transference" (Jung 1946, para. 358). From the conceptualisations of participation mystique, the psychoid unconscious, and the interconnected world or *unus mundus*, analytical psychology offers a particular and fertile foundation for developing new forms of intervention which can process these psychological events which are otherwise impossible to symbolise. "I believe that engagement at the level of the psychoid unconscious is essential for processing unsymbolised experiences", writes Lagutina (2021, p. 20).

According to Giles Clark (1996), the psychoid dimension of the unconscious does not correspond only to the intrapsychic level of life but encompasses those areas of experience that include very primitive sensations that are experienced internally and, simultaneously, relationally, for example through participation mystique (Clark 1996, p. 354).

Facilitating the opening to the psychoid level of the unconscious in the process of analysis requires certain conditions in the therapeutic relationship. Authors such as Lagutina (2021) and Winborn (2022), among others, consider, for example, the importance of the function of affective attunement, the capacity for reverie, and the resonance to subliminal perceptions. In previous work, I postulated, along with these factors, the inclusion of the somatic dimension in clinical work (Fleischer 2023). I think that fostering unconscious-to-unconscious communication also requires, on the part of the analyst, an embodied understanding of what Jung stated so long ago: that "the symbols of the self arise in the depths of the body" (Jung 1949, par. 291).

There are nowadays – and within the context of analytical psychology – several ways, practices, and approaches that allow the development of a somatic listening that, in my opinion, broadens and deepens what we understand by analytical attitude. In the second part of this chapter, I want to share one of these possible modes.

The Somatic Dimension in Jungian Clinical Work

From the beginning of his work, Jung had emphasised the correspondence of mind and body, understanding that the distinction between the two was an artificial one.

> The difference we make between the psyche and the body is artificial. It is done for the sake of a better understanding. In reality, there is nothing but a living body. That is the fact; and psyche is as much a living body as body is living psyche: it is just the same.
>
> (Jung [1934–1939] 1988, p. 396)

In recent decades, new studies in experimental psychology, neuroscience, trauma theory, and research on early development had contributed to scientifically ground this correspondence – stated by Jung a hundred years ago – describing how mental process are initially grounded in somatic experience (Beebe & Lachmann 2002; Cozzolino 2002; Fonagy et al. 2002; Gallese 2009; LeDoux 1996; Panksepp 1998; Schore 1994; Damasio 1994, 2000, 2010). "In the beginning" writes Damasio, "there was no touching, or seeing, or hearing, or moving along by itself. There was rather a feeling of the body as it touched, or saw, or heard or moved" (Damasio 1994, p. 237). With different words, Gallese (2003) explains this same correlation when he says that these multimodally signals (received initially through hearing, vision, and touch) are integrated into the sensorimotor system, and that during development, these sensorimotor mechanisms acquire new roles in the construction of concepts, language, and imagination. In consequence, feelings and thoughts are formed not only *in* the body but also *between* bodies.

Following this trend, there has been in contemporary Jungian – and also psychoanalytic – literature increasing interest in including the somatic embodied dimension in clinical work, from the perspective of the analyst and also of the patient (among others, Eulert-Fuchs 2020; Fleischer 2020, 2023; Fogarty 2018; Godsil 2018; Kalshed 2020; Lagutina 2021; Martini 2016; Merchant 2015; Schellinski 2009; Sidoli, 2000; Sletvold, 2014; Wilkinson 2017; West 2016; Zoppi 2017).

Within these contemporary perspectives, and in accordance with the seminal ideas proposed by Jung regarding an embodied mind, I would like now to turn to one possible way of working with this intercorporeality in Jungian analysis when we are faced with clinical circumstances and conditions for which a solely verbal/explicit approach remains insufficient.

Embodied Active Imagination – Authentic Movement

As I pointed out in previous work, Authentic Movement as a form of active imagination invites us to bring our attention within the body/psyche, attending to the signals – sensations, images, and emotions – that emerge from there, and allowing them to express themselves and take shape through the body (Fleischer 2020, 2023).

When working with Embodied Active Imagination, in an analytic session, it is necessary to have an adequate space for it. To begin, the patient, after finding a comfortable position – either lying on the floor or on a couch, standing, or sitting – closes the eyes, bringing the attention inward, and waits for an impulse to emerge. In this inter-relational context, somatic experiences are explored as emergent phenomena.

In his description of the process with active imagination, Jung spoke of the need *to leave the lead to spontaneous fantasies* to facilitate a state of openness to the contents of the unconscious. Mary Whitehouse, the creator of Authentic Movement, described such a state as staying in an open waiting until something happens. The person experiences a letting oneself into the body, a descending into the sensations, allowing them to come alive in the movement. Thus, through an increasing receptivity to somatic experience, these gestures, postures, and sensations of being moved by something often begin to emerge. At other times, there may only be stillness, and yet it may contain barely perceptible somatic expressions, such as a sudden sensation of coldness, warmth, or changes in breathing. In the presence of an attentive and resonant listening attitude by the analyst, these first sensory experiences and expressions usually – after some time – give rise to patterns of movement that tend to repeat themselves until a new sense emerges.

In a second stage, and after reopening the eyes, and with the experience still present in the body, the person lets it continue to express itself through other art materials, such as drawing, painting, sculpture, or clay, thus facilitating this function of *translation from the sensory-motor to the visual image.*

In this passage, experiences may emerge which, encoded in the implicit memory system, never reached a representational level, thus re-establishing a connection between the ego and early aspects of the self. The analyst's trust in the time of the unconscious is fundamental, as well as to not fill the void of uncertainty and not-knowing with some form of interpretation (Fleischer 2023).

In the next subsection, I would like to show how, by including the somatic dimension through Authentic Movement in the analytic context, it became possible both to recover the translation function and to repair the rupture in the ego–Self axis, making possible a new development of the transcendent function. It is important to consider that the piece I am about to share is only one element of a long process. Thus, the restoration of these functions was not due to a single event, but may be illustrated by this.

Clinical Vignette: Vera and the Wind

A patient, whom I called Vera in previous works (Fleischer 2022, 2023), tells me in our first session that she is the daughter of disappeared people. Her parents were taken to a clandestine centre when she was 3 years old. She has only one black-and-white photo of them, showing a smiling couple. She was then raised by other people with whom, in her words, "It was not possible to talk about what had happened". In addition to denying her knowledge about those important years of her early history, she was also denied the right to a life of affection, support, and care.

At the end of this first session, I felt deeply and emotionally touched; I wrote in my notes that it had been difficult for me to continue with another patient. An almost immediate feeling of mutual empathy had taken place. I think I somehow identified with those wounds of history and with the abandonment at such a young age, which stirred in me the need to go to the rescue of this little girl. Gradually, I was able to realise that there also was a little girl in me in need of rescue and acknowledgement, a little girl who had over adapted to what she had lived through during those same years.

Vera had tried several ways, first to maintain and then to recover the memory of her parents: as a child, in her play in nature, she built houses where she kept waiting for them; in her youth, through her participation in human rights movements, she directly sought Memory, Truth, and Justice (I have capitalised these words, as they are terms related to the Human Rights movement in Argentina). The absence of symbolic representations of these early intersubjective experiences, as well as the concealment of the truth about the life and death of both, acted as a force holding her to the past, making it impossible for her to build her own life. When she asked herself about her own wound, she recognised it in this permanent waiting, this permanent search, this "not having a place . . . not even to locate the dead".

Experience With Embodied Active Imagination

In the months prior to this experience, after a large number of traumatic dreams in which persecutions and assassination attempts followed one after another – testimonies of the efforts of the psyche to be able to elaborate something of those denied deaths – Vera realised, as she put it, that "the burden of searching for the bodies could continue all her life and that, when she carried it unconsciously, it acted as a force that kept her clinging to the past", not allowing her to live her own life or follow her own path.

In this context, and during a session, the following experience took place:

Vera sits down and takes a blanket. Later she will say that with the blanket she felt the presence of the wind. After a while her body begins to tremble. With her eyes closed, her body slides through the space between us until it comes close to me. There she remains in stillness, in a posture of recollection. Almost imperceptibly, at a certain moment, her hands come so close to my feet that they seem to touch them and remain so until the end of the experience.

When she opens her eyes, and allowing the necessary time to do so, the experience continues to take form through a drawing, in which she begins to create a place for a symbolic burial, a rite that will continue to manifest itself throughout several sessions (Figures 5.1 and 5.2).

After drawing and before the end of the session, she tells me:

With the blanket I felt the presence of the wind that brought me the connection to death. The trembling was very physical – of fear. I needed to get close to you, it was like getting close to my mom again, knowing that she was okay, and that it had been real. When I felt your feet close, I felt a cord all the way down my spine, something that reminded me of my mom's existence. I needed to be able to remember her like that so I wouldn't have to carry her in my body anymore. It was only then that I felt the need to mourn, and these words came to me: "Because I love you, I bury you".

Figure 5.1 Vera's drawing of a symbolic ritual.

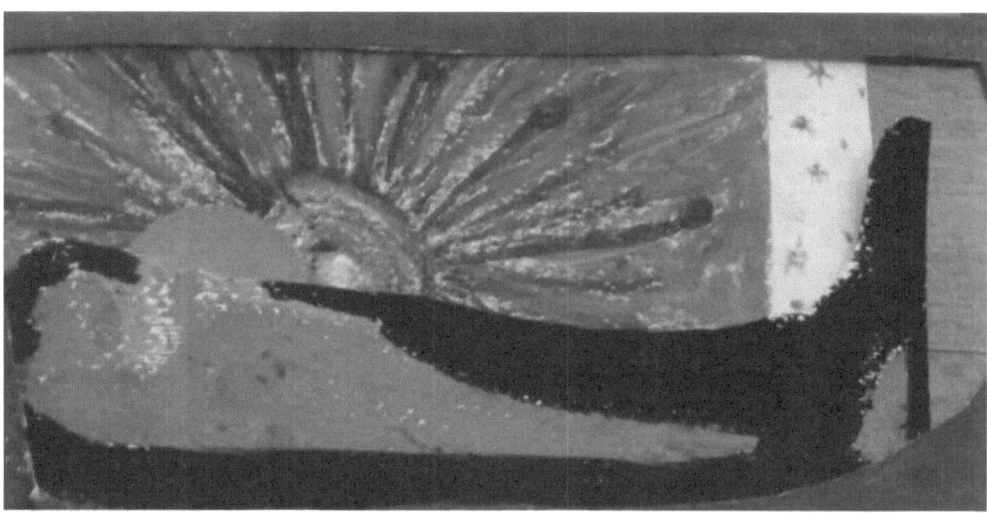

Figure 5.2 Vera's drawing of a symbolic ritual.

Some Concluding Thoughts

In the Hebrew and Arabic traditions, the word *Ruh* is also wind, breath, and spirit. The wind in Vera's somatic experience opened the possibility of a connection with that which until then had been denied: the power to know about the origin of both life and death.

Initially the experience emerges, bringing with it a sensation in the body but not necessarily an explicit knowledge of what is happening there; there is not yet an understanding of the meaning. Being able to remain in contact with the somatic sensation may give rise to emotions that can then continue to be expressed through another medium – in this case, drawing. Thus, a first translation from the sensory-motor-affective to the imagistic becomes possible. Then, the subsequent translation from the symbolic image to the verbal expression happens as a natural continuity when the appropriate context is provided. Words do not have to be offered from the outside, by the analyst; what is offered is the containment and the necessary time to wait for them.

In this experience, a mother's absent body comes to life in Vera's somatic experience, contained and mediated by the link with another body now present – that of the analyst. Fear can be embodied; trembling no longer destroys. Holding in proximity an absent body of mother and the trembling of fear creates a new symbol – that of the cord in the spine, linking the origin of life with the possibility of death. A new funeral rite – a symbolic burial – becomes possible, thawing a long-suspended mourning.

When there is a shift from the ego–Self axis to the ego–traumatic complex axis, an implicit and unconscious phantasy of sharing the surface of the analyst's self may emerge in the patient. This relational state, which Luci (2023) calls adhesive, allows the patient to feel supported, protected, and contained through this experience of continuity with the analyst's psychic skin.

When this implicit mode emerges in the transference, it may be challenging for one or both members of the dyad, and may even generate somatic countertransference responses that are difficult to understand. I believe that the relational-bodily context of the Authentic Movement form favours the symbolic and somatic manifestation of this unconscious phantasy of psychic continuity. When it is consciously shared by the analyst, through a participation mystique experience, it enables the opening to that early dissociated traumatic affect, as well as to a new experience of deep interconnectedness. At the same time, it implicitly communicates to the patient the deep sensation of being sheltered, which in turn allows the patient to experience these affects – the trembling and the fear of death – without the threat of destruction, giving rise to the acceptance of a loss hitherto impossible to symbolise.

Figure 5.3 Vera's composition of her journey that she created with her paintings and art forms that have emerged during embodied active imagination experiences.

I also consider that the same experience represents a new and incipient operation of the transcendent function, of Vera being able to sustain trembling/fear and maternal presence as opposite feelings related to the same experience.

And finally:

Shortly thereafter, Vera has a dream. It should be noted that she was familiar with some Jungian notions and terms that emerged in the process of analysis and also from her interest in working with the body and active imagination.

In the dream, Vera writes a composition and at the end signs her name, the initial of her paternal surname, her maternal surname, and then "Self".

The dream – and her emotion when remembering it – express not only the possibility of a conscious return to those personal origins that were denied to her, but also to her own Self, the psycho-somatic-affective matrix, which is archetypal in nature, and the foundation of a singular life that could begin to recover a sense of its own temporality which had been for a long time suspended (Figure 5.3).

References

Beebe, B. & Lachmann, F. (2002). *Infant Research and Adult Treatment*. Hillsdale, NJ: The Analytic Press.

Bianchi, M. G. (2023). "Enforced disappearances and torture today: A view from analytical psychology. Victims of enforced disappearances: Absent bodies, inner presences." *Journal of Analytical Psychology*, 68(2), 327–336.

Bion, W. R. (1962). *Learning from Experience*. London: Heinemann.

Bion, W. R. (2006). *Volviendo a pensar*. Ediciones Hormé.

Cavalli, A. (2020). "Noah's Ark. Technical and theoretical implications concerning the use of metaphor in the treatment of trauma." *Journal of Analytical Psychology*, 65, 5.

Cavalli, A. (2023). *Transforming Infantil Trauma in Analytic Work with Children and Adults. The Clinical Writings of Alessandra Cavalli*. Edited by Martha Stevns and Lucinda Hawkins. London: Routledge.

Chevance-Bertín, M. P. (1987). "Memoria para lo impensable. El caso de los hijos de desaparecidos argentinos robados por militares o policías." En *Restitución de Niños, Abuelas de Plaza de Mayo. 1997*. Eudeba: Buenos Aires.

Clark, G. (1996). "The animating body: Psychoid substance as a mutual experience of psychosomatic disorder." *Journal of Analytical Psychology*, 41, 353–368.

Cozzolino, L. (2002). *The Neuroscience of Psychotherapy: Building and Rebuilding the Human Brain*. New York: W.W Norton & Company.

Damasio, A. R. (1994). *Descartes' Error*. New York: Avon Books.

Damasio, A. R. (2000). *Sentir lo que sucede. Cuerpo y emoción en la fábrica de la consciencia*. Chile: Editorial Andrés Bello.

Damasio, A. R. (2010). *Self Comes to Mind*. London: William Heinemann.

Davoine, F. & Gaudilliere, J. M. (2011). *Historia y Trauma: La Locura de las Guerras*. Fondo de Cultura Económica.

Eulert-Fuchs, D. (2020). "The other between fear and desire – countertransference fantasy as a bridge between me and other." *Journal of Analytical Psychology*, 65(1), 153–170.

Fleischer, K. (2020). "The symbol in the body: The un-doing of a dissociation through embodied active imagination in Jungian analysis." *Journal of Analytical Psychology*, 65(3), 558–583.

Fleischer, K. (2022). "At the train station: The self, suspended in collective trauma. Symbolic analysis with victims of childhood trauma caused by state terrorism." *Journal of Analytical Psychology*, 67(1), 130–144.

Fleischer, K. (2023). "Collective trauma, implicit memories, the body, and active imagination in Jungian analysis." *Journal of Analytical Psychology*, 68(2), 395–415.

Fogarty, H. W. (2018). "Emergent somatically embedded memories." Paper presented at the VIII Latin American Congress of Analytical Psychology, in Bogotá, Colombia.

Fonagy, P., Gergely, G., Jurist, E. J. & Target, M. I. (2002). *Affect Regulation, Mentalization, and the Development of the Self*. New York: Other Press.

Fordham, M. (1976). *The Self and Autism*. London: Karnac.

Fordham, M. (1985). *Explorations into the Self*. London: Academic Press.

Freud, S. (1917/1993). "Duelo y Melancolía." In *OC-Tomo XIV*. Buenos Aires: Amorrortu Editores.

Gallese, V. (2003). "The roots of empathy: The shared manifold hypothesis and the neural basis of inter-subjectivity." *Psychopathology*, 36, 171–180.

Gallese, V. (2009). "Mirror neurons, embodied simulation, and the neuronal basis of social identification." *Psychoanalytic Dialogues*, 19, 519–536.

Godsil, G. (2018) "Residues in the analyst of the patient's symbiotic connection at a somatic level: Unrepresented states in the patient and analyst." *Journal of Analytical Psychology*, 63(1), 6–25.

Jung, C. G. (1916). "The transcendent function." In *CW 8*. Princeton, NJ: Princeton University Press.

Jung, C. G. (1946) "The Psychology of the Transference." In *CW 16*. Princeton, NJ: Princeton University Press.

Jung, C. G. (1949). "The psychology of the child archetype." In *CW 9/1*. Princeton, NJ: Princeton University Press.

Jung, C. G. ([1934–1939] 1988). *Jung's Seminar: Nietzsche's Zarathustra*. Edited by William Mc.Guire.

Kalshed, D. (2020). "Opening the closed heart: Affect-focused clinical work with the victims of early trauma." *Journal of Analytical Psychology*, 65(1), 136–152.

Knox, J. (2010). "Responses to Erik Goodwyn's 'approaching archetypes: Reconsidering innateness'." *Journal of Analytical Psychology*, 55(4), 522–533.

Lagutina, L. (2021). "Meeting the orphan: Early relational trauma, synchronicity and the psychoid." *Journal of Analytical Psychology*, 66(1), 5–27.

LeDoux, J. (1996). *The Emotional Brain: The Mysterious Underpinnings of Emotional Life*. New York: Simon & Schuster.

Luci, M. (2023). "Enforced disappearances and torture today: A view from analytical psychology 2. Torture survivors and the unthinkable: A hyper-present body in the therapeutic process." *Journal of Analytical Psychology*, 68(2), 337–347.

Martini, S. (2016). "Embodying analysis: The body and the therapeutic process." *Journal of Analytical Psychology*, 61(1), 5–23.

Merchant, J. (2015). "Foetal trauma, body memory and early infant communication: A case illustration." *Journal of Analytical Psychology*, 60(5), 601–617.

Panksepp, J. (1998). *Affective Neuroscience. The Foundations of Human and Animal Emotions*. Oxford University Press.

Schellinski, K. (2009). "When psyche mutters through matter. Reflections on somatic countertransference." *Spring Journal*.

Schore, A. (1994). *Affect Regulation and the Origin of the Self: The Neurobiology of Emotional Development*. Hillsdale, NJ: Erlbaum.

Sidoli, M. (2000). *When the Body Speaks: The Archetypes in the Body* (P. Blakemore, Ed.). London & Philadelphia: Routledge.

Sletvold, J. (2014). *The Embodied Analyst. From Freud and Reich to Relationality*. London & New York: Routledge.

Stern, D. (1991). *El Mundo Interpersonal del Infante. Una perspectiva desde el psicoanálisis y la psicología evolutiva*. Buenos Aires: Paidós.

Stern, D. (2017). "Unformulated experience, dissociation, and Nachträglichkeit." *Journal of Analytical Psychology*, 62(4), 501–525.

Trevarthen, C. (1998). "The concept and foundations of infant intersubjectivity." In S. Braten (Ed.), *Intersubjective Communication and Emotion in Early Ontogeny* (pp. 15–46). Cambridge: Cambridge University Press.

Volnovich, J. C. (1992). "Las abuelas, entre dioses y ausencias." En *Restitución de Niños, Abuelas de Plaza de Mayo. 1997*. Buenos Aires: Eudeba.

West, M. (2016). *Into the Darkest Places. Early Relational Trauma and Borderline States of Mind*. London: Karnac.

Wilkinson, M. A. (2017). "Mind, brain, and body. Healing trauma: The way forward." *Journal of Analytical Psychology*, 62(4), 526–543.

Winborn, M. (2022). "Whispering at the edges: Engaging ephemeral phenomena." *Journal of Analytical Psychology*, 67(1), 363–374.

Zoppi, L. (2017). "Chilled to the bone: Embodied countertransference and unspoken traumatic memories." *Journal of Analytical Psychology*, 62(5), 701–709.

Chapter 6

Dream With the Heart, and the Heart of Dream

Heyong Shen

Abstract

Based on Jungian analysis and the psychology of the heart, the chapter investigates dreams related to war in China and paintings from artists associated with war. At the third level, the chapter studies images of war in Chinese culture. For instance, the message in Chinese characters of "military" (Wu) carries the meaning of finding a way to stop fighting; the character of "auspicious" (Ji, lucky) conveys the significance of withdrawing weapons and troops. This essay uses Jung's "the knowledge of the heart" as methodology to reflect on the current predicament, including a survey on the attitude to war, dreams of war, and data from artists, including their drawings and paintings.

In his later years, C.G. Jung expressed his deep concern for the fate of mankind, and the danger of the world. He said: "The world hangs by a thin thread" (Wehr 1989, pp. 438–439). Jung went through World War I and World War II, and had been working with the related deep wounds. For the needs of patients, he draws inspiration from Chinese culture, such as *I Ching* or the *Book of Changes*, and Tao, and transforms them into methods and practices of depth psychotherapy.

The origin of Jung's *Red Book* is linked to the wartime atmosphere, affecting him deeply with continuous nightmares and panic imagery. In the Chinese context, the *Red Book*, or *Liber Novus*, holds the meaning of the book of the heart. Jung himself stated that he wanted to find "the knowledge of the heart" (Jung, RB, p. 233) and the "method of the heart" (intuitive technique) (Jung, CW8, para 863) in his most arduous exploration. So, the world hangs by a thin thread, and that thread is the human heart and soul.

Richard Wilhelm came to China in 1899, experienced World War I in China, and left behind a far-reaching book: *Soul of China*; which contains the inspiration to me for this chapter. A Chinese philosopher, Liang Shuming, known as the last Confucianist in China, used a conversation with his father in 1917 as a clue to leave behind his last book, *Will the World Be Good*? Readers who have read this book will naturally contemplate how to make this world a better place. Liang Shuming left his own answer in his psychology: *Human Heart and Life*.

The Cultural Imagery of Chinese Characters Related to War and the Contained Meaning

According to the *Art of War* by Sunzi, attacking the city is at the bottom, and attacking the heart is at the top: "Know your enemy and know yourself, and you will not lose a hundred battles" (Sunzi 1999, pp. 22–23).

DOI: 10.4324/9781003390039-22

In Chinese cultural context, people's understanding of war is mainly manifested in the imagery of the following Chinese characters: Zhan (戰), Dou (鬥), Zheng (爭), Wu (武), Bing (兵), and Ting (鼎). These images contain a cultural attitude and wisdom, and they convey the meaning which can help us for further understanding and reflection. As C.G. Jung said, after he studied *I Ching* and Chinese characters, the image of *I Ching* and Chinese characters are readable archetypes.

Zhan (戰), the Chinese character for "war" (battle, fighting), combined with two parts: Ge and Dan. Ge is an ancient weapon, and Dan means one and big. The symbol of Dan is related to cicada; Dan (or Chan) is determined by the pronunciation of Zhan.

Ge 戈, the symbol of an ancient weapon, later almost doubled of it, forming the Chinese character for ego (Wo, Me): 我. According to the classic Chinese dictionary, Wo (Me), those who follow the Ge (Weapon), take the Ge and hold on to themselves.

The classic dictionary interpreted this character of Zhang (War) as Dou (鬥: War, Battle, Fighting): two fighters face to face, with weapons behind them. Then, the classic dictionary quoted *The Zuo Zhuan* (*The Zuo Tradition*): for Zhan (War), the sage takes it very carefully.

Therefore, the image of "war" in the Chinese context has rendered to the overdevelopment of ego. The overdevelopment of ego, losing respect for divinity, and violating nature and Tao, will inevitably cause disasters; thus, war is extended to fear: Ju. The left part of the character for fear in Chinese, 忄, is the symbol of the heart, and the right part is a symbol of an ancient weapon. So "war fear" (Zhanju), a term with the image and meaning, contains a vivid psychological picture.

Usually, the word of war (Zhan) comes with Zheng (as Zhan-Zheng) and Dou (as Zhan Dou) together. For Zheng, 爭, the symbol is two men struggling, competing for something, with their hands. For Dou, 鬥, the symbol is two men fighting face to face, with weapons.

In ancient times, Wu was a more formal word for war, usually translated into modern English as military, martial, and warlike. The Japanese Proficiency Testing Association at Kiyomizu Temple in Kyoto selects a Chinese character as the symbol to represent the social emotion of the year, and Wu became the symbol for 2022, as reported by Japan *Asahi Shimbun*.

The earliest representation of the Chinese character of "Wu" in the oracle bone inscription appeared in the Wu Ding period (1250 BC). The primary meaning of wu, 武, with Ge (Weapon) as the upper part, and Zhi (Stop, Halt) below, contains a symbolic revelation that to stop using the weapon, conveying the message that the purpose of war is to stop or end using the weapon. This became the basic cultural attitude to war, especially related to Taoist tradition. As Lao Tsu said: "Weapons bode not well. All creatures under heaven fear them. Who follows the Way honors them not, and adopts them only at the last" (Laozi 1999, pp. 62–3).

So, from the Chinese character for Ji (lucky, propitious, good), 吉, we can read the meaning from the symbol – to put the weapons away, don't use them – as lucky and good. There is a legend that Dayu (Yu the Great), founder of the Xia Dynasty (about 2070–1600 BC), melted the weapons and cast them into nine tripods. Ting (Tripod) was used for cooking and worship, and later became the most important symbol for a country, the imperial power.

I Ching has a hexagram named Ting (Tripod, Cauldron, the 50), ䷱; above is the trigram of Li, the symbol of fire; and below is the trigram of Xun, the symbol of wood. According to Richard Wilhelm's (Wilhelm and Baynes 1967) translation and understanding, the image of the hexagram Ting suggests the idea of nourishment. When one causes wood to penetrate the fire, food is cooked. The holy man cooks to sacrifice to God, the Lord, and cooks feasts to nourish the holy and the worthy. As Ting after Ko (49, Revolution/Molting), the hexagram is structurally the inverse of the preceding one; in meaning also, it presents a transformation. The meaning of Ko is more like revolution through war, so Ting shows the correct way of going about social reorganization.

> The two primary trigrams move in such a way that their action is mutually reinforcing. The nuclear trigrams Ch'ien and Tui, which mean metal, complete the idea of the *ting* as a sacred ceremonial vessel. These old bronze vessels – as still occasionally found in excavations – have been connected throughout all time with the loftiest expressions of Chinese civilization.
>
> (Wilhelm and Baynes 1967, p. 1149)

In the culmination of civilization, the *ting* serves to sacrifice the highest earthly values to the divine, as expressed by Richard Wilhelm. The supreme revelation of God occurs through prophets and holy men, and veneration of them is true veneration of God. Accepting the divine will through humility brings inner enlightenment and genuine understanding of the world, leading to great good fortune and success.

While finalizing this chapter, I visited Jerusalem and engaged in a discussion with Dvora Kutzinski about war and shadow issues. As a gift, I presented her with an ancient Chinese tripod model, along with insights into the Ting, the image, and the significance of the *I Ching* hexagram. Dvora mentioned that the Book of Isaiah aligns with the imagery of Ting in the *I Ching*: "They shall beat their swords into plowshares, and their spears into pruning hooks."

Regarding War, Current Attitudes, Dreams, and Expression

On February 25, 2022, the International Association for Analytical Psychology (IAPP) expressed profound concern on its website about the dire situation in Ukraine and conveyed sincere sympathy and solidarity with the affected group members and their communities: "We cannot accept the military invasion and hope that the situation will not get any worse, and that the peace of land and soul will return to the region very soon" (IAAP 2022). I wrote a letter to IAAP officers and the executive committee, support the statement, and participate in related actions.

Soon after that, we started two investigations: about the attitude to war, and dreams related to war.

The questionnaire for the attitude of war included the following measurements.

War Effectiveness: 1, 7*.
The moral justification for war: 5*, 8*.
Economic consequences of war: 3, 10
Social consequences of war: 2, 11
The positive humanitarian impact of war: 4, 6
Negative humanitarian consequences of war: 9*, 12*.
* – reverse declaration.

There were 98 participants who answered the questionnaire. Participants were recruited online. One was removed because almost every item (11 of 12) was answered with the same value. Thus, the participants consisted of 97 participants.

Table 6.1 Results for the KMO and the Bartlett Test of Sphericity for the Exploratory Factor Analysis

Kaiser-Meyer-Olkin		.74
Bartlett test of sphericity	χ^2	340.92
	Df	66
	Sig.	.000

The number for the KMO and the p value for the Bartlett test met the typical standard (KMO >.6; Bartlett p <.05)

Table 6.2 The Component Matrix and the Variance Contribution Rate for the Exploratory Factor Analysis

Element	Initial Eigen Values			Rotation Sums of Squared Loadings		
	Total Number	Variance Rate (%)	Cumulative Rate (%)	Total Number	Variance Rate (%)	Cumulative Rate (%)
1	3.97	33.04	33.04	2.41	20.05	20.05
2	1.44	11.97	45.01	2.29	19.10	39.15
3	1.32	10.98	55.98	1.53	12.79	51.93
4	1.04	8.67	64.65	1.53	12.72	64.65

Exploratory Factor Analysis Results

Extraction Method: Principal Component Analysis

The cumulative variance rate met the typical standard (cumulative variance rate >60%).

Thus, the raw data can be divided into four factors. The item items contained in these four factors and the names of the factors are as follows:

Factor 1 (Self-Enhancing Effect of War): (1) sometimes, war promotes national progress; (2) war develops the best abilities in human beings; (3) participation in war is a special experience that leads to a deeper understanding of life; and (4) war promotes the economic development of the country involved.

Factor 2 (Self-Protective Consequences of War): (1) sometimes, war is necessary to resolve international conflicts; (2) Sometimes, war can protect one's own citizens and enforce human rights; and (3) war helps protect those who suffer from injustice.

Factor 3 (Negative Consequences of War): (1) war brings out the worst in people; (2) war makes people cynical and takes away their faith in goodness and justice; and (3) international conflicts should be resolved through negotiations, not through the use of arms.

Factor 4 (Moral Reasons Why War Is Incorrect): (1) waging war is a sign of weakness, an inability to achieve one's own goals; and (2) 5here is no excuse for waging war.

The Result of Difference Comparison

The mean difference comparison of the four factors showed that Friedman test, $\chi^2 = 187.78$, p <.001. Pairwise comparisons were performed using the Wilcoxon signed-ranks test. Key results: the self-improvement effect of war < the self-protection effect of war (z = 5.41, p <.001,

Table 6.3 Mean data for each factor in the sample

Factor Type	Mean (Standard Deviation) (a)
Self-enhancing effect of war	2.55 (1.25)
Self-protective consequences of war	3.44 (1.60)
Negative Consequences of War	5.82 (.75)
Moral reasons why war is incorrect	5.77 (.87)

(a) 1–7-point Likert scale, 1 = totally disagree, 4 = generally agree, 7 = completely agree

phi =.55); self-protective effects of war < negative consequences of war (z = 7.82, p <.001, phi =.79); self-improvement effect of war < moral justification for war being incorrect (z = 8.39, p <.001, phi =.85); negative consequences of war ≈ moral justification for war being incorrect (z =.18, p =.86).

Therefore, the results show that the self-enhancing effect of war < the self-protective effect of war < the negative consequences of war ≈ the moral justification for the incorrectness of war.

In addition, in the choice of war as the top and war as the bottom, among the 97 subjects, 96 people think that the war is the bottom (99%), whereas one thinks that the war is the top. This is consistent with the negative evaluation of the war by the subjects in the previously mentioned difference comparison results.

Around March 12, 2022, we conducted a questionnaire to explore the content of recent dreams related to war. Out of 315 responses, 120 valid replies remained, all of which reported dreams associated with war. We worked on the emotion in the dreams using Nvivo 12 software automatic coding. Total codes: 120; negative sentiment: 55 (45.8%); mixed sentiment: 40 (33.3%); neutral sentiment: 19 (15.8%); positive sentiment: 6 (5%).

Chi-square test results: no significant difference between negative mood group and mixed mood group (2 = 2.13, p =.14); negative mood group > neutral mood group (2 = 20.99, p <.001, phi =.29); mixed mood group > neutral mood group (2 = 9.91, p =.002, phi =.20); neutral mood group > positive mood group (2 = 7.55, p =.006, phi =.18); therefore, judging from the pairwise comparison results between the above groups, the distribution of emotions in dreams is as follows: negative emotions ≈ mixed emotions > neutral emotions > positive emotions (see Figure 6.1)

Examples of dream texts for various emotional groups in the auto-encoded results follow.

Negative Emotions: dreaming of a terrorist attack, almost being killed by others.
Mixed Emotions: the dream was at sea, near the South China Sea, submarines and warships were fighting, China and the United States were at war, I was driving a warship; and found that we could not defeat the United States, and then I was about to drop the atomic bomb, but my conscience suddenly realized that I couldn't drop the atomic bomb, and finally, I became friends with the people on the U.S. warship.
Neutral Emotions: I am standing alone on the roof of the school dormitory, and all I can see are gray concrete buildings. There is news of a war with Taiwan. My mind is spinning very fast, wondering which city I am in. To where should I flee? Should I go back to my hometown in Northeast China to find my parents?
Positive Emotions: I dreamed of waking up to have breakfast in the morning and turning on the news broadcast as the background sound according to the habit. It turned out that we sent troops to Taiwan. At that time, I wondered why it was so sudden and whether it would take a long time. As a result, around the afternoon, people from both sides of the strait began to celebrate the family reunion.

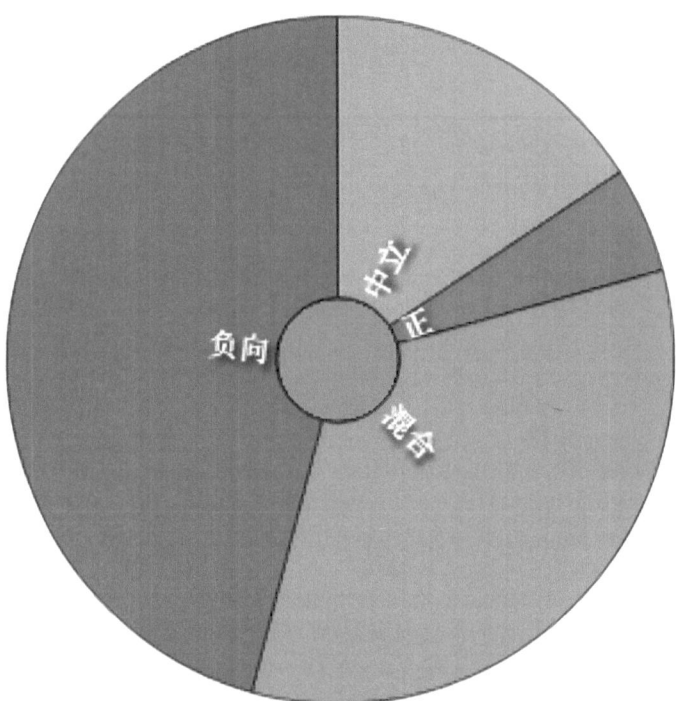

Figure 6.1 The emotion in the dreams.

I have a friend who is a painter. On the morning of Feb 25, 2022, he recorded the following dream and sent it to me.

In a very modern large shopping mall, relatively clean and brand new, in a European style, there are tourists and parents with their children wearing European-style autumn and winter clothes inside. They are looking at the map on the big screen, paying attention to and discussing the issue of the Ukrainian war.

The second scene of the dream was early in the morning. I saw a special car passing by the gate of Fang-zhou Court where I live. It was a very standard sphere as a whole, with a metallic silver-gray color, and two wheels sticking out below. It can be seen that it has been professionally modified from a motorcycle. On both sides of it, there is a large camera lens with a diameter of about 50–60cm. It seems that it also has the function of taking pictures along the way. Half asleep and half awake, I realized that it was a ball cart, and the homonym is a prison cart.

Then I realized that maybe I was very lucky, not born in the era of creating the group sculptures of "Rent Collection Yard," and I didn't have to do this kind of sculpture like the sculptors and authors back then. Born in an era with creative freedom, I can use my works to sort out the historical trauma left by that era.

Most of the rest of the dream is forgotten.

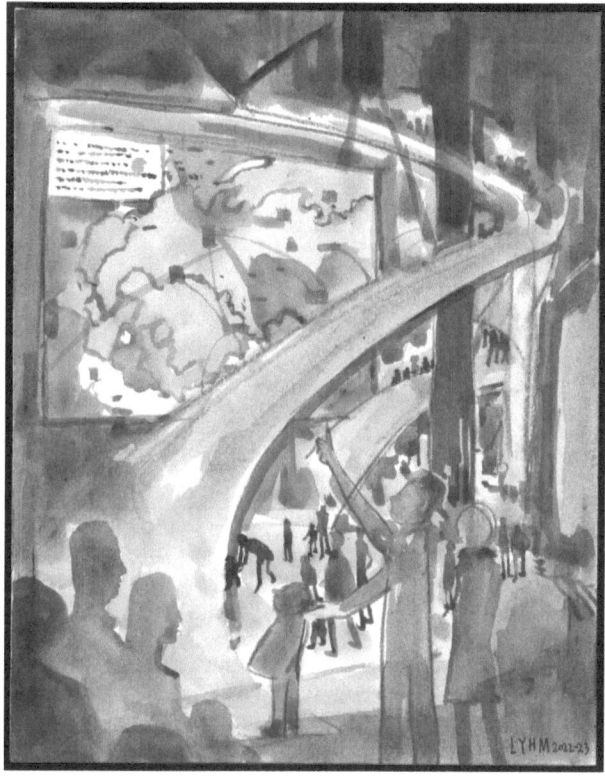

Figure 6.2 Liyang Huameng, European shopping malls, people who pay attention to the Russian-Ukrainian war; watercolor on paper.

My friend the dreamer, named Liyang Huameng, painted the dream scenes (Figure 6.2 and Figure 6.3).

I interviewed the painter, Liyang Huameng, about his dream, his paintings, and his feelings. He mentioned that for the dream, he felt that the influence of the war is not just on Ukraine but the whole of Europe. Liyang Huameng finds the appearance of Europeans in his dream curious. Being in China, he wonders why they appear when the war is far away. Upon further discussion, he notes that the dream establishes a passage of time, connecting the peaceful life on his side with the hidden crisis under the abundance on the Europeans' side. Inside, there's a mix of excitement and uneasiness, reminiscent of ancient times when – despite being thousands of miles away – one could sense the distant rise of warlike signals. The energy, though not directly affecting the current situation, subtly permeates the surroundings.

The Shi Hexagram of *I Ching*: Discussions and Reflections

The seventh hexagram of *I Ching* is named Shi (䷆), combined by earth and water, and contains the understanding and the meaning of war. Richard Wilhelm translated this hexagram as The Army.

Figure 6.3 Liyang Huameng, ball cart passing by the gate of Fang-zhou Court; watercolor on paper.

The hexagram before the army is Sung (䷅): above the trigram is Qiang, and below is water; its translation is Conflict. As Richard Wilhelm interpreted: The upper trigram, whose image is heaven, has an upward movement; the lower trigram, water, in accordance with its nature, tends downward. Thus, the two halves move away from each other, giving rise to the idea of conflict.

Therefore, there are Sung and conflict, and the coming Shi and army, to prepare for war. The basic meaning of Shi (the Army), is following the Sung (conflict), for the use of military, united to resist aggression. The hexagram's image suggests that self-protection is challenging without assistance from various forces. Richard Wilhelm further elaborated on this interpretation: This hexagram is made up of the trigrams K'an, water, and K'un, earth, and thus it symbolizes the ground water stored up in the earth. In the same way, military strength is stored up in the mass of the people – invisible in times of peace but always ready for use as a source of power. The trigrams indicate internal danger and external obedience, reflecting the essence of an army – internally perilous, yet requiring discipline and obedience externally.

As the *Tuan zhuan*, or *Commentary on the Judgment*, says: "Shi" means army and "zhen" means perseverance in the right way. If a man can command an army and persevere in the right way, he is capable of being a king. The nine at the second is firm in the middle position of the lower trigram and responds to the six at the fifth. He engages in dangerous tasks and follows the right way. Because of this, he leads his army to conquer the land under heaven and all people follow him. There is good fortune. What harm there can be? The image of Shi signifies that in the middle of the earth is water, the representation of the Army. Consequently, the superior man augments his masses through generosity toward the people. Richard Wilhelm further eluci-dated that the vast expanse symbolizes the attribute of the earth, representing the masses. Water embodies serviceability, as everything flows towards it.

Following my response to the IAAP officers and the executive committee regarding the state-ment, I corresponded with Jungian colleagues and friends, engaging in discussions about the situation and our shared concerns. Together, we explored avenues to offer support to those affected by and suffering due to the war.

In my letters, I mentioned the project of "Jung's *Red Book* for Our Time: Searching for Soul under Postmodern Conditions" by Murray Stein started in 2016, considering the uncertainty of the world and the shadow issues, that threaten any sense of coherent meaning, personally and collectively – and now, we have encountered the epidemic and pandemic, and the Ukraine crises. I want to know how my Jungian colleagues see and feel the situation and the future, and the significance of Jungian analysis for today. As Jung said, "The world hangs by a thin thread, and that thread is the psyche of man" (Wehr 1989, pp. 438–439). Jung also mentioned that "The only real danger that exists is man himself. He is the great danger. We know nothing of man, far too little. His psyche should be studied" (Jung, 1977, p. 436). Then, what we can and should do now, for the psyche and the heart of man. As I mentioned to my colleagues and friends, I con-sulted with *I Ching* and got a response; even in such a difficult time, we still have confidence in Heaven and Humanity.

Dennis Merritt, a senior Jungian analyst – we have known each other since 1997 – wrote a letter back as follows.

Dear Heyong,

Interesting: I have been thinking of you a lot recently as well and look forward to the possibility of talking with you. I've had some thoughts of doing a paper together on the *I Ching* and would like to run some ideas by you.

In so many ways the world is at a **Turning Point**-Hexagram 24. Very pleased with your sup-port of the IAAP statement. The Psychosocial Wednesday's group had a 1½hour Zoom meeting yesterday with a Ukrainian Jungian MD living outside Kyiv: I will send you the link when avail-able if you are interested. Last Sunday a dear friend put together a conference about the urgency of the times and I gave a half hour presentation. I can send you the link as soon as it is available. I came up with 10 points to guide us at this time.

Please tell me what question you put to the *I Ching*, if you can, so I can ponder the hexagrams. Surprise! I have some ideas about them and it would be interesting, and important I think, to discuss them. I'm glad I'm a Jungian at a time like this where we can analyse a situation at the deep archetypal levels, including with a little help from our friend – the *I Ching*.

Have to get back to my big presentation to the Chicago trainees tomorrow on Jungian ecops-ychology, especially how to work with spirit animals and sacred landscapes in dreams. They will be watching the "Seasons of the Soul" film I showed in China as part of their preparation.

Wishing you all the best,
Dennis

Dennis, a Jungian analyst, *I Ching* scholar, and professional ecologist, provided valuable insights during our discussion. He highlighted the significance and inspiration inherent in the turning point, especially within the context of the 24th hexagram. As Richard Wilhelm explained in his translation of the *I Ching*, "Return leads to self-knowledge" (Wilhelm and Baynes 1967, p. 911); turning back to one's inner light. There, in the depths of the soul, one sees the Divine, the One. "In the hexagram of Return one sees the 'heart'-mind of heaven and earth." (Wilhelm and Baynes 1967, p. 912).

Christa Robinson, former president of the Eranos East and West Foundation and a senior Jungian Analyst in Switzerland, became acquainted with Rudolf Ritsema during our meeting in 1995 in Ascona. Christa visited China frequently, attending International Conferences of Analytical Psychology and Chinese Culture, notably participating in the Oriental Roundtable for *I Ching* and Jungian Analysis.

Dear Heyong,

I appreciated your message of this morning deeply, especially your sharing of the very clear answer of the *I Ching*.

From my point of you, it agrees with Jung's statement, that the destiny of the world will depend on the amount of people trying to learn about the psyche, becoming more conscious. Shadow confrontation would be an important part of it, (not snapping at people, Hex. 10, and the illuminating 2nd line), but there people often shy away, as it is inconvenient. – My experience with the 2nd hexagram showed me, that it can be the result or outcome of the 1st one or accompanying the process of the 1st one as a background energy. I have the impression, that in this case both perspectives apply.

So, your trust in heaven and humanity is supported and underlined, but it says also, that we have to check where we should get rid of outworn attitudes and ways of thinking and strip them to arrive at a self-renewal. Hexagram 10 reminds us that individuation is a solitary process, and again, people in the West often shy away from it, because it is inconvenient. In a collective society like China, it is even more challenging. But to have a clear look at ourselves and trying to develop ourselves like we are taught in hexagram 15, may be a tiny but significant contribution in these troubled times.

I slowly start to understand the difference of the concept of individuation in East and West, why it cannot be the same. I do not particularly like to give lectures, but here I have the impression that it might be important for ISAP students to have some more knowledge about the fundamentally different social systems that govern our world They are the front and backside of the same coin, but we have to learn about the differences to arrive at more understanding and a potentially fruitful exchange. It may be my way to kindle some light.

This war is overshadowing me on a personal level, as my daughter-in-law has relatives in the Ukraine and her parents in Belarus. I see her deep suffering and the impossibility to do anything, apart from offering the last bed in their house to a refugee woman.

I send you warm thoughts and greetings, best wishes and thank you much for supporting the standpoint of IAAP.

Christa

The 10 hexagram that Christa mentioned is Lü (☰ Treading/Conduct). The Judgment of Lü says: trending; treading upon the tail of the tiger; it does not bite the man; success. (Wilhelm and Baynes 1967, pp. 165–166). When Richard Wilhelm interpreted the second line of Lü, he paid attention to the "dark man": the perseverance of a dark man brings good fortune. "He is central and does not get confused. This line is light, but occupies a dark place; hence, the image of a dark man. However, since he walks in the middle of the road – the line is central" (Wilhelm and Baynes 1967, pp. 796–797). The Chinese word for "Dark Man" is "You Ren" (secluded man), the character of You: 此; two symbols of Xuan in the deep mountain. "Xuan," translated into English as the "darker than any mystery" which Lao Tzu used as "the Doorway whence issued all Secret Essences" (Laozi 1999, pp. 2–3)

John Beebe attended the first International Conference for Analytical Psychology and Chinese Culture in 1998 on behalf of IAAP, and participated in the development of Jungian analysis in China since then. I first met him in San Francisco on August 31, 1997, at the birthday party of Dr. Joe Henderson.

Dear Heyong,

It is good to hear from you. We are at the very hinge of history, I believe, and no one knows which way the door will open. I too consulted the *I Ching*, on January 1, asking about the world, and I received gua 26 ("The Taming Power of the Great" in Wilhelm's translation) with the unusual combination of change lines first and top, changing to 46 ("Pushing Upward"). Reading this as the movement of the year, it predicted starting the year with yao 1, the man who would like to make an advance and finds himself firmly held back (this I associated with Putin) and also with yao 6 (the man who having "attained the way of heaven" would have "success") and this I associated with Biden. By years end, the "superior man" will have heaped up small things in order to achieve something small and great, and that I think refers to the world at large, which is rarely united in condemning what is not fair play and trying for a measure of compassionate cooperation, at least with regard to committing uncalled for atrocities. That would be pushing upward, spiritually, even though into the darkness (Yao 6 in Gua 46) since the victory for the light is not of the kind that banishes evil altogether. For me, however, it is enough to be confident that the world is discovering the Psychology of the Heart as a way of dealing with the terrible love of war (Hillman) that this time will not succeed in seducing the integrity of the world.

Please know I appreciate your reaching out to me at a time when you have already grasped these matters. Though I feel that you know that I also feel them along similar lines, it is very good to take your hand.

As ever,
John

John Beebe is a senior Jungian analyst, sinologist, and *I Ching* scholar. Like C.G. Jung, he is also "a jealous lover of the *I Ching*" (Jung 1973, pp. 154–155).

Tom Kelly is a senior Jungian analyst in Canada, and past president of IAAP.

Dear Heyong,

Many thanks for your very touching email. These are indeed very difficult times and your quote from Jung about "the world hanging on a thin thread" describes it so painfully well. It is remarkable to see how destructive personal complexes can become and how powerful and devastating the effects can be. To me, the answer to the question you raise as to what we can learn from this now for the psyche and heart of man is to listen carefully to the dream images. This is where we will get direction as to what each one of us can do. I have been in touch with some of our members both in Ukraine and in Russia. It is striking how both are suffering, though in different ways, because of this crisis.

I intend to spend some time in reflection on the two hexagrams you mention in your email . . . in the hope that th*ey can help maintain confidence in Heaven and Humanity, as you mention.*

I very much appreciate your email and you sharing your thoughts and concerns this way, Heyong. Let us hope and pray for the best in this very dark and difficult period.

With my very best and warmest wishes to you and to the members of the group in China,
Tom

Tom Kelly contributed to the International Conferences for Analytical Psychology and Chinese Culture, providing instruction and supervision for candidates at the Oriental Academy for Analytical Psychology and subsequently the China Society for Analytical Psychology.

John Merchant, a senior Jungian analyst in Australia, wrote the following letter for his response.

Dear Heyong,

It was encouraging to see your letter on the IAAP page as I think that's one of the things we can do for Ukraine at this time, support for our colleagues who are there helping people. . . .

Apart from that, if we can also find ways to encourage our politicians to provide humanitarian relief that would also be good.

As you say, Jung was very aware of the tenuous situation the world faced with the advent of the atomic bomb. Interestingly, I have been recently re-reading C. G. Speaking: Interviews and Encounters. In the interview for his 85th birthday with Georg Gerster, he commented on the Delphic oracle above the entrance to his Kusnacht house saying "religious phenomena are to be met with everywhere . . . It only needs an emergency, a serious emergency . . . under very emotional conditions." And further he said, "situations in which a man feels defeated very often give rise to religious phenomena." It will be interesting eventually to hear from our Ukrainian colleagues what numinosity their emotional emergency has activated. As you say, we can still have confidence in Heaven and Humanity.

And I'm hoping in the not too distant future to be able to see you again in person. My memories of the Xi'en conference are constantly on my thankfulness list.

With every best wish,
John

John Merchant referred to the 8th Xi'an conference in 2018, themed "Enlightenment and Individuation: East and West," where international Jungian analysts engaged in profound discussions with Chinese scholars in *I Ching*, Taoism, Confucianism, Buddhism, and traditional Chinese medicine.

Dyane Sherwood, a Jungian analyst, and Sandplay Therapist, wrote the letter for her consideration.

Dear Heyong,

Thank you for writing at this moment when much hangs in the balance.

A powerful image came to me of the gravity of this time and of how unknown is the outcome.

I made a small "blurb" on the Press website, and I hope you will be able to read it and to share it if you feel so inclined: https://www.analyticalpsychologypress.com/post/kyiv-before-the-war.

I very much appreciate your devotion to China and dedication to providing care for those in China who have suffered from natural disasters and other difficulties, using both heart and mind.

You are an inspiration to your students and to many others!

With my warm regards,
Dyane

Dyane Sherwood came to China for the fourth International Conference for analytical psychology and Chinese Culture: "The Image in Jungian Analysis, Active Imagination as a Transformative Function in Culture and Psychotherapy" (Fudan University 2009); She is the guest professor of the Orient Academy for Analytical Psychology.

George B. Hogenson, a senior Jungian analyst, is the past vice-president of IAAP. He was on the faculty of the Yale School of Management, and a consultant to many organizations on leadership and strategy.

Dear Heyong,

. . . Concerning your question, I think it is fairly self-evident that we are entering a period of dramatic change, certainly in the international order and politics, but also in the more spiritual or psychological way. Jung commented on how systems of cultural symbolism which at one time shaped a culture can exhaust themselves and become merely signs that no longer carry the significance they once had. I have felt for a long time that this was definitely the case in the West, because of the significance of the Christian tradition for the development of the culture. Jung, as early as the Zophingia lectures, was pointing out that the system of symbols that had shaped the Christian tradition had given way to only a system of practices with little symbolic depth – to use a term I coined in one of my papers. In Asia I think things are somewhat different, but change is undoubtedly happening there's as well. Overall, however, we are certainly seeing a compounding of global issues that do not respond to normal political or social responses, and for which we do not have a sufficient symbolic system to mediate the experience. When you link the [COVID-19] pandemic to the impact of global climate change and now warfare that everyone thought we had gotten past the impact of the combination exceeds the mediating power of the traditional symbolic systems, and requires something new . . . .

These are just some preliminary thoughts. I would be curious what your point of view is, and how your work with the *I Ching* is leading you to new ideas. Please stay in touch, and also please greet all my friends in China. I often think of our wonderful times together hope that we may be able to meet again in person one of these days.

Warm regards,
George

George's insights prompt me to contemplate the importance of cultural symbolism. Through his writings on archetypal theory, synchronicity, and the nature of symbols, I discern a resonance with the principles found in the *I Ching*, the Book of Changes, the Book of Living Symbols, and the Book of Readable Archetypes.

Tom Singer is a senior Jungian analyst. He developed the cultural complex theory, and published several important books in the field.

Dear Heyong,

It is good to hear from you. We are all suffering great confusion and distress/horror at the Russian invasion of Ukraine . . . .

There is no doubt that the world is shifting on its axis right now and it is profoundly disturbing because Putin's aggression (he would say it is the West's and Ukraine'[s] aggression against him) is so brutal and ruthless – it horrifies all of us and seems so unnecessary.

Anyway, global alliances are in the midst of deeply significant changes.

I did write an article regarding cultural complexes and unconscious forces in international relations at the time of the US invasion of Iraq following the bombing of the World Trade Center.

We manufactured that war, although the 9/11 bombing was certainly a huge precursor. I will attach that article.

It might interest you in terms of how I have thought about cultural complexes and what I call "archetypal defenses of the group spirit." I am thinking about how the cover image for the article might be upgraded or reframed in terms of today's realignment and conflict. I will also attach a link to a video that a young Russian with a Jungian orientation used the cultural complex idea to think about Russia – some of it is a bit garbled in terms of his understanding and at time it is hard to make out what he is saying, but it is certainly a good faith attempt to use the cultural complex idea to understand Russia . . . .

This is a list of readings/links that I recently put together for seminars I have been giving.

Warm greetings,
Tom

I have been working with Tom Singer on the research of cultural complexes. I agree with him that the cultural complex theory and related cultural unconscious and cultural archetypes are still crucial issues today.

Andrew Samuels, senior Jungian analyst and professor of psychology, is well known for his work at the interface of psychotherapy and politics.

Dear Heyong,

I admire you for your sense of responsibility and care.
I believe therapists and analysts have a role to play in offering support and insight.
In terms of policy and actions, my theory has always been we need to work with other people and groups. On our own, we are useless.
I work with the UK Stop the War Coalition.
The *I Ching* reading is most interesting.
I look forward to seeing you again someday. I was supposed to come to China in Aril 2020!

Best wishes,
Andrew

Andrew Samuels played a crucial role in promoting analysis and activism. He supported our Garden of Heart & Soul initiative, providing psychological aid for Chinese orphans and disaster relief during the COVID-19 epidemic.

Ann Casement is a senior Jungian analyst and a senior member of the British Psychotherapy Foundation and the British Jungian Analytic Association.

Dear Heyong,

You are, and have been, much in my thoughts over these past weeks and I was warmed to receive your email. A book that is relevant to the war currently being raged is James Hillman's "A Terrible Love of War." James and I became close in his later years so I commend his book to you if you are not familiar with it already. I saw your entry on the IAAP website and was so happy to see China represented there as I have been an ambassador for China over many years now. I love working with the many practitioners and students there who are keen to develop their knowledge of psychoanalysis and psychotherapy, including the current participants from all over China who are attending the ethics seminars . . . .

With best wishes,
Ann

I delved into James Hillman's profound exploration, *A Terrible Love of War*. His intriguing perspective questions why we fear war if it's deemed a "normal" part of our existence. He challenges the possibility of tempering the impulse for war and underscores the importance of understanding and imagination to mitigate its horrors for life to move forward.

Murray Stein, a senior Jungian analyst, is a past president of IAAP. He started the development of analytical psychology in China officially with Tom Kirsch in 1994.

Dear Heyong,

Thank you for your thoughtful message. The question you ask is crucial: how can Jungian analysts and scholars contribute TODAY to the desperate need in the world for a new myth to live by? I applaud your approach in consulting the *I Ching*. I will study the hexagrams you mention. Maybe they will inspire some thoughts. At the Symposium at Eranos this will definitely be a topic for discussion. Maybe the Red Book offers some hints for how to bring greater consciousness to the world by working on ourselves. It is an indirect approach but I feel the bet and most authentically Jungian approach. Hope to see you in Ascona soon!

Warmly,
Murray

Murray Stein invited me to join the project: "Jung's *Red Book* for Our Time: Searching for Soul under Postmodern Conditions" in 2016. He felt and predicted the uncertainty of the world then, and the related threats and dangers. Similar to C.G. Jung, Murray is an enthusiast of *I Ching*. His significant contribution lies in exploring the connection between Richard Wilhelm and C.G. Jung.

The image of "Shi" represents both fighting and the teacher, with its original purpose being to ensure livelihood, benefit human relations, increase morality, and contribute to national well-being. Reflecting on its meaning, "Shi" is intended for protecting the country and resisting insult, not causing harm. Its fundamental principle stems from Tao, virtue, and a restrictive function, akin to the combined trigrams of water and earth, gathering power to resist invasion.

Conclusion

C.G. Jung studied the *I Ching* and Chinese characters, and he once expressed his special feeling that they are "readable archetypes." When the archetypes become readable, we can have the chance to appreciate the symbolic meaning within, as well as the thread, the way, and the inspiration for action.

So, the image and meaning of the hexagrams of *I Ching*, as well as the related Chinese characters, is not only in the book but is just what is happening in real life. Ji-Xiong-Hui-Lin, "good fortune and disaster, regret, and stinginess," are the four characters used very often in *I Ching*. As the Great Treatise says: The sage sets up hexagrams to observe the phenomenon, attached judgments to understand the good fortune (Ji) and disaster (Xiong), and alternation of firm yang and yielding yin gives rise to endless changes. Therefore, good fortune and disaster are images of loss and gain, regret (Hui) and stinginess (Lin) images of worry and caring.

I once believed that Jung foresaw the World War II in dreams and images, pondering the significance of this prediction. As the war unfolded, it became a disaster for mankind. Later, I considered that Jung's foresight led to the creation of analytical psychology, incorporating Eastern wisdom like the *I Ching*, dedicated to developing depth therapy for traumatized souls. Now, I also believe in Jung's expression that the world hangs by a thin thread – the human heart and soul. What Jung did, and what Jung could do, is his depth psychology in accordance with the Tao, the Way, a kind of knowledge of the heart,[1] the intuitive method (method of the heart);[2] only this can meet the needs of the moment, whether it is therapy or healing or individuation and transformation.

Notes

1 注释，荣格红书，黑书
2 注释，中国方法

References

IAAP (2022) https://iaap.org/support-letters-for-our-colleagues-in-ukraine/
Jung, C.G. (1973) *Letters of CG Jung: Volume 1, 1906–1950* (G. Adler and E. Jaffe, Ed.). London: Routledge.
Jung, C.G. (1977). *Speaking: Interviews and Encounters* (W. McGuire and R.F.C. Hull, Ed.). Bollingen Series XCVII. Princeton, NJ: Princeton University Press.
Laozi. (1999) *Laozi* (A. Waley, Trans.). Beijing: Foreign Languages Press.
Sunzi. (1999) *The Art of War* (W. Rusong and L. Wusun, Ed. and Trans.). Beijing: Foreign Languages Press.
Wehr, G. (1989) *An illustrated Biography of C.G. Jung.* Boston: Shambhala.
Wilhelm, R. and Baynes, C.F. (Trans.) (1967) *The I Ching, or Book of Changes.* Bollingen Series XIX. Princeton, NJ: Princeton University Press.

Chapter 7

The Sacrificial Murder of Palestine

Grinding Bones to Dust

Heba Zaphiriou-Zarifi

Abstract

This chapter was written by a Jungian analyst of Palestinian descent ten months into the devastating assault on Gaza. In it, the author asks which definitions of war are applicable to the Palestine–Israel conflict. She questions whether analytical psychology should be employing C.G. Jung's understanding of the problem of opposites to articulate the dynamics of this catastrophic conflict as if a reconciling third might enable the world to transcend it without first calling it out. Inquiring whether it is ethically right to equate the two parties as opposites, she asks her fellow psychotherapists whether they can afford even as human beings to ignore how they feel about having the carnage of Gazans presented as a needed Israeli defence. She asks whether it is psychologically healthy not to condemn the murder of thousands of Palestinian civilians, when maiming, amputations, rape, starvation and abduction reveal issues other than simple self-interest on the part of world powers that have silently permitted these atrocities. She notes the psychological dangers inherent in disavowing the element of sacrificial murder in this kind of retaliation. Finding a prophetic analogy in the Book of Genesis, the author reminds us that the earth itself raised a cry of outrage at Cain's crime against his brother and suggests we might do better to recognise that when fratricidal genocide starts to occur, the horror that arises is calling not for denial but rather for atonement.

Preface

This chapter was written ten months into the devastating assault on Gaza.

By 'Israel', I mean the Israeli occupying state and its military, not its citizens, Jews or Judaism as a religion and way of life. Anyone analysing the comparative map of Palestine from 1948 to the present time cannot but be shocked by the shrinking Palestinian geographic presence. It is in this context that we need to understand the agony of the current catastrophe. It is regrettable, but experience tells us that there will be those who insist on interpreting my words as an expression of hostility to the Jewish Israeli people, and to Jews in general. I do not wish to stir hatred or revenge and certainly not the dehumanisation of any of the peoples entrapped in the hideous conflict; rather, this is my own appeal for solidarity among the international Jungian community to help us all face this dark reality in order to change it.

I feel it incumbent on me as, so far as I know, the only Palestinian analytical psychologist that I should press the case of an occupied people against the might of a modern militarised, ideologically-driven state. It is in this context that I emphasise Palestinian suffering from a *de facto* genocide in Gaza that includes an explicit process of ethnic cleansing of the West Bank and East Jerusalem. This situation has increasingly deteriorated over the last 30 years and recently with unrestrained violence under the direct control of the most extreme elements of Israeli Prime Minister Benjamin Netanyahu's cabinet. The genocide must be recognised for what it is and stopped.

DOI: 10.4324/9781003390039-23

Most importantly, I would like to stress my understanding and that of my fellow Palestinians of what Israel and world Jewry went through after the 7 October attack in 2023. My hope is that there could be a growing chorus of mutual compassion. May conscious grief and loss make way for the two-winged dove of justice and peace to fly over the Holy Land and into the hearts of her people, respecting one another and living with dignity.

Then the Lord said to Cain,
'Where is Abel your brother?'
He said: 'I do not know.
Am I my brother's keeper?'
What is war?

Before we define (and dismiss) the Palestine–Israel conflict as 'war', we should reflect on the choice that implies. War is generally defined as violent conflict between nation-states. War may erupt between opponents of opposing factions within one nation (civil war), or between distinct states who *declare* war against each other, engage in prolonged and intense military combat and have recourse to armed forces. Palestinians may have splinters of armed resistance; they certainly do not have war weaponry, let alone an army. Furthermore, as for the *de jure* Palestinian state, it is left on the back burner by international consensus with the entire 'hob' in full view but no political will to engage it. But if we insist on describing the conflict as 'war', then it has to be recognised as the Israeli state's war on *de facto* state-less Palestinian people.

Jungian Theory vs. Reality

Numerous analytical psychologists, in seeking a *coniunctio oppositorum*, may view Palestine and Israel as a pair of opposites at war. This perspective, under the veil of neutrality, betrays a justifiable ignorance of the (Christian) Zionist plan of establishing a Jewish (only) state of Israel, not only *in* but more actively *instead of* Palestine. This replacement project is a typical example of settler-colonialism. Political correctness limits intellectual honesty when it insists that we equate the conflicting parties with one another. The two were not conceived even-handedly; rather, one at the detriment of the other. For opposites to interact, 'this as well as that' have to exist. Over the last century, the gradual erosion of historical Palestine from geography, now dramatically encapsulated in the total devastation of Gaza, precludes the possibility of any transcending conjunction of the opposed forces. It is the human experience of opposites that are evenly matched, which Nicholas of Cusa called 'contrariety', that enables us to appreciate the sublime union of contraries that we find in God.[1] The abstruse impartiality that posits both sides, the occupier and the occupied, as equals has blindly condoned a military occupation by an apartheid regime that has profoundly diminished every aspect of the indigenous people's lives so that there can be no possibility of their becoming contrary.

Sentimental concepts, borrowed from John Gower's *Confessio Amantis,* may have allowed Jung, discussing the difficulties of the analytic transference, to summon the erotic promise in the idea of a 'warring peace, a sweet wound, a mild evil' as his epigraph for *The Psychology of Transference* (1966),[2] but they certainly do not apply to facts on the ground. To favour Jungian concepts over embodied reality is to misapply Jungian theory. Theory (θεωρία *theôria*) demands an intelligible explanation founded on observation – a contemplation based on a careful viewing. Such unexamined views regarding the pairing of Israel with Palestine treats theory as absolute. Whether it pares down reality to accommodate theory or bends reality to conform to theory, it destroys both theory and reality in one mental backflip. The evasion of reality on the ground poses a serious ethical dilemma; yet, speakers elaborate arguments, even heartfelt truths that appear valid but are factually wrong. Not only is this false equation an outright misunderstanding of analytical psychology, but more seriously a perversion of Jungian theory into dogma.

There is no equating a colonised with a coloniser. The Palestine–Israel conflict is not played out between agonistic opponents of similar or even comparable strength, 'a war by algebra', to use Carl von Clausewitz's metaphor.[3] Rather, it is a cruel and calculated destruction of the occupied by the occupier that justifies the centrality of war as enabler of politics. If anything, what we witness is an *aggression* against the Palestinians' inalienable right to freedom and self-determination. When sociology professor Baruch Kimmerling argued that Israel's domestic and foreign policy is 'largely oriented toward one major goal: the politicide of the Palestinian people' (2003),[4] he was only stating what we have learned to expect from colonialism. By politicide he meant 'the process that has, as its ultimate goal, the dissolution of the Palestinian people's existence as a legitimate social, political, and economic entity'. The assassination of numerous political figures is a hallmark, a long pattern, of Israel's aversion to Palestinian self-determination. Moreover, the 2018 bill, passed by the Knesset, stipulating a nation-state 'for Jews only', has precipitated further expansion of its illegal settlements beyond the state's internationally recognised borders, as it has most recently – whilst the barbarous mass murder of Gazans climaxes to absolute insanity – and annexed the largest West Bank land grab in 30 years.

Israel vs. Palestine: Two Opposites?

War is reciprocal, 'nothing but a duel on a larger scale', writes von Clausewitz.[5] But how can a duel be fought when one party, armed to the teeth and enjoying a sanctioned use of violence, forcibly removes at gunpoint their preyed-upon opponent who, unbacked, is not given even the slightest right to self-defence or means to withstand a duel, and moreover is tagged as an 'absentee' by their opponent? *Absentee* is Israel's euphemism for the 1947–1949 brutal expulsion of 800,000 Palestinians to neighbouring countries. Palestinians' voices resonate with Mahmoud Darwish's poem (2008), mourning 'our weight has become light like our houses / in the faraway winds'.[6] War is an act of force to compel the enemy to do one's will, in this case for Palestinians to disappear. But absentees the Palestinians are not. The deceptive narrative, concocted to uphold Israel's *terra nullius* fictional myth of 'a land *without* a people', has usurped and induced ignorance of Palestinian history, culture and economy in order to proceed with the Zionist plan to illegally steal and occupy by force the land, without her people. Paying no heed to the existence of a people on their land further blinds the ignorant 'to his capacity for evil, hence making possible for this evil to be projected, and depriving him of his capacity to deal with it' (Jung 1976).[7] Ignorance is also the means for international erasure not only of the Palestinian people, but of the true nature of the relationship between the Israeli regime and the indigenous population. A drone operator or a warplane bomber is shielded from hearing the horrifying screams of parents scooping up the shredded bodies of their children, or from seeing the hideous mutilations of bodies amputated by the blast, or uncovering those buried alive under collapsed concrete buildings, or those eviscerated to a horribly painful death. The bomber's willful ignorance deprives him from perceiving himself as a mass murderer. Ignorance is indeed, as Rabelais noted, the mother of all evils, especially when it obfuscates one's own.

Pressing Questions

In an informed world, where facts and figures are made widely available through a large corpus of scholarship, how can anyone persist in ignoring not only the specific injustices inflicted upon the Palestinian people but also the horror of their intentional erasure? Do we find acceptable the use of the Bible as a means of oppression? Can journalism be ethical in the face of occupation politics? How will the world recover from the genocidal massacres that have dislodged – in order to wipe out wherever possible – an entire Palestinian population for the sake of establishing a swelling Israeli state on a 'clean' slate? How is it that we abdicate knowledge, evade responsibility, abandon morality, deny the scale of mass murder, circumvent the immense psychological and existential suffering, to benefit the governing and obfuscating colonial *zeitgeist* that must deny its own evil in order to exist? How is it that we come to accept and even normalise Israel's sheer military brutality that eclipses the political and the moral? How can we pretend to live in a decolonised world when an unstoppable colonial war that abides by no rules of engagement is ferociously waged on Palestinians? Have we forgotten that *hubris* is followed by *nemesis*? Surely analysis began, with Freud, to teach us to recognise the psychopathology in defences like denial, displacement and projection. Can Jungians not also see the dark side of the Self when it enacts itself on behalf of the integrity of individual nation-states, without regard for its brother and sister nations? As God himself asked, 'What have you done with Abel?' Has cowardice, branded by Michel de Montaigne 'the mother of cruelty',[8] become the implacable neurosis of our time?

7 October 2023 and Beyond

The current history did not begin on 7 October 2023, but what the attack provoked, however callous and shocking, induced a tearing-off of the mask of concealment from the Zionist project over the Middle East with Palestine as its linchpin. Western imperialism continues to tolerate Israel's own imperialism, yet there has been a paradigm shift and victimhood can no longer hide victimisation. The upheaval is filtering through Western universities and media, and a collective uproar among youth the world over is rising independent of their governments' policies. Does the collective uprising swelling up from the freshly released information herald a new aeon? The Aquarian Age transitioning into the collective consciousness is like a tidal wave gradually building in intensity. An enantiodriomia is inevitable; there is only so much the earth can take.

'When a nation loses control of territory, resistance is applicable'.[9] Resistance to occupation is a constant in history, even when the occupied has no recourse to conventional forces. Resistance against the oppressor describes not only the character and spirit of the occupied, but it also manifests as the direct consequence of the occupation itself and its normative power. The length of time, as well as the historical context and the nature of the occupation, define the kind of resistance that an occupation will provoke. It is widely understood that 'occupiers that victimize the occupied population will face greater resistance' (Collard-Wexler 2013).[10] This is especially true with a prolonged 'occupation', a euphemism for colonial plunder and ethnic extermination, as in the case of Israel's of the Palestinian Territories. As long as Israel imposes occupation, Palestinians will seek liberation. In that respect, the colonial entity will have to bear the intrinsic moral and political responsibilities for the kind of resistance its occupation engenders.

'War as Reset' is Israel's agenda, fully endorsed by hefty financial, military and political cover from the USA's American Israel Public Affairs Committee (AIPAC). Replacing the Sykes-Picot lines drawn by the Anglo-French pen in the scramble for colonies, the lines of the

Middle East map are now redrawn with Netanyahu's knife. Netanyahu's map, recently publicised with nonchalance, geographises the Zionist Eretz-Israel land grab as though it were only a matter of better-defined boundaries. It was in fact in retaliation for this 'reset' that 'Operation Upset' was unleashed by Hamas. The 'upset' is Palestinian exposure of Israel's face from behind the mask of innocence; the Zionist project is, effectively, incompatible with a *unio oppositorum*. In fact, it abhors opposition even from within its Jewish population.

This is not to deny the evil in war. But this war was meant to upset, and Palestine's resistance – armed and unarmed – has shattered the world's willingness to ignore a longstanding pre-existing persecution. If it is at all a war between two peoples claiming their right to exist, then the unevenness is evident. There is no political Hamas in the West Bank, yet the state has armed Israel's fundamentalist settlers to carry out further vicious ethnic extermination. Gobbling up even more of Palestine raises serious doubts as to how much Israel can digest, given what it has already filled itself with, as if that were wholesome. Water wells and olive trees are not Hamas, yet Jewish settlers destroy them with messianic zeal.

Present-day Israel's proficiency in lying about its intention to destroy not just Hamas but all Palestinians – and in having even U.S. presidents lie for it – is legendary. State-backed racial profiling has gained in popularity, and the Israeli soldiers' shameless posting of their horrors in Gaza, which reel with bestiality, including the brutalising of both men and women by Israeli soldiers, are often popular tropes within Israel, celebrated by people who ought to know better. Israel's occupation forces in Gaza mockingly pulled out of a torched bedroom drawers containing the lady of the house's lingerie, jeering and sneering with satisfaction during ecstatic orgies of looting. Carpets, jewellery and money were stolen, the soldiers' lupine avidity rapturous with indecency. The desecration of holy sites and the sanctity of homes, churches, mosques, schools, music conservatory, hospitals, mental health centres – nothing stood in face of the inebriated lust for evisceration. Children's toys were smashed to pieces in a frenzy of sadism. Doctors, nurses, professors, writers, poets, teachers and artists were slain in an unprecedented degree of attacks on learning and healing; an explicit intent for a scholasticide, augmenting an already entrenched memoricide. That the murder of journalists has been followed by an imposed state-imposed silence is symptomatic of the occupier's control of the narrative. Sewage and electricity infrastructures were exploded, for they too are Hamas! Far-right Zionist protestors stormed Israeli military facilities, pressing for the release of soldiers who gang-raped Palestinian prisoners, one to the point of paralysis. Torture and rape are war crimes. They are not a new phenomenon, as Dr Samah Jabr attested: 'We've seen these crimes before, behind closed doors of interrogation rooms. The difference is that they are now doing it, shamelessly, in front of cameras, and more frequently with more soldiers engaging in these acts'.[11] Israel – represented by its soldiers, in the eyes of the caring world – has become a criminal, evil state. Released Palestinian hostages reveal 'shocking allegations of physical, psychological and sexual abuse'. The West Bank and now Gaza serve as Israel's laboratory for a morbid trade in stolen bodies and body organs.[12] One wonders if giving Israel a loose rope to indulge in countless crimes with no accountability is truly a sign of the USA's or Germany's care for Israel?

The Victimised

The world has failed to clear the fog of misconceptions, let alone rectify the injustices Palestinians have endured for more than 76 years. It is on the back of its victims that Israel has etched in bold scars its founding crimes, each mark cutting deeper into the flesh. A cruel immolation was forcibly imposed upon Palestinians. No Israeli book has the spine to tell the tale of those who were murdered under its colonial state, or to uphold the stories of those who struggled and cried what it means to live and die for a cause. In a Reuters interview in 2020, Benny Morris considered disappearing Palestinians from view an ideal solution for Israeli Jews.[13] Palestinian bones are ground to dust and, with them, any hopes and dreams of co-existence vanish.

Fortunately, this is no longer an attitude that has world Jewry's support. U.S. Jewish orthopaedic surgeon Dr Mark Perlmutter describes the 'distinct signs of [overwhelming] genocide' in Gaza. 'All of the disasters I've seen, *combined* – 40 mission trips, 30 years, Ground Zero, earthquakes, all of that combined – doesn't equal the level of carnage that I saw against civilians in just my first week in Gaza'.[14] A child shot twice in the head and abdomen is not a collateral damage, nor is shredding the bodies of young elementary school children or worshippers at prayer with cluster bombs an accident. Charred children walk the pummelled streets with open burns. A man drags to hospital his brother whose exploded limbs hang by a shred of skin. A toddler, her small body covered with blood and nostrils smouldering with grey dust, cries in absolute panic and terror at the sight of her parents' body parts flung all around her. She looks around her in despair, yearning to meet friendly eyes, but no one of her family is left alive. She is now another orphan, adding to the 19,000 already orphaned. 'With their bulldozer they ran several times over my husband's body, eviscerating his organs; his brain poured out of his nose, blood gushed out of every pore of his skin, his last breath of life was a bloodcurdling scream. I will never forget'.[15] A young mother played dead over her daughter's injured body, then crawled in the midst of blood and dismembered corpses to a corner where the bulldozer could not reach. It was two days later that she managed to escape. Her daughter died in her arms.

Essentialising Palestinians as monsters to be obliterated evokes the Nazis' attitude towards European Jews and sets the tone for what follows. Subsequent to the unexpected 7 October incursion and the hostages taken, which the world immediately condemned, the Israeli Minister of Defense, Yoav Gallant, declared Palestinians to be 'human animals and we will act accordingly'.[16] The savagery of the retaliation leaves no room for doubt as to the bestial vengeance he was intending and the degree to which the playbook for that revenge had already been written. David Ben-Gurion's 1938 statement of a 'compulsory transfer' of Palestinians, for which he openly declared he had no 'moral compunction',[17] was certainly both echoed and enacted by Netanyahu's current war of attrition on Gazans.

Israel's policies of superpower betray an *übermensch*, inflated mindset – above God and above law – that regards the Palestinians as subhuman, *untermensch*, pushed down to the very bottom of Israel's racialised classification ladder. Israel's portrayal of 'heroic' conquest of land completely overshadows the forced disappearance of Palestinians. It describes Israel's wars against a 'primitive, uncivilized, savage and backward' population (Halperin et al. 2010) as 'moral', a well-known colonial trope.[18] Elaine Schwager identified the dualistic splitting between the good self and the evil other as a clear drawing of lines that in part contributes to a 'genocidal mentality' (2004).[19] Eliminating the lowly, evil, other becomes an irreversible necessity in such a context. Because this projection of evil obscures the humanity of the recipient, the latter is made more easily the target of unconstrained sadism, and also an object of intense paranoid terror, further promoting the drive towards their destruction. No guilt, no shame, no compunction, no contrition, but a lethal desire to murder more. The extermination of the Palestinian other becomes the illusionary conveyer of Israel's safety. Whilst committing crimes against demeaned others, perpetrators perceive themselves as 'good'. The use of self-preservation as an incentive for eliminating others is more of an avowal than a justification of crimes against the population held under the perpetrator's control. The projection that has underpinned Israel's genocidal policies of destroying 'them' before they destroy 'us' splits oneself from one's own shadow.

The rationalisation of 'evil' for 'good' reasons seems to justify the archetypal projection of 'Amalek' onto Palestinians. Palestinian victims of the Crusades were also called Amalekites. White settlers used the name for Native American Indians. Professor Philip Jenkins (2012) noted that, historically, Christian extremists have variously labelled Protestants, Catholics and Tutsis as Amalekites to justify their genocides.[20] The late Yasser Arafat was called 'the Amalek and Hitler of our generation' by 200 rabbis (Horowitz 2018).[21]

Loss of Humanity

The casting of collective shadow on Palestinians as the archenemy of the state dehumanises them to the level of demonisation. It betrays a dark and uncontained shadow within the occupying state and the collective it represents, virulently influencing racial policies and their indiscriminate implementation. Driven by unscrutinised beliefs, projection empties oneself of one's own humanity. With this loss of humanity, any evildoing is overlooked; the archetypal defences of the self (Kalsched 1996) take over whilst the thinking-feeling capacities shut down.[22] Jung would agree with Hannah Arendt (1976, p. 459) that 'Suffering . . . is not the issue, nor is the number of victims. Human nature as such is at stake'.[23]

The recent finding of the International Court of Justice affirmed the plausibility of Israel's committing genocide in Gaza. The ongoing war *on* the Palestinian people seems to be possessed by a pervasive need to reduce Palestinian lives to dust.

Nations That Fall Have Wings

The kind of new ammunition employed calcinates bodies to such an extent that there is nothing left, not even dust to return to dust. Instead, the body is altered radically to disappear absolutely except for leaving a residue, like a film that evaporates into thin air. The message is clear: there are no bones to bury in mother earth, no embowelment of her child in her womb, and therefore no renewal or rebirth. This return to genocidal war is the most direct attack on the sacred, the sacredness of life and on everything that sustains it. The invoked Muslim prayer 'God suffices me, He is the disposer of affairs' (حسبي الله ونعم الوكيل) evokes the only consolation, commiseration, left for the dispossessed, a redress for the unilaterality of immoral scales. The prayer mirrors the biblical understanding that blood and life belong to God alone, and that destroying either is a direct attack on God.

'Men never do evil so completely and cheerfully as when they do it from conviction', Blaise Pascal noted,[24] and religious conviction is the most lethal of weapons when in the hands of power-worshippers. Garbed in virtue, some of our world's political and spiritual leaders exhibit a piteous loss of authenticity. Their lack of courage in addressing Jewish supremacy over the entirety of the Holy Land is a disgrace to the cloak they wear. Their collusion with those in power exudes messianic impetus. Prepared to sacrifice almost anything on the altar of elections for the money and the power at stake, construed religious supremacy and political ideologies turn men into murderers. The inaudible weeping of God behind the clamour of wars based on biblical scriptures cries: Not in my name. 'A voice is heard in Ramah, weeping and great mourning, Rachel weeping for her children and refusing to be comforted, because they are no more' (Matthew 2:18).

And the Lord said:
'What have you done?
The voice of your brother's blood

is crying to me
from the ground'

Through my commitment to universal humanism by which, along with Arendt's view, all peoples are recognised as integral members of the human family, I see a peculiarly naked and terrible fratricide taking place in Palestine. The maternal earth has been made to drink the slain brother's blood. She will in her time reject Cain, for he has harmed her so profoundly that she can no longer house him. To offer my own response, I should like to insist that we are our brother's keeper. To murder our brother is to shatter the link of brotherhood and carry the mark of Cain for many generations to come. We become the fugitive, the wanderer till the end of time for stubbornly attaching to 'the wrong attitude', as Jung would say. The unconscious will take to us the attitude we take to it. For every colonisation, there is a dark shadow of dehumanisation over both occupier and occupied, and the shadow will perpetuate itself through more and more unredeemed killings.

There is more than one God and more than one altar on which to offer sacrifice. This brings us to the real meaning of Jihad: a purification of the heart, the turning of the sword against the killer within, not without. Christ immolated himself and lived his Jihad. Not as Christians, as Jung writes in *The Red Book*, but Christ the *Anthropos* within, as destiny; a unity of opposites between coloniser and colonised, for a humane consciousness transcending the prevailing *zeitgeist*. The sacrifice that pleases God is not the sacrificial murder of one's courageous brother, but a measure of sacrifice of one's own one-sidedness. 'Where is your brother?', your other, God asked Cain, for Self is not enough. The other, your brother, is the other face of God with whom we learn the meaning of love. 'For love is the Mystic flower of the soul. This is the centre, the Self' (Serrano 1966).[25] Until Cain comes to terms with the spilling of his brother's blood, the earth will raise a cry of complaint, *vox oppressorum*, the voice that appeals for legal protection. The uncontrollable carnage in Gaza has sent cries across the globe calling for lawful protection of the slain. Utterly bewildered, despair on their faces, children roam their bombed-out homes in search of survivors, whilst the dead haunt our days and nights calling for redemption, for they have disappeared before their time. Cain will have to suffer the pain of consciousness, of utter restlessness until he comes to terms with his crime. For this reason, his life is protected. It is protected because God needs to come to terms with his duality. Cain's death is withheld, under God's tutelage, for he cannot continue to spill more blood. 'The punishment that God inflicted on Cain is not to be the occasion for barbarism among men' (Von Rad 1972).[26] He is rejected from the community and, as Jung would add, he has to suffer duality – meaning to learn to open his heart to his brother, his other.

'When Men Are Cast Down, Then Thou Shalt Say, There is Lifting Up' (Job 22:29)

'We are Job': I heard many Gazans repeatedly praying: 'Ya Sabrak Ya Ayyoub' ('Job, grant us your patience'). I heard them cry: 'Everything has been taken from us: our freedom, our country, our land, our homes, our children, our families, our friends, our neighbours, our clothes, our health, even the flesh on our bones. Thank you, God'. They do not curse, nor abandon the love of God. Grinding Palestinian bones to dust, then sweeping them away, by a future Israel leaves no space for imagination of an *imago dei* in formation. It is indeed through Job's suffering that God repents. The demonic fourth is in full visibility and will not transform if we do not engage the humanity of those who suffer it. The divisive God needs humanity to become conscious of its darkness, its absolute opposite within. Palestine, today, forces upon us a reflection of the necessity for human consciousness to transcend the God/unconscious and consciously suffer otherness. It faces us with shame for our avoidance of responsibility in the face of the killing of our brothers. The favouring between the sons and the envy it engenders provokes the murder that manifests God's unconscious duality that cannot be reconciled except in the wholeness of the human heart.

Zionism turned Jewish arrivals in Palestine from the immigrants they had been into settlers perverting the need for refuge with the 'wrong spirit' and with the 'wrong attitude'. Many Palestinian families took Jewish refugees under their wing, fed them, housed them and found them jobs. In return, the newcomers took hold of the property and flung the original owners out. Isn't it time to right the wrong? Isn't it time for a unified Jewish population to thank the host who welcomed you, whom you have envied and sought in every way possible to destroy his very existence which is now yours? Isn't it time to share that piece of earth between both the tiller of the ground and the keeper of sheep? The hot resentment that has risen in Cain's heart has distorted even his body! Hasn't Zionism distorted the body of Judaism enough? Isn't it time for Jewish values to take hold of the State and *trans*mutate it existentially in the cauldron of the heart? Cain might himself be the victim of many of his descendants, but through him, the brotherhood of man can be reawakened. In effect, Cain needs to learn that the love of God is demonstrated by the quality of love of each individual towards his brother.

The enormous suffering of the Palestinians and the hostages' families has opened the heart of the world for those who connect with the heart. Every consideration of mercy, justice and expediency is required involving a huge international effort to save countless lives, whilst Gaza is bled to its last gasp. The killing of hostages and the aborting of cease-fire efforts are but a parcel of the chain of evil that we witness when brother decides to turn against brother in the name of bad faith.

'If you do well, will not your countenance be lifted up?'[27]: lift your face, Zion, to your God, whose face you see in your brethren. Let the beast of prey pray at the door of the heart. Evil eagerly takes possession of the human mind, yet only the human heart can curb it. The geographical and historical place where the biblical story of Cain and Abel takes place is in Palestine, a Palestine which is in each of our conflicts and killings. The upheaval that Palestinians have recently conjured, if supported in its cry for freedom, has a chance for a reset towards wholeness of brotherhood. Each of us is a sacred vessel for achieving greater consciousness if we bring the other to be an opposite – not an occupied, nor a besieged, nor tortured other, but an opposite. The terrifying responsibility of integrating both the good and evil of a previously projected duality requires a fundamental restructuring of the relationship between the Jewish Israeli and Palestinian national communities, a process of decolonisation, such as liberated both whites and Blacks in South Africa. Palestinians' right to exist must be recognised alongside that of Jewish Israelis if we are to emerge towards a new humanity for a new aeon. The precepts on which Zionism has erected its state have failed, and punishing Palestinians for it is another failure. Rather, Israel has to face another 'culture' for an appropriate gardening, for, as Jung clarifies, God is in his 'oppositeness'.

Nations, when they become inflated by their own nationalism, need to atone by accepting the suffering of reconciliation with the brother and the nation to which they belong. I would therefore like to conclude these remarks, which would otherwise seem sympathetic only to the suffering of Palestinians, with a prayer that shows what I am looking for Palestine to give to

Israel and the world, and not just what it can belatedly get by way of understanding. A reconciliation between Israel and Palestine would be a coming together, at last, of brothers.

May Palestine be the garden of 'divine conflict' where mankind, rather than through wars enacting the gods of power, shows it has endured suffering them, so that the gods may suffer mankind; and from this learning may reconciliation between the two augur a reconciliation within the *imago dei* radiating back unto humanity.

Notes

1 David Henderson (2010) 'The coincidence of opposites: C G Jung's reception of Nicholas of Cusa'. *Studies in Spirituality* 20: 101–113.
2 Carl G. Jung (1966) *The Practice of Psychotherapy: Essays on the Psychology of the Transference and Other Subjects*, vol. 16, 2nd edn., trans. R.F.C. Hull. Princeton University Press, NJ, p. 167.
3 Carl von Clausewitz (1984) *On War*, eds. and trans. Michael Howard and Peter Paret. Book 1, Princeton University Press, NJ, p. 76. http://slantchev.ucsd.edu/courses/ps143a/readings/Clausewitz%20-%20On%20War,%20Books%201%20and%208.pdf
4 Baruch Kimmerling (2003) *Politicide: The Real Legacy of Ariel Sharon*. Verso, NY.
5 Carl von Clausewitz, *On War*, p. 75.
6 Mahmoud Darwish (2008) Excerpt from 'Who Am I, Without Exile?' from *The Butterfly's Burden*, © 2008 Mahmoud Darwish. Trans. Fady Joudah © 2008. Reprinted with the permission of The Permissions Company, LLC on behalf of Copper Canyon Press, coppercanyonpress.org.
7 Carl G. Jung (1976) Excerpts from *Collected Works of C.G. Jung*. Princeton University Press, Rockville, MD, pp. 10, 19.
8 Michel de Montaigne (2012) *The Essays of Michael Seigneur de Montaigne, in 3 vols,* Chapter 27, Vol. II. 8th edn, ed. and trans. Peter Coste. Gale ECCO, Farmington Hill, MI, pp. 489–501.
9 Maj. Gen. Kirk Smith, US Air Force. Commander, Special Operations Command Europe (2019) Foreword to *Resistance Operating Concept*, Otto C. Fiala, Ph.D., J.D. COL, USAR(R). © Special Operations Command Europe SOCEUR. Swedish Defence University, Stockholm, p. 9. https://www.diva-portal.org/smash/get/diva2:1392106/FULLTEXT01.pdf
10 Simon Collard-Wexler (2013) *Understanding Resistance to Foreign Occupation* (submitted in partial fulfilment of his PhD for the Graduate School of Arts and Science). Columbia University, NY.
11 Dr. Samir Jabr (2024, 13 August) 'Psychiatrist EXPOSES PERVERTED Israeli war tactics'. *One Path Network*.
12 Channel 4 (2024, 19 August) 'Palestinian detainees allege torture and sexual abuse by Israeli captors'. https://www.youtube.com/watch?v=GWqvDYbpbIg
13 Stephen Farrell, Dan Williams and Maayan Lubell (2020, 29 September) 'Palestinians out of sight, almost out of mind for Israelis seared by 2000 uprising'. *Reuters*. https://www.reuters.com/article/israel-palestinians-intifada-anniversary/palestinians-out-of-sight-almost-out-of-mind-for-israelis-seared-by-2000-uprising-idUSKBN26J2B3/
14 Dr. Mark Perlmutter (2024, April) 'Dr. Mark Perlmutter describes finding "distinct signs of genocide" in Gaza'. YouTube video. *Middle East Eye*. https://www.youtube.com/watch?v=2QdalGDpCIU; Tracy Smith (2024, 21 July) 'Children of Gaza'. *CBS News*. https://www.cbsnews.com/news/children-of-gaza/
15 Survivor describing to the author her experience. Name withheld for professional confidentiality.
16 Sanjana Karanth (2023, 10 September) 'Israeli Defence Minister announces siege on Gaza to fight "human animals"'. *HuffPost*. https://www.huffingtonpost.co.uk/entry/israel-defence-minister-human-animals-gaza-palestine_uk_65245ebae4b0a32c15bfe6b6
17 David Ben-Gurion speaking to the Jewish Agency Executive, June 1938.
18 Eran Halperin, Daniel Bar-Tal, Keren Sharvit, Nimrod Rosler and Amiram Raviv (2010) 'Socio-psychological implications for an occupying society: The case of Israel'. *Journal of Peace Research* 47(1): 59–70. DOI:10.1177/0022343309350013
19 Elaine Schwager (2004) 'Transforming dualism and the metaphor of terror, Part II: From genocidal to dialogic mentality: An intergenerational struggle'. *The Psychoanalytic Review* 9(4): 347–393. DOI:10.1521/prev.91.4.543.48748

20 Philip Jenkins (2012) *Laying Down the Sword: Why We Can't Ignore the Bible's Violent Verses*. HarperCollins Religious US. ISBN 978-0061990724

21 Elliott Horowitz (2018) *Reckless Rites: Purim and the Legacy of Jewish Violence*. Princeton University Press, NJ.

22 Donald Kalsched (1996) *The Inner World of Trauma: Archetypal Defenses of the Personal Spirit*. Routledge, London.

23 Hannah Arendt (1976) *Origins of Totalitarianism*. Harcourt Brace Jovanovich, NY.

24 Blaise Pascal (1909 [1670]), in *Pensées*, ed. L. Brunschvicg, no. 895.

25 Miguel Serrano (1966) *C.G. Jung and Hermann Hesse: A Record of Two Friendships*. Routledge & Kegan Paul, London.

26 Gerhard Von Rad (1972) *Genesis: A Commentary. Old Testament Library Series*. Westminster Press, Philadelphia, PA.

27 Genesis 4:7.

Chapter 8

The Northern Ireland Conflict

From IRA, to Sinn Fein, to Peace Ireland's Cultural Complexes Transformed

Kathleen Kirgin

Abstract

Drawing on the theories of complexes and the shadow, as set forth by C.G. Jung, and the cultural complex theory, as developed by Dr. Thomas Singer and Samuel Kimbles, this research explored the psychological dynamics of division and healing between Republicans and Unionists in Northern Ireland, and the significance the war named The Troubles and the Good Friday Agreement had on the Irish collective psyche. Examining Ireland's evolution through a depth psychological lens, this analysis sheds light on the ongoing process of recovery from the cultural trauma sustained over the course of Ireland's history of oppression and violence. This study highlights Ireland's reclamation of its culture, and the transformation of a once colonized nation to a sovereign land whose people, from both Northern Ireland and the Republic of Ireland, are dedicated to a continued peace and reconciliation for all who live on the island nation.

On April 17, 2023, the people of Northern Ireland celebrated the 25th anniversary of the Belfast Agreement, also known as the Good Friday Peace Agreement. A three-day conference was held at Queens University in Belfast, called Agreement Twenty-Five, featuring global leaders from the past and present, offering their reflections on the conflict in Northern Ireland. They discussed what was learned from the peace process and the progress made in the previous 25 years in the region. Although the event was a celebration for the achievement of peace in Northern Ireland for over two decades, the mood in the room was solemn. Each participant spoke of the horrors of war, the fragility of peace, and the humility and courage one must have when trying to reconcile two waring communities. Listening to each speaker, one could not help but realize how rare such a peace treaty is, considering the complicated history of the Anglo-Irish relationship. Those who fought for peace, from both sides, had to endure within themselves the intensity of centuries of hatred, grief, and loss so a new beginning could arise. The Good Friday Agreement offered a potential of a shared future – one of peace, respect, and tolerance.

One could not argue against the fact that Ireland – a once colonized and oppressed nation which has been at odds with its colonizers since its colonization – has entirely transformed its identity and its reality. Ireland's transformation occurred over centuries, but it was the Northern Ireland peace agreement that became the apex of a metamorphosis that reached across the entire island. Whether it was the political will, an inclusive negotiation process, or the tireless dedication to a peaceful resolution, the agreement shifted a deeply wounded Anglo-Irish relationship forever.

The agreement was signed in 1998, ending a war named The Troubles, a three-decade secular conflict between two communities: the Catholic Nationalists, also known as Republicans, and the Protestant Unionists, also known as Loyalists. The conflict between these two groups originated centuries prior when England first colonized Ireland in the 16th century. This historical moment was followed by centuries of subjugation of the Irish people. Eventually, due to the

DOI: 10.4324/9781003390039-24

continued strife between the Republicans, Loyalists, and the British forces, the country was divided, resulting in a 26-county Republic of Ireland, a free state in the south, and a six-county region in the northern province of Ireland named Ulster, which would become a part of the United Kingdom. Various cultural themes emerged from the Anglo-Irish relationship over time, most notably themes of victimization, division, revolution, and martyrdom-motifs that cycled through each generation for centuries. The era of The Troubles and the signing of the peace agreement shifted these themes, perhaps healing them, at least in part. Exploring Ireland's evolution through a depth psychological lens can offer a distinct understanding of the arc of transformation during this moment in Irish history. From an island with a history steeped in sectarian conflict and suffering, it has become a culture that has shed its colonized identity and is now a multicultural nation poised to participate in a new discourse within the global community as a culture whose once deeply divided and volatile history continues to move to higher thresholds of reconciliation and healing.

My Ireland

My Irish ancestry can be traced to both southern and northern Ireland. In the early 1980s, my father, a second-generation Irish American born in Bronx, New York, bought a home in Donegal, a small coastal town located at the tip of the northern territory. My father had always been interested in Irish politics and was a supporter of the Republican movement in Ireland. Upon his move to Donegal, he began to attend local Sinn Fein meetings. Sinn Fein is the Irish Republican/ nationalist political party in both Northern Ireland and the Republic of Ireland. It was also the political wing to the nationalist-driven paramilitary group named the Irish Republican Army, better known as the IRA. Eventually my father was introduced to Gerry Adams, the president of the party at that time, and Siobhan O'Hanlon, a Sinn Fein activist organizer and a close confidant of Adams. O'Hanlon introduced my father to her husband, Pat Sheehan, an ex-hunger striker, and to Martin McGuiness, the head of the IRA. My father began to spend time with Adams and O'Hanlon, and when our family visited Ireland, I found myself in conversations with them, listening to the stories of their Ireland. I was not that connected to my lineage then, always finding myself leaning away from the Americanized version that I experienced growing up. But these people introduced me to a culture and a history that I never really knew – one that was captivating but also very dark. When visiting, our family were exposed to what the people in the north lived with every day; the checkpoints, soldiers patrolling areas of the city with military weapons, and the subtle tension that an armed struggle brings to daily life. Gone was the whimsical Ireland I was taught as an American – a land of fairies, leprechauns, enchantment and song. I was introduced to and got to know a much more complicated and compelling culture. Decades later, when I decided to research and write about Ireland, I discovered a nation that was able to evolve out of its dark history of colonization, and – despite its trauma – truly transform as a nation.

Depth Psychology and the Cultural Complex Theory

Depth psychology, cultural complex theory, and Jung's theory on the shadow can contribute a valuable psychological viewpoint when examining long-standing cross-cultural conflict. Depth psychology not only explores the unconscious aspect of a culture but also considers its fundamental layers – its history, religion, politics, literature, art and mythology, and the accumulative information and disinformation that is passed down to each generation (Singer, 2010). Critical to understanding what lies at the core of cross-cultural conflict is the cultural complex theory, a theory developed by Dr. Thomas Singer and Samuel Kimbles. Cultural complex is a term from depth psychology, the definition of which is partially derived from Carl Jung's original theory on complexes. For Jung, complexes are feeling-toned representations in the unconscious mind, centering around patterns of perceptions, emotions, and memories that are formed by experience and the individual's response to that experience. Singer and Kimbles apply Jung's theory to their theory of cultural complexes, viewing them as a global anthropological condition in order to examine the collective psyche – specifically, those cultures that have experienced long-engaged hostility and struggle.

Singer (2010) describes cultural complexes as "emotionally charged aggregates of ideas and images that tend to cluster around an archetypal core and are shared by individuals within an identified collective" (p. 234). His theory extends Jung's work on personal complexes and Joseph Henderson's theory of the cultural unconscious. Henderson described the cultural unconscious as the mediating layer between the personal and primordial unconscious (Henderson, 1990). Just as personal complexes emerge from the personal unconscious, cultural complexes can be viewed as arising from the cultural unconscious, a process in which personal and archetypal realities interact with the broader world – archetypal meaning inherent universal patterns and images. According to Singer, cultural complexes are made up not only of cumulative experiences that validate a certain biased point of view but also of ancestral memories rooted in actual historical events residing within individuals and the group psyche of the community (Singer, 2004).

The cultural complex theory and Jung's theory on the shadow help explore a possible transformation of trauma within the Irish culture. Jung's theory on the shadow illuminates the need to recognize the darker aspects within ourselves. Jung considered the personal shadow as "the other in us . . . the reprehensible inferior, the other that embarrasses us or shames us" (Abrams & Zweig, 1991, p. 3). He applied his theory of the shadow to the individual psychological process, but arguably it could be used with a culture when exploring the deep roots of its historical trauma, specifically those wounds from colonization, and the significance of exposing what was rejected or repressed within its collective psyche.

The Cultural Complexes of Ireland

The Victim Complex

The colonization by England became the first and most critical cultural wound for the Irish culture, followed by the subsequent repetitive trauma of subjugation experienced by the Irish natives. Colonization is a story of asymmetrical cultural dynamics between the colonized and the colonizer, where the colonized eventually become dependent on the colonizer, no longer having a voice in their own fate. In his publication, *The Colonizer and the Colonized,* Albert Memmi (1965) writes, "The most serious blow suffered by the colonized is being removed from history" (p. 91). He continues, "He carries its burden, often more cruelly than others, but always as an object" (p. 92). Colonization altered each layer of the Irish culture dramatically. As in other colonized nations, it created a psychological dynamic of the oppressor and the oppressed, fortifying an enduring master–servant relationship between the Irish natives and the British. What made Ireland's colonization so catastrophic was the refusal by the English to assimilate at all with the culture. Instead, the Irish were considered bestial and lesser then. Historically, for a colonist to successfully colonize another culture, the colonizer must dehumanize and victimize that culture, not only to control that society, but also to legitimize the occupation and oppression to themselves

(Memmi, 1965). Prior to Ireland's colonization by England, there had been a series of foreign invaders on its shores, most notably, the Vikings in800 BC. Although at the time, they caused great suffering for the Irish people, the Vikings eventually became integrated with the culture, even adopting their Christian faith. However, the English refused to integrate into the Irish culture, and reimagined the Irish identity to that of subhuman, or other, shaping Irish history for the next millennium (Bartlett, 2010). For Ireland, their "othering" was to become the initial psychological split that became rooted in their collective psyche.

Seen through the lens of a Jungian perspective, the reimagining of the Irish identity by the English allowed for a collective shadow projection to take place. Projection means the unconscious process of attributing what is unconscious or unknown within ourselves onto another (Jung, 1978). When the superior thinking culture – England and the settlers – projected the notion of inferiority onto the opposing culture, the Irish natives, a transference of blame as a justification of the oppression on to the Irish as other took place. This behavior is incited by unexamined fear of the other by the self-proclaimed superior culture (Kimbles, 2003). In his essay "Cultural Complexes and Collective Shadow Processes," Kimbles (2003) writes, "When fear becomes the primary affect that organizes their world, it leads to scapegoating. Collective shadow projections become a kind of contagion" (p. 227).

In the 18th century, Ireland experienced a devastating event, which came to be known as the Great Famine, a mass starvation primarily caused by a disease called *blight* that decimated the potato crop, a food source that sustained the poor peasantry of Ireland. This tragic crisis resulted in the loss of one million Irish souls and eventuated a vast emigration of more than two million natives. This exodus scattered the Irish diaspora around the world. The famine, more than any other historical event, catalyzed the victim complex. It left an indelible scar upon the Irish psyche, carving a deep reservoir of loss and grief, and inflicted upon the Irish people a profound sadness. The grief was not only for those who died from the famine, but for the two million natives who left the country, never to return.

The Martyr Complex

At the beginning of the 19th century, the victim complex began transmuting into its opposite. The retaliatory revolutionary to some and others, the terrorist, started to appear in cities throughout Ireland. Kimbles (2003) describes the bipolar nature of victimization and retaliation, claiming that it typically emerges in cross-cultural conflict. This bipolarity becomes apparent when the cultural complex is constellated.

The revolutionary/terrorist became prominent figures in the Easter Rising of 1916 (a battle by Irish nationalists against English rule over Ireland) and in The Troubles, which began in 1968. In 1916, the Irish nationalists were suffering great discontent in the urban areas and began challenging the old British political order. The IRA fought against the British security forces in the rebellion. Ultimately, their efforts failed and a new treaty followed that resulted in the division of the country. The theme of division became literalized and an actual partition was created between the north and the south. The split left the north cordoned off in a kind of alembic container, shifting centuries of cultural trauma to this small region of the island. Over time, simmering tensions between the Republicans and the Loyalists intensified. By the late 1960s, tensions escalated and conflict returned to the Ulster region. The IRA re-emerged in the north and became what they considered to be freedom fighters, whose purpose was to not only end the oppression of their community but to ultimately force the British government to leave the north, uniting Ireland once again. Memmi claimed that for the colonized to truly free themselves, they must begin with their oppression. He writes, "In order that his liberation may be complete, he must free himself from those inevitable conditions of his struggle" (Memmi, 1965, p. 152).

By the spring of 1968, Republicans organized peaceful marches in the town of Derry, north of Belfast. The protests were to call attention to religious discrimination they experienced that caused severe unemployment and insufficient housing in their community – two sectors controlled by Loyalists. These demonstrations triggered old fears in the Protestant community. To disperse the marchers, local Loyalist police used batons and water hoses. The Republican community escalated the violence by throwing rudimentary petrol bombs in retaliation. From this moment, the violence increased, marking the beginning of a violent and bloody 30-year battle. As the victim complex shifted to the retaliatory freedom fighter/terrorist, the martyr complex was also activated and re-emerged in a very profound way.

By the 1980s, the IRA rose in power and popularity, and the oppressed expression of the victim complex began to transform. In a sense, the IRA embodied the possibility for liberation as they became a legitimate military force. Gerry Adams, a Catholic from Belfast, became the leader of Sinn Fein. Adams was a controversial figure. He had been interned in Long Kesh prison, also known as the H-Blocks, alongside many accused of being in the IRA in the 1970s. He never disassociated himself from the IRA but was vocal in claiming he was not a member of the paramilitary group. Once in leadership, he defended the Republican armed struggle as he worked toward a negotiated peace. He was politically astute, articulate, and shameless in his pursuit for a united Ireland. With Adams at the helm of the Republican movement, the IRA killed over 1,700 people, including 600 civilians.

From a Jungian depth psychological standpoint, Adams can be viewed as the personified representation of the shadow element within the cultural psyche, and the IRA seen as the archetypal defense system for the Republican community. Singer (2006) writes, "These archetypal or daimonic defenses are ferocious and inhuman. The daemonic defenses often direct their primitive aggression back onto the wounded spirit of the group" (p. 209). What was seen as terrorism in the Loyalist eyes was seen as the protection of Catholic neighborhoods by the IRA. Ultimately, the freedom fighters/terrorists may represent an evolution of the victim complex on a deeper level. In his book, *Bloodlines: From Ethnic Pride to Ethnic Terrorism*, Vamik Volkan (1998) explores the possible driving force of ethnic terrorism. In an interview with political psychologist Jeanne Knutson, he asks her what was the common element in the personal history of IRA volunteers and leaders that motivated them to commit terrorist attacks. Knudson claims the common theme was that they or their families were victims of the terrorism inflicted upon them. Volkan (1998) cites her claiming, "One never erases the identity of a victim. A life preserving, primitive belief in personal safety has been breached. Once having been terrorized, a victim thus simultaneously grieves over the past and fears the future" (p. 161).

A galvanizing moment in the conflict emerged out of the hunger strikes of the 1980s. The hunger strikes were Republican prisoner protests in the H-Block prison. Those IRA members jailed considered themselves to be political prisoners rather than criminals, and they demanded to be treated as such, refusing food and prison attire until their dictates were met. The first strike took place in December of 1980, and after 53 days, the prisoners gave up the strike without any gains. The second strike occurred in March 1981, ending in ten deaths, including Bobby Sands, the IRA's commanding officer in the H-Blocks. It was the death of Sands that captured the attention of the world. Before his death, photos of the emaciated Sands and other imprisoned hunger strikers – all wearing blankets instead of clothing in protest – were leaked to the press, unleashing intense emotions and reactivity, specifically for the Catholic population in Ireland and around the world. Several images of Sands depicted as Christ on the cross were painted as murals on the separation walls in and around Belfast and Derry. Separation walls had already existed since the 1920s and 1930s in the north, dividing the Loyalist and Catholic communities. They became more prolific after the outbreak of civil unrest in the late 1960s early 1970s. By the 1980s, these walls became canvases where local artists from either side told the story through murals of the conflict as it unfolded. Sands and his fellow martyrs became eternal symbolic images of persecution, sacrifice, and the eventual resurrection of the Irish people. Sands seen as a Christ figure gripped the global Catholic community. It also re-activated the martyr complex.

Many Republicans have died by martyrdom for the cause of a free Ireland throughout Ireland's history. Prior to Sands, some of the infamous martyrs include the 17th-century Protestant barrister Wolf Tone and 19th-century nationalists Padrick Pearse and Terence McSweeney. But it was Sands who activated something deep within the cultural psyche. Conceivably, mural images of Sands depicted as Christ figures evoked the Christian myth. Jungian analyst Edward Edinger (1987) writes, "The death and resurrection of Christ as an archetype lives itself out not only in the individual but also in the collective psyche" (p. 119).

The images of Sands and his fellow hunger strikers also linked The Troubles to the Great Famine. Irish historian Joseph Lennon (2014) associates the profound reaction of Sands' death with unprocessed collective grief that resulted from the famine. Lennon (2014) writes, "The ideological cornerstones of political and cultural leaders continued to reflect the trauma of a lost generation, a fractured culture" (p. 59). In Lennon's view, the famine was never grieved by the culture and the hunger strikers ripped open the old trauma stimulating "a communal need to mourn, revisit, or act out the past" (p. 59).

Negotiating With the Shadow Element

By the mid-1980s, the British government realized steps needed to be taken to bring peace to Northern Ireland. The IRA's campaign expanded to the British mainland, bombing various targets throughout London. Margaret Thatcher, Prime Minister of the United Kingdom from 1979–1990, also became a target in the Brighton Hotel bombing in the fall of 1984. It was then she realized a political solution was needed. In November 1995, the Anglo-Irish agreement was established, a treaty between the United Kingdom and the Republic of Ireland to help put an end to the violence. Although it failed to decrease the violence in the north, some would argue that since the IRA was not a part of the talks to find a solution, the conflict would not end. Others would say that although the agreement did not reconcile the issue between the two communities, in retrospect, it did become the foundation for future peace talks.

In the mid-1990s political parties in the north and Britain quietly began to talk to Adams. Northern Irish politician John Hume, the leader of the Social Democratic and Labor Party (SDLP), became critical in the peace negotiations. Although Hume, a Catholic, was anti-IRA, he understood the importance of bringing them to the negotiating table. He understood that without including Sinn Fein and IRA representatives in the peace talks, the conflict would continue. From a depth psychological viewpoint, it was Hume who first allowed the shadow element – personified by Adams – to emerge into the light of the day.

Before his election in 1992, U.S. President Bill Clinton was also engaged in the peace process in Ireland. Once elected, Clinton appointed a peace envoy for Northern Ireland, spearheaded by retired U.S. Sen. George Mitchell. Clinton did what few other heads of state would do: authorize a visa for Adams to visit the United States to enlist support from Irish Americans in government. Clinton gave Adams a restrictive 48-hour visa, a controversial move that angered the British immensely. However, his exposure in America, specifically with Irish American politicians such as U.S. senators Ted Kennedy, John Kerry, and Daniel Patrick Moynihan, as well as the American media, empowered Sinn Fein. His visit became a landmark moment in the peace process. One could surmise that Adams' visit sparked something in the Irish American psyche – that hidden component that carried their instinctive devotion and unconscious drive for atonement for Ireland and retribution for their own ancestors.

The Good Friday Agreement: The Beginning of Peace

By the 1990s, the Republican movement became less about retaliation and inflicting terror and more about negotiation and the accumulation of political power. They shifted from an ethnos stance – a position of tribal loyalty – to a polis position, one that embraces a legal, political process. In the dim light of the us-versus-them dynamic and the projection of the shadow onto the other, the Irish saw themselves in the English, who reflected back something they had repressed or never realized: a certain kind of political savviness. The Troubles mobilized the Irish people's capacity to negotiate on the global stage. In 1998, Irish and British governments and the political parties in the north came together to sign the Good Friday Peace Agreement. But it was the UK Prime Minister at the time, Tony Blair, who invited the IRA leaders to join publicly in the negotiations. Blair's bold action was thought by many to be the impetus for what would become a lasting peace. In 1997, Blair met with Adams in Northern Ireland at Stormont, the parliament building for the Northern Ireland Assembly. It was the first meeting between a Sinn Fein leader and a British PM in 76 years.

One could argue that Blair withdrew England's projected shadow (i.e., colonization and tyranny over the Irish) by including the once disenfranchised representatives to sit at the table and engage in negotiations. From a practical standpoint, all sides were weary of the war and needed a solution to bring about its cessation. From a Jungian perspective, it was Tony Blair who withdrew the English shadow and thereby increased the light of consciousness. His contribution was integral in ending one of history's most contentious, long-standing cross-cultural conflicts. Blair assisted in helping the Irish to reclaim some of the dignity and wholeness lost to colonialism, whether he was aware of it or not.

Kimbles (2000) writes, "The opening of a traumatic past in a public way is only the first step toward connecting with ongoing individual and social processes" (p. 167). He adds, "Real healing of a cultural complex requires a collective effort which should include addressing the perpetrators . . . as well as the victim" (p. 168). Kimbles suggests that the path towards the healing of cultural complexes, specifically between two opposing groups, is to mutually consider the roots of the hostility they have for one another. He writes, "Bringing a cultural complex to consciousness requires a real inquiry into what each group means to one another and how they have functioned within the us/them dynamic to carry each other's shadows" (p. 168).

25th Anniversary of the Good Friday Agreement

In the morning session of the Agreement Twenty-Five conference on April 17, 2023, the first day of the 25th anniversary of the Good Friday Agreement, Hillary Clinton, the now chancellor of Queens University of Belfast, commenced the conference by recalling her memories of that time during her husband's presidency and his dedication to seeing peace in Northern Ireland. She paid tribute to all who assisted in the hard-won peace in Northern Ireland, claiming the path to peace often looked unattainable, but all parties endured knowing peace was the only option. She states, "Northern Ireland stands as an example of even the most staunches adversaries can work together for a common and greater good" (Queens University Belfast, 2023). On the panel was former President Bill Clinton, former British Prime Minister Tony Blair and former Irish Taoiseach Bertie Ahern. They each spoke about their experience and their roles in establishing the agreement. President Clinton discussed how the peace process was never linear and that, most of the time, there was no plan or vision of how to move forward. O'Hearn complimented Blair for his dedication to find a peaceful solution to the conflict. Blair discussed how peace was still not guaranteed in the last few days before the signing. In every meeting with the Republicans and Unionists, he made sure he and O'Hearn were present, showing a force of unity between the British and Irish governments. Although at times, he felt the entire process could fall apart, he also thought that a long-lasting peace was attainable. He felt a specific determined energy amongst all parties to find short-term and long-term solutions. Blair reflects, "You can have whatever technical agreement you like, but if the spirit is not sitting alongside it, it will never work" (Queens University Belfast, 2023).

Blair emphasized the need to understand where all sides were coming from in the conflict to find a resolution. He stated, "I don't think anything could have happened unless we were prepared to sit down and to talk to people and to understand why these conflicts come about, because there are two distinct narratives in opposition to each other." He continues, "The only way you will ever make peace is if one side understands how the other side feels. They don't have to agree with it, [they] don't even have to sympathize with it, but they have to understand it" (Queens University Belfast, 2023).

Kimbles (2003) insists that in order to heal cultural complexes that involve two warring nations, both groups must confront the core of their hostilities toward the other. For example, he writes, "Bringing a cultural complex to consciousness requires a real inquiry into what each group means to one another and how they have functioned within the us/them dynamic to carry each other's shadows" (p. 168).

A Shared Space and a Unified Ireland

Since the signing of the accord in 1998, peace – albeit fragile – has remained in Northern Ireland. Arguably, the north and the Republic have experienced an arc in transformation. Socio-economic and political issues plagued both regions. In the 1980s in the south, controversy within the Catholic Church began to emerge. Allegations of the sexual abuse of children in Catholic institutions came to light. Some would argue this moment was the beginning of the end of the grip the Church had on the Irish people and their government. In the 1990s, the Republic of Ireland evolved from being one of the poorest countries to one of the most successful in just one generation. Due to its economic tour de force within the European community, the Republic was given the title of the Celtic Tiger. Ireland had found a certain swagger, leaving behind the reputation for needing financial assistance from the country's former parental lords and queens. The Irish were once again empowered with full ownership of their future. Fewer people needed to be "on the dole," the Irish welfare system for the unemployed. Ireland found herself in the unfamiliar territory of prosperity, no longer identifying itself with poverty, joblessness, and desperation. Old dealings of secular violence and poverty were replaced by increasing affluence and global respect. The old wounds of colonization seemed to have disappeared, as the Irish had finally found the elusive sovereignty for which they had yearned. Regrettably, in 2008, the boom ended abruptly, leaving the Irish once again victimized – this time by Irish bankers, property developers, and politicians. Eventually, foreign investors and tourism returned to the south, and as of spring 2023, the Republic of Ireland was touted as having the fastest-growing economy in the European Union (EU) (KPMG, 2023). However, the Republic is not problem-free. A lack of affordable housing, homelessness, and wealth inequality now challenges the south. Societal issues many face in the modern world.

In the north, there is a continued commitment to healing the wounds of the past, which has helped to dilute the themes of division, victimization, and inequality. Adams believes the violence will never return to the north. In a recent interview, he argued, "It will never go back to what it was before, and that is because the will of the people is behind the peace . . . the peace is here and here to stay." (Adams, 2023). Watching him speak, I could not help but notice a shift in his demeanor compared to when I met him several decades ago. The tone of his voice was gentle, his facial expression serene as he spoke about the future of Ireland. His focus on a united Ireland and an end to English rule has not faded, his devotion to unity is as steadfast as it was. Although acknowledging that there is a difference in opinion on who should rule in the north, Adams emphasized the continual process of working together, highlighting inclusivity and respect for all as the country moves forward. Whether one is politically aligned with him or not, Adams was and continues to be instrumental in shifting the course of Ireland's future, perhaps towards its own wholeness once again. He reminds the interviewer that it is the signed agreement itself that will unify the country due to a referendum that states if the majority of people no longer wants to be a part of the United Kingdom, then a border pole – the term for a referendum on Irish reunification – must occur. As of 2023, 63% of the people in the north believe that Brexit – the term for the UK's departure from the EU – will result in unification (Pogatchnik, 2023). Most believe it is not if the country will unify, but when. Undoubtedly, the Good Friday Peace Agreement has changed Ireland forever. It is now a living symbol of a culture once held in the grip of ancient sectarianism, perpetual adversity, and division, evolving into a nation that has truly reconciled with its own darkness for the sake of a sustained peace upon their land.

References

Abrams, J., & Zweig, C. (Eds.). (1991). *Meeting the shadow: The hidden power of the dark side of human nature*. Los Angeles, CA: Tarcher.

Adams, G. (2023). *Gerry Adams: Peace is the will of the Irish people, IRA will never return (times radio)* [Video]. YouTube. https://www.youtube.com/watch?v=6tqNPwq1SCE

Bartlett, T. (2010). *Ireland: A history*. Cambridge, United Kingdom: Cambridge University Press.

Edinger, E. F. (1987). *The Christian archetype: A Jungian commentary on the life of Christ*. Toronto, CA: Inner City Books.

Henderson, J. (1990). *Shadow and self: Selected papers in analytical psychology*. Wilmette, IL: Chiron.

Jung, C. G. (1978). Archaic man. In H. Read et al. (Series Eds.), *The collected works of C. G. Jung* (Vol. 10, 2nd ed., pp. 50–73, R. F. C. Hull, Trans.). Princeton, NJ: Princeton University Press. (Original work published 1946)

Kimbles, S. (2000). The cultural complex and the myth of invisability. In T. Singer (Ed.), *The vision thing: Myth, politics and pscyhe in the world* (pp. 157–170). Routledge.

Kimbles, S. (2003). Cultural complexes and collective shadow process. In J. Beebe (Ed.), *Terror, violence and the impulse to destroy* (pp. 211–233). Daimon Verlag.

KPMG (2023, March). *Ireland's economy: Continued growth, comes with caveats*. https://kpmg.com/ie/en/home/insights/2023/03/irelands-economic-outlook.html

Lennon, J. (2014). The starvation of a man: Terence MacSwiney and famine memory. In O. Frawley (Ed.), *Memory Ireland: The famine and the troubles* (Vol. 3, pp. 59–90). Syracuse, NY: Syracuse University Press.

Memmi, A. (1965). *The colonizer and the colonized*. Boston, MA: Beacon Press.

Pogatchnik, S. (2023). A united Ireland looks more likely thanks to Brexit, new study finds. *Politico*. https://www.politico.eu/article/united-ireland-look-more-likely-brexit-study-uk-belfast/

Queens University Belfast (2023, April 18). *Agreement twenty-five: Marking the 25th anniversary of the belfast/good friday agreement* [Video]. YouTube. https://www.youtube.com/watch?v=hR4xbvt3D04

Singer, T. (2004). The cultural complex and archetypal defenses of the group spirit. In *The cultural complex: Contemporary Jungian perspectives on psyche and society* (pp. 13–34). New York, NY: Routledge.

Singer, T. (2006). The cultural complex: A statement of the theory and its application. *Psychotherapy and Politics International*, 4(3), 197–212.

Singer, T. (2010). The transcendent function and cultural complexes: A working hypothesis. *Journal of Analytical Psychology*, 55(2), 234–241.

Volkan, V. (1998). *Bloodlines: From ethnic pride to ethnic terrorism*. Boulder, CO: Westview Press.

Chapter 9

When Our Shadow Makes Us Blind and Deaf to Suffering

Elana Lakh

Abstract

When facing information about life under continuous trauma caused by humans, there is a part of the psyche that resists knowing. Various mechanisms of denial and disavowal are used in order to avoid knowing. Drawing on the concepts of collective shadow and cultural complex in the Israeli context, this chapter examines the obstacles to seeing, hearing and acknowledging the human experience of trauma suffered by Palestinian people. This chapter offers a discussion of the threat that knowing presents to the conscious mind, and asks under which conditions we will be able to face this knowledge. How is it possible to agree to know, when the suffering is caused by our own group, and in our name? In light of Erich Neumann's idea of "New Ethics" based on taking responsibility for one's shadow, the chapter looks at shadow projection processes that compromise the possibility of acknowledging the other's suffering. This paper describes the archetypal themes and historic processes that are at the basis of the Israeli cultural complexes of victimhood and power, and tries to explain how the collective trauma of the past is used in the service of denying and justifying the suffering perpetrated in the present. It also offers a description of the effect that socio-political processes in contemporary Israel regarding the ongoing occupation of Palestinian lives has on the ability to take responsibility for our shadow and acknowledge the other's suffering.

This paper was written in 2022 and published first in May 2023 before the devastating events of October 7th.

In October 2015, I was asked by Amnesty international to comment on a documentary film screened in the Jerusalem cinematheque, named *Shivering in Gaza*. The film tells the story of a trauma treatment intervention conducted in Gaza by a Dutch therapist, Jan Andreae. "Shivering in Gaza" as a headline elicited extreme reactions in the Israeli public, though the documentary is mainly human and psychological. Screening of this documentary had been banned in the south of Israel and was seriously threatened in Jerusalem.

I would like to relate to the events around this screening as an example of a collective psychological process of dealing with shadow contents, in the Israeli context.

C.G. Jung's concept of the shadow relates to a part of the psyche which consists of the tendencies that the ego considers to be unwanted (Jung, 1959, 1968, p. 284, para. 513). These are the parts that we judge to be bad and shameful. On a collective level, the culture's unwanted characteristics develop throughout the history of the group, and are expressed in the cultural canon (Neumann, 1990). The archetypal shadow is the dark side of culture and of human nature, repressed and rejected from the light of consciousness, and considered evil (Jung, 1959, 1968, p. 322, para. 567). This evil is what the ego experiences as hurtful and damaging to its continuity and sense of safety.

The shadow contents are alienated from consciousness because they form a threat to a person's or a group's self-perception of being good and living according to a moral code with high

DOI: 10.4324/9781003390039-25

values, and so the shadow contents tend to be projected on the other (Jung, 1959, p. 9, para. 15), who is perceived as evil and thus dangerous. Moreover, when confronted with our own shadow acts, various means are taken in order to avoid awareness of our shameful and hurtful deeds.

In the spirit of these ideas, I will try to suggest here an understanding of the reasons that make the screening of *Shivering in Gaza*, and the human experience that it brings to our awareness, impossible for the Israeli consciousness to tolerate.

I see the attempts to prevent the screening as a clear example of what my psychoactive colleague, Dr. Sharona Komem, has termed "actively wanting not to know". It is different than not wanting to know, which can be a rather passive process. Wanting not to know forces us to make an active effort not to let information reach our consciousness or denying information that evades this attempt at complete avoidance.

In Israel today (referring to 2022), with all the available evidence about the consequences of the occupation, it becomes more and more difficult not to know, and requires stronger psychological measures of denial and disavowal of the Israeli acts, and of their effect on Palestinian lives.

On the way to the cinematheque, the taxi driver told me that the documentary must not be screened as it portrays Israel in a shameful way, and when I said that the film is about trauma treatment in Gaza, he told me that Palestinians do not suffer from trauma because they are not human. He was serious, and his statement reflects a point of view shared by many Israelis.

Jerusalem municipal council member tried to prevent the screening of the documentary in Jerusalem cinematheque, and threatened to sanction the place: "I will do anything in my power to prevent screening this false documentary about our soldiers!" she wrote on her Facebook page (Anderman, 2015). It is obvious that there is something about the documentary that consciousness cannot tolerate and therefore must be eliminated by any means.

The city council member also wrote: "In these days, when Israeli security forces work day and night to defend our security, terrorism assisting organizations such as Amnesty are raising their heads again" (ibid.).

The council member's statement, and the taxi driver's words reflect the psychological processes of denial and projection of the shadow, characteristic of a state of mind that prevails in Israeli society today. This state of mind is rooted in the cultural complexes of extinction versus redemption, as well as power and aggression as opposed to weakness and fear, described by Erel Shalit in his *The Hero and His Shadow* (2004).

So why cannot a documentary about trauma treatment in Gaza be screened in Jerusalem? What is so dangerous to the consciousness that needs to be annihilated? And what is the danger of the "terrorism assisting organization" Amnesty?

The danger is of connections and relatedness. Relatedness opens the way to complexity, and complexity compromises the clear distinction between good and bad, between ego and shadow, and between "us" and "them". Thus, the active effort not to know is meant to protect consciousness against information that is intolerable.

This is a dissociation mechanism that is characteristic of traumatic states. Donald Kalsched describes the dissociation as "a trick the psyche plays on itself" (1996, p. 13) – a trick that divides the unbearable experience, and makes part of it inaccessible to consciousness, thus leaving elements of the traumatic experience in the unconsciousness, where they become shadow parts. The traumatic events and relationship turn into an autonomous complex in the psyche, forcing itself on the conscious mind, as Jung described it (Jung, 1954, pp. 131–132, para. 266–267). Thus, trauma has its own life inside.

Israeli society lives in an ongoing traumatic mode of experiencing reality. This mode is based on a cultural complex of a constitutional trauma that is alive in our collective unconscious, and is kept alive in the collective consciousness by means of public discourse that begins in kindergarten and is sustained throughout life. The complex influences consciousness and prevents adjusting to inner and outer reality, impairs judgment, and compromises human contact (Jacobi, 1959).

In traumatic states, consciousness is possessed by the sense that one's life is threatened, and becomes entirely survival-oriented. Time, space and context cease to exist, and the person re-lives the traumatic events again and again. In such a state of mind, there is no space for complexity, and thinking processes are blurred. There is no distinction between kinds and levels of threat and danger. In such a state of consciousness, a teenager stabbing a fully armed soldier at the checkpoint is perceived as a total annihilation threat to the state of Israel.

This state of mind does not allow any distinction between the current events that are happening within a specific political context that has causes and possible solutions, and the constitutional Israeli narrative of victimhood and surviving attempts to obliterate the Jewish nation, since the Biblical Amalek, through the Holocaust and until today. Thus, in situations dominated by the experience of a life threat there is an attack on thought processes and on linking (Bion, 1959), and the notion of cause and context cannot be remembered. The Palestinian attacks are thus experienced as antisemitism and hatred of Jews just because of being Jewish, and are dissociated from the current context of decades of Israeli occupation of Palestinian lands and lives.

This context is denied and forgotten, and only the belief in a threat to the Jews prevails, timeless and placeless. Fear dominates reality perception. In such an ever-traumatic mode of thinking, there is no place for two subjects, and it is impossible to see the suffering of the other. There is clear good and clear bad, victim and perpetrator. We can easily recognize them and distinguish them. In such a line of thinking, Amnesty becomes a terrorism-assisting organization, and Breaking the Silence activists become traitors, for confronting the Israeli public with the details of everyday life under occupation.

By telling what they have done as soldiers sent by the state, Breaking the Silence activists are complicating the clear distinction between good us and evil them. By screening *Shivering in Gaza*, Amnesty forces us to see the people in Gaza, the people who suffer trauma, and does not allow the Israeli viewers not to know. Watching this documentary, seeing the faces of Gazan people, their homes, their children and their pain, might compromise the dissociation that keeps the shadow away from consciousness.

Seeing the humanity of the Gazan people makes it impossible to deny their suffering, and raises profound questions about our responsibility for this suffering. The Israeli cultural complex of victimhood makes it very difficult to acknowledge our perpetrator shadow part. As the Jerusalem city council member said: "The film stains our soldiers' morality" (Anderman, 2015). Not that the soldiers' acts stain morality, but the film. If the documentary is not screened, perhaps the clear-cut distinction between moral and immoral can be preserved.

The shadow is always visible when projected on the other (Jung, 1959, p. 9, para. 16), thus preserving self-perception, and the split between good and evil (Stein, 1995). In this way, the Palestinians are perceived as violent, murderous and immoral, keeping the Israeli self-perception of morality and goodness intact. Encounter with our shadow parts is a painful one, since it forces us to see the dark and shameful parts in ourselves, and not in the other that we project upon.

This is the danger we face watching *Shivering in Gaza*. It is almost impossible for the collective Israeli consciousness that is based on a well-preserved narrative of victimhood. For such a consciousness, acknowledging the Palestinians as victims of trauma is dangerous because it means that we are the perpetrators causing their trauma, and therefore acknowledging our own shadow, in our acts.

Jung stated in many of his writings that no one is free of evil within, and no one can be entirely good: "None of us stands outside humanity's black shadow" (Jung, 1970, p. 297, para. 572). He thought that denying the shadow is dangerous, because "negligence is the best means of making man the instrument of evil" (ibid.). Therefore, Jung considered the task of owning one's own shadow to be a crucial one: "One does not become enlightened by imagining figures of light, but by making the darkness conscious. The latter procedure, however, is disagreeable and therefore not popular" (Jung, 1967, para. 335).

Jung called for self-knowledge, as the way for an individual to deal with the problem of evil. He related to this need for self-knowledge as a moral imperative, in response to cultural and mass processes of projecting evil on the other (Jung, 1959, p. 8, para. 14). This imperative is also valid on a collective level, as described by Erich Neumann in his *Depth psychology and New Ethics*.

Screening *Shivering in Gaza* in Jerusalem for Israeli audiences calls for such acknowledgment and owning of the shadow, but is very difficult to achieve, as the case of the reaction to the screening shows. Nevertheless, individuals and groups that insist on owning the shadow and bringing it into the collective's consciousness, have a very important role of promoting moral responsibility and development of individual and collective consciousness (Neumann, 1990), despite the attempts to silence them.

Bibliography

Anderman, N. (2015, October, 17). The film 'Shivering in Gaza' will be screened in Jerusalem cinematheque today: City council member is acting to cancel it. *Haaretz Gallery*. https://www.haaretz.co.il/gallery/cinema/2015-10-17/ty-article/0000017f-e276-d75c-a7fffeff69ba0000. (in Hebrew)

Bion, W.R. (1959). Attacks on linking. *International Journal of Psycho-Analysis*, 40: 308–315.

Jacobi, J. (1959). *Complex, archetype and symbol*. Bollingen series. Princeton, NJ: Princeton University Press.

Jung, C.G. (1954). *The practice of psychotherapy* (2nd ed., R.F.C. Hull, Trans.). Bollingen Series XX. Princeton, NJ: Princeton University Press.

Jung, C.G. (1959). *Aion* (R.F.C Hull, Trans.). London: Routledge & Kegan Paul.

Jung, C.G. (1967). *Alchemical studies* (Translator: R.F.C. Hull). Bollingen Series XX. Princeton, NJ: Princeton University Press.

Jung, C.G. (1968). *The archetypes and the collective unconscious* (2nd ed., R.F.C. Hull, Trans.). Bollingen Series XX. Princeton, NJ: Princeton University Press. First published in 1959.

Jung, C.G. (1970). *Civilization in transition* (R.F.C Hull, Trans.). London: Routledge & Kegan Paul. First published in 1964.

Kalsched, D. (1996). *The inner world of trauma: Archetypal defenses of the personal spirit*. London and New York: Routledge.

Neumann, E. (1990). *Depth psychology and a new ethics* (E. Rolfe, Trans.). Boston and Shaftesbury: Shambala.

Shalit, E. (2004). *The hero and his shadow: Psychopolitical aspects of myth and reality in Israel*. Carmel, CA: Fisher King Press.

Stein, M. (editor). (1995). *Jung on evil*. Princeton, NJ: Princeton University Press.

Insight Into an Analysis With a Patient Who Became Frozen in Fear Because of the War

Marianne Meister-Notter

Abstract

The author presents the crippling effects of the "special operation", as Russia's invasion of Ukraine was euphemistically called by the Russian rulers, on a 40-year-old analysand – called Lisa in the chapter – who had already found herself in a vulnerable life situation during the COVID-19 period and thus before the beginning of the war in 2022, but had remained stable so far. The onset of destruction in Ukraine fostered a severe regression in which feelings of powerlessness in the face of this situation affecting the collective took possession of Lisa: A deep-seated fear began to spread, overshadowing her entire life day and night, and as feelings of hopelessness and futility began to accompany this fear, she felt increasingly paralyzed and frozen. Several factors favored this negative development that the outbreak of war exerted on the analysand: the interplay of typological predisposition with the stressful family history (multigenerational perspective) that shaped her, and the current personal and professional situation. The author shows both the descent into absolute darkness and the arduous journey back into the light by including dream material and images.

Introductory Thoughts

It is probably known to everyone that different people react differently to the same or a similar situation. It is the same in the case of this Ukraine war. Some find it merely unwise and unnecessary, and also differentiate in the attribution of blame for the cause of the war: it is then not simply the "evil Russians" or the "deranged," "power-hungry," "unscrupulous" [Russian President Vladimir] Putin who bears the responsibility for this war and who would go to the last; others think that the West – namely the North Atlantic Treaty Organization (NATO) states and above all the United States – also bear at least as much responsibility for it.

My analysand – an intelligent, sensitive, talented and lovely woman in the approximate middle age – could not relativize for various reasons, however, but experienced the Russian president with his nuclear weapons as an overpowering danger for the world and for her own life. Lisa's unconscious will reflect this theme in one of the dreams. Perhaps this paralyzing fear was also reinforced in part by the fact that she, living in a country neighboring Russia, also perceived the threat of a new war as much more real than a person living farther away from all the destructive action. During the day, her thoughts circled only around the war events, preventing access to distraction, diversion, or concentration on her own goals. The carousel of thoughts did not allow her to sleep at night. She finally lapsed into a torpor that was dangerously close to catatonia. Temporary drug treatment became necessary to prevent catatonia.

In Lisa's case, men and women relate very little to each other throughout the family history, which is set in a rural–rural context, because the men behave towards their female relatives

DOI: 10.4324/9781003390039-26

either disinterested or then either openly aggressive and devaluing or passively aggressive and thus thwarting them. Lisa's grandmothers, aunts, and female cousins are also clearly lower in their own evaluation than the male family members who dictate the rules of the game with a rigid hand. The observed effect of the men's everyday disrespect for the women meant that the female family members lacked the ability to make effective decisions and take action, which meant that they repeatedly ended up in the role of victims.

In Lisa's case, other aspects come into play that began to cut off her life through increasing anxiety. One of them is her high sensitivity and openness to collective processes, which is accompanied by a weakness of demarcation against external influences. This has a typological component, which I cannot go into in detail about in the context of this short chapter, but I would like to mention that Lisa is predominantly – but not unilaterally – introverted. In addition, she had become unemployed, which fueled her existential fears. The many unsuccessful job applications also contributed to her becoming increasingly hopeless and depressed.

Another issue is the lack of social contact with peers: The few contacts she had were older relatives – parents, uncles, and aunts – but Lisa generally experienced their influence as weakening rather than strengthening, because the majority of them had a very pessimistic view of life. Why? As mentioned previously, in this family – as far as Lisa knows – there is a history of devaluation of the feminine going over several generations, by which my patient and her mother were directly massively affected. Thus, Lisa also received little attention from her biological father. He also did not support his wife and daughter in their fight against a male relative who tried to cheat Lisa's mother out of her considerable inheritance and who, because of his aggressiveness, had an easy time of it, all the more so because neither mother nor daughter could count on a battle-hardened and battle-tested animus.

This lack of positive support and encouragement from the close male relatives, respectively the suffering from their persistent aggressiveness, caused mother and daughter to try to help each other in an ambivalent symbiotic bond, which, however, did not succeed satisfactorily due to the insufficiently developed self-esteem and self-confidence of mother and daughter. The resulting feelings of powerlessness in the face of many everyday situations triggered anxiety and feelings of hopelessness. This was despite the fact that both mother and daughter were intelligent and had many abilities and talents.

The considerable existential fears, which grew as unemployment continued and the war progressed, led Lisa into an actual paralysis that prevented her from finding creative ways to make her abilities fruitful even as a hobby, let alone professionally.

What Does It Look Like in Lisa's Unconscious?

Following are Lisa's dream images in crisis.

1. Here is a dream in which Lisa enters a cramped little apartment that has no windows.

   Scene Change: Suddenly, Lisa faces a friendly older woman who tells her that she has seen a nice man who could be a good match for Lisa. Lisa, however, replies that many have said that, but it has never happened yet. The woman then says the word "openness" to Lisa.

The image of the windowless, small, cramped apartment reflects Lisa's feeling of confinement and hopelessness; these rooms without windows do not allow views of the outside world and thus prevent fruitful impulses that could penetrate from the outside to the inside. It is an image of being trapped and of darkness and fits the feeling of not being able to access the world.

The unconscious now gives Lisa an archetypal positive mother figure in the form of a friendly older woman who counters this constricting darkness with hope: Namely, she has seen a man who could fit Lisa. For people who are familiar with Jung's theories, this unknown person is an animus figure – that is, the male energy to which Lisa has too little contact, which is also clearly expressed in her concrete life, because she neither has a partner whom she ardently desires nor can she sufficiently energetically stand up for herself and her wishes and needs. Lisa, however, remains skeptical; at that moment, she is not amenable to the hopeful prospect of a desirable future. The woman, however, still says the word "openness" to Lisa. Exactly this openness is missing in her conscious attitude, but Lisa is not yet able to open herself for new things.

The dream fragments of the next day lead into a fear-filled scenery, which I reproduce here in the words of Lisa:

2. A heavy dark sky, deeply overcast with heavy dark clouds; storm clouds move from left to right (from west to east) across the sky above me (multi-layered clouds), for I am standing outside in front of a house, on the porch, looking up at the sky; a thunderstorm departs without discharging; I think to myself "lucky again"; but the fear remains, for the heavy dark sky with the heavy dark clouds remains and the next black thunderstorm clouds are already lining up from right to left (from east to west)."

    Scene Change: "At [the] house, there are at least human-sized angel sculptures on the porch; one of the two sculptures is not standing, but lying on its side; both sculptures are made of light-colored stone, [and] seem like sculptures that have been forgotten and deposited here; scene not appealing."

Lisa commented on this dream that she had dreamed a lot and extensively in the previous week, and could hardly remember these dreams, but that in each case only one feeling remained, and that was a dark, oppressive one.

The sky as a symbol of the mental and also spiritual sphere – where, in the imagination of many religions, angels and gods dwell – is here covered by dark heavy clouds; it is threatening, frightening, because in this dream experience there is no light in the sky: neither sun nor moon nor stars are visible. These sky lights are hidden behind dark nebulous formations that are impenetrable and intangible, and can change their shape in advance. They symbolize Lisa's experience of an uncontrollable event, for she as an individual feels at the mercy of the arbitrary decisions of those in power. The figures of light, the angels, are also petrified, still bright but no longer alive, and only one is left standing, the other lying carelessly; these figures of light seem to have been forgotten. This absence of light and clarity, hope, and protection symbolize here in an impressive way the depressive attitude into which Lisa had slipped, an absence of transcendence.

Lisa redundantly describes the clouds obscuring the light as "heavy" and "dark." These clouds are multi-layered; Lisa spontaneously interpreted these black, heavy clouds as a symbol for the war that had broken out in Ukraine and could put the whole of Europe in great danger.

The multi-layeredness of the dark clouds could be due to the unconscious intuitive grasp of the complexity and multi-layeredness of the conflict between West and East.

Very interesting are also the two directions of movement of the heavy, dark clouds: first they move namely from west to east over Lisa, afterwards Lisa observes how now in the east new heavy, dark thunderclouds build up and are ready to move to the west.

Here, the compensatory function of the dream clearly comes into play: In Lisa's unconscious perception, the dark, heavy clouds first come from the West (NATO with the United States and Europe) and move to the East (Ukraine, Russia); in her conscious assessment, however, Russia clearly corresponds to the intangible, dark, threatening element that is primarily responsible for the conflict. In the dream, however, the dark, heavy storm clouds in the east (Russia) build up only secondarily, namely after those from the west (NATO) had moved eastward.

However, the thunderstorm has not yet unloaded over Lisa and thus over Europe: "Lucky." At least there is still the possibility of an omission of the complete annihilation. But the ability to act lies with others, not with Lisa. She feels at the mercy of the changing rules of the game and events around her in many ways, not only towards collective powers such as the political rulers who could decide to bring war or peace upon countries, but also with regard to her work situation and thus also towards the labor market in her country.

Is there no hope for Lisa to get out of the situation and into motion?

3. A first fleeting glimmer of hope in the form of a colorfully dressed Roma woman who appears fleetingly in a dream, disappearing from a confining, dark building in which the dreamer finds herself.

Lisa cannot get in touch with her, although she would have liked to, because she associates very positive qualities with her: the Roma woman has something mystical, she can be clairvoyant and is a fortune teller; in the word "gypsy" resonates something hidden, mysterious; she is on the move, free, wandering; she is emotional, intuitive, can improvise; unfortunately, the Roma are not respected in society – they are stigmatized.

It is easy to see that this Roma woman is a very positive shadow figure of Lisa, who is optimally connected with her feelings and instincts, with her creativity, and can thus live all the qualities and abilities that are currently inaccessible to Lisa.

This is because Lisa is still in a state of depression and blockage, which is shown in another dream image in the same night.

4. Looking at herself in the mirror, she has dressed all in black; black pants, black patterned blouse, and black leather jacket; she tries to have the blouse sleeves stay in the jacket sleeves.

Lisa reflects her clothing, which is all black. In the black, all the colors are swallowed, comparable to the emotions full of life, which are suppressed in depression. The blouse sleeves must also be quite correctly tucked inside the sleeves of the jacket; they must not fall out of it loosely and playfully. In reality, she always wore colorful and also bright clothes. She associates these black clothes with the black, heavy cloud layers in the sky from the dream a few days before, an impressive image also for her depression. These black clothes, in which the dreamer or the ego complex sees herself, reflect this depressive attitude towards life, which was mainly triggered by the war that broke out.

Nevertheless, there are counterforces to the blackness in the unconscious: Lisa, in fact, finds herself in nature in the next dream, a liberating contrast to the confinement and darkness in the dream with the house.

5. So, there are resources, because here now also appears a light-brown horse, a symbol for motive power and thus vitality that is widespread in literature. This horse, however, is only able to jump up from a standing position, but not to move forward, moreover, after each jump up, it lands on its lower back; however, this does not seem to bother the horse, nor does it hurt itself: the life energy comes into place. This corresponds in the external life to the undaunted search for a job, where she did not find a possibility to get a job in spite of

good qualifications, but had to experience repeatedly that in each case, another person was preferred by the decision-makers and she fell by the wayside: it did not go forward.

6. In the next dream now, the dream-I stands opposite an oversized king:

"Outside in nature; in the mountain world; I am walking there with a man at my side; suddenly the king and his queen are standing with us (his parents); the king is very tall, almost oversized; he speaks to me and places expectations on me that are contrary to the agreements between the man at my side and me; it is about the sole responsibility beyond me that is supposed to rest with me [context no longer remembered]. I see that the king does not have a living face, but that of a bronze figure (frosted); disagreeing, I decide at a fork in the road to take the 'right' path and leave the man, king and queen behind. It goes uphill; besides me, there are also many hikers, both on the right and on the left path.

"At some point, I decide to leave the group of hikers and to continue my way alone; however, the destination seems to be the same for all of them. It occurs to me that if I continue my way alone, I will also have to spend many nights alone in nature until I reach my destination; thus, I decide to shorten the way and then to take the way to Paris. Having arrived at the rural station, I learn that my mother is looking for me. She sits with others at a large round table and tells her worries about me; do I speak to her and continue my way?"

This dream contains many aspects that could be discussed, but the dreamer's associations are essentially as follows: The king as a rigid, exaggerated, overbearing superego, dominates both his son – the animus/companion of Lisa and his wife, the queen, both of whom behave passively. It is interesting to note that he is made of bronze and therefore artificial, meaning that he is not human. A bronze figure is not movable, but rigid. In this dream, apparently, the face is made of bronze; the body is not explicitly mentioned. The face now primarily guarantees the identity of a person (or a god); therefore, also on all identity cards of a person the head or the face of the person is shown. It concerns here obviously an inhuman or superhuman archetypal father figure, which has superhuman demands on the dreamer. The latter does not accept this and separates from the archetypal parents – as well as from their son, who had been her companion.

Now she is traveling alone, without reference to an animus. But since she is traveling in nature, Lisa is in contact with the Great Mother Earth.

With Lisa, the dream ego symbolizes the ego complex; there are many wanderers on the way whom Lisa does not know and who thus symbolize subject-level, unconscious parts. Finally, she also separates from them because she wants to shorten the hike – this could speak for a certain impatience of Lisa towards her process. She wants to be able to relate directly to the self as quickly as possible, as indicated in the image of the (fast) train to Paris. But without support from like-minded shadow parts, the ego is left to its own devices and hereby falls into a regression to the parental complexes. In this dream, the mother complex comes into play: it does not prove to be strengthening and supporting, because "the mother" looks at her daughter as someone who worries her instead of trusting her to make her way. Lisa understood this part of the dream as her mother weakening her through her fears, which show themselves in a false "caring"; the mother complex as one of the pillars for the ego complex slows down the dream ego with its lack of confidence in its ability to set out, in its impulses to set out.

What about the father complex?

7. The dream ego Lisa crosses a huge unplowed field and then sees her father standing on the opposite side, passive, at a great distance.

She experiences her father as intangible; in fact, they have no real relationship to each other, for the father has shown little interest or commitment to her in Lisa's memory. This lack of a father in Lisa's experience led to a lack of constructive masculine energy in Lisa's unconscious, and then manifested itself in another dream in the following way.

8. Lisa tried in vain to drive forward in her cars in a narrow cellar (progression), while driving backward (regression) was always possible without any problems. Suddenly, darkly dressed men appeared and shot everyone dead with machine guns. Lisa closed her eyes, pretended to be dead, and hoped that those who had not yet been discovered by the killers would survive.

This is unrelated masculinity: On the archetypical level, the oversized king with the bronze face and his inhuman demands on the dream ego and these killing animus figures that populated Lisa's inner self were projected by her onto Putin and his soldiers, and these projections unleashed the same overpowering force over Lisa as the underlying inner male figures and plunged Lisa into feelings of powerlessness coupled with extraordinarily great fear that paralyzed her. Psychologically, Lisa was under the absolute dominance of a negative, destructive male energy against which she was unable to rise in her weak female identity.

How could Lisa find her way back into life? By increasingly beginning to (re)connect with her own instinctual and emotional side. I would like to mention here two dream images with animals that had a strengthening effect on Lisa and supported her to write applications for a job again and to expose herself to qualification interviews.

9. "I am running with a golden retriever across plowed farmland, the soil of which is dark and moist with sporadic young grass blades. The farmland is slightly uphill, and we run up (and down?) across the field."

This dream is experienced as beautiful. Lisa feels earthy in it and has a strong sense of freedom. This dog as a friend and companion symbolizes an instinct side that is reliable, gives Lisa a good "nose" and security.

The last dream I want to attach here is about a cat.

10. Lisa is walking along a street in the dark of night past a cat crouching at the edge, of which she is only semiconscious. As she continues walking, Lisa suddenly becomes aware of what she has seen, goes back, and calls the cat's name several times into the darkness. The cat runs towards Lisa and jumps into her arms.

Now they continue their journey through the night together.

This is a wonderful image of a re-encounter with one's own capacity for autonomy, since the cat symbolizes independence, autonomy – precisely what Lisa had lost through the shock of the outbreak of war and which had plunged her into a deep regression. Now the turn towards progression, towards a departure which has come with the rediscovered cat as a symbol of her own female independence and strength, an important step has been taken towards overcoming the paralyzing fear.

However, one must remain aware that, due to the family history alluded to previously, Lisa still has a long way to go to become the self-confident, strong young woman she has the potential to become.

Chapter 11

Destructiveness, Complexity, and Archetypal Epistemology

Critical Reflections[1]

Renos K. Papadopoulos

Abstract

In this chapter, destructiveness is approached as a multi-dimensional phenomenon whereby the mental health perspective addresses only one of these dimensions. An attempt is made to locate this phenomenon in the context of epistemological and societal considerations. Critical of mono-dimensional explanations based on causal-reductive epistemology, the chapter instead proposes the idea of an 'ecology of destructiveness', according to which mental health professionals cannot possibly continue to assume the role of detached observers. The ordinariness and archetypal fascination of destructiveness are proposed as indispensable pre-conditions for a proper understanding of this phenomenon. In addition, it is suggested that 'destructiveness may be a tragic facet of the human condition', without this implying any justification of it. The importance of terminological precision is emphasised and, accordingly, the term 'Severe Forms of Collective Adversity' is introduced and developed, to replace the inaccurate expression 'traumatic events'. Finally, it is clarified that, in effect, this chapter elaborates on what is termed 'Archetypal Epistemology', which is the exploration of the impact archetypal influences have on the way we conceptualise events and experiences.

Destructiveness, in its various forms and variations, has occupied a central place in both psychoanalysis and analytical psychology. In the main, psychoanalysis seems to have approached destructiveness through the notion of 'aggressive impulses', drives, or instincts, whereas analytical psychology seems to have placed its emphasis on the notions of 'shadow' and 'the problem of evil' in its treatment of the same subject. Both schools have accepted that destructiveness is one of the most fundamental issues in considering human nature.

However, any consideration of destructiveness and violence is inevitably influenced (consciously or non-consciously) by a host of other factors, from philosophical views, cultural and theological positions, and moral and ethical perspectives, to the pragmatics of the socio-economic, political, and historical realities; that is why destructiveness is a multi-faceted and multi-dimensional phenomenon that cannot and should not be limited to the realm of any single discipline. Yet, given that its very nature is so highly emotionally charged, this tends to cloud even further its complexity. Therefore, it is imperative to explore the very way we approach destructiveness before we embark on any further elaboration. In short, I argue that it is essential to seriously reflect on the epistemology that we imperceptibly employ in approaching the subject of destructiveness.

Epistemological Concerns

Regardless of how one defines destructiveness (that is, in terms of 'aggression' or 'the evil', etc.), there are three over-arching factors that tend to interweave within the way we conceptualise it. The first has to do with the position in which we locate ourselves during the process of

DOI: 10.4324/9781003390039-27

considering destructiveness; the second refers to the way we approach destructiveness (implicitly or explicitly), either as a personal characteristic or a collective phenomenon; and the third factor concerns our ideas about the origin and aetiology of destructiveness. As with any logical and philosophical pursuit, one has to ask the question: 'Who' is the person who is addressing the topic of destructiveness, and 'under what circumstances', 'within which context', and 'for what purpose' is this pursuit taking place? In answering these questions, what emerges is that our very position and identity as analysts (that is, as mental health professionals) decisively affect the way we approach destructiveness.

At first glance, one would claim that there is nothing unusual in the idea that society recruits mental health professionals to assist with understanding and explaining destructiveness by providing psychological insights about this phenomenon. Viewed from this perspective, mental health professionals are mere commentators who supply their particular expert opinion. This type of assistance is based on the formula that is used widely in society, and which I have called 'the Societal Discourse of the Expert' (Papadopoulos 1998). This discourse governs the rules and regulations, the expectations and obligations, and the positions, identities, and relationships of everybody involved in the totality of the sets of actions and interactions, belief systems, and networks, which are activated whenever we turn to experts for help in areas where specialist expertise is sought. The Societal Discourse of the Expert creates a certain type of inter-dependence among its actors that, at best, facilitates smooth and easier interactive conditions and, at worst, fosters exploitation and manipulation (that is, authoritarianism on the side of the expert and impotence on the side of the consumer). In the context of destructiveness, the Societal Discourse of the Expert develops intricate ramifications that should make us re-examine the idea that what is happening is simply a recruitment of professionals to explain an area 'where specialist expertise is sought'.

At the outset, it is important to observe that as mental *health* professionals, we are *de facto* located on the side of health and, perforce, against non-health or pathology. This means that the lenses we use in approaching any issue are coloured by this oppositionality.

Whatever phenomenon we address, this dichotomous narrative is active and forms a certain predisposition which is unavoidable, regardless of whether we are aware of it or not. This means that when, for example, we look at human pain and suffering, we cannot but formulate our observations and theories (whatever these may be) within this binary narrative, and therefore the idea of pathology or health will, inevitably, affect any conceptualisation of these phenomena (albeit in a crude or a subtle way). Similarly, in examining the very same phenomena, other professionals will view them through their own lenses; for example, theologians will be predisposed to formulate their comments in the context of the language of good and evil, and sociologists in terms of societal considerations. Thus, whenever we address destructiveness, as mental health professionals, we are bound to locate it in the context of the pathology–health polarity. Moreover, insofar as destructiveness is something that is humanly abhorrent, it will invariably end up being pathologised; that is, destructiveness will not be associated with health and, thus, will be located on the pathology end of the polarity.

Elsewhere (Papadopoulos & Hildebrand 1997), I compared the mental health professional's narratives of destructiveness with those of divorce; both are widespread phenomena which demand our attention and understanding. With reference to divorce, society seems to have achieved a rather remarkable feat, by avoiding its pathologisation: until not long ago, we used to refer to divorce using terms such as 'broken homes', implying that there is something psychologically wrong with people who get divorced. Thus, our narratives used to suggest that the divorced partners were 'unable to form relationships'. However, insofar as the phenomenon of divorce is so widespread, society changed its narratives and now we talk about 'one-parent families' and, with reference to the divorced partners, we talk about 'incompatibilities', emphasising the 'effects of circumstances', etc. In other words, 'we have managed to normalise the experience of divorce without minimising its disruptive effects' (Papadopoulos & Hildebrand 1997, p. 208) or the pain involved.

However, society seems to have failed to do something comparable with reference to destructiveness and violence. Destructiveness is an equally widespread phenomenon, in its various shapes and forms, and yet we tend to find it difficult to get a proper grip on it. By simply condemning it, we do not get closer to the phenomenon – and thus we cannot begin to understand it. We seem to get caught up in a debilitating conundrum: on the one hand, we cannot afford to 'normalise' destructiveness, whilst on the other hand, we cannot delve more deeply into it to understand it more fully, unless we adopt a less judgmental stance. This impossible situation has multifarious repercussions; one of them is exemplified by our attitudes towards the outbreak of new wars. After each eruption of violent hostilities or war, we believe that we shall never forget it – that we shall learn from the horrible experience and thus avoid any future repetitions. Yet, when the next outbreak occurs, we react with remarkable dismay as if we are witnessing something outrageously unfamiliar.

> It seems that there is a protective function in human beings which enables us to 'forget' painful memories of war and react with the wrath of naïve ignorance when conflict recurs. It is as if humanity needs to keep cleansing itself from the horrors of war by constantly 'forgetting' them, and thus renewing its virginal innocence.
>
> (ibid., p. 208)

A simple explanation of this complex phenomenon may be that, as human beings, we cannot afford to be neutral about destructiveness; inevitably, moral considerations influence our perception and judgement of destructiveness and violence. This means that there will always be an epistemological confusion (in mixing definitional precision with moral considerations) whenever we approach the subject of destructiveness.

Limitations of a Linear-Causal Epistemology of Destructiveness

Insofar as destructiveness is a multi-faceted phenomenon, it is impossible to develop one precise and mono-dimensional understanding of it. Moreover, as we have seen, an epistemological confusion prevents us from having a clear perspective on destructiveness due to our position as mental health professionals with all the resulting predispositions (societal and theoretical). Nevertheless, destructiveness – being such an important facet of our lives – imposes pressing demands on us to develop a clear understanding of it. It is then that we tend to resort to a familiar methodology, the predominant one that we use in our lives: to understand a phenomenon by reducing it to one or several clear causes. For example, we think that we 'understand' anger when we believe that we have identified what has 'caused' someone to be angry. This process is based on a linear epistemology according to which causes have clear effects which, in turn, become causes for other effects. In this way, for example in a family, when we approach an angry father, a nagging mother, and a disruptive son, we tend to order them in a causal and linear way, trying to discover 'the' original reason which causes the rest of the chain to unfold; then, depending on our own perspective as observers, we order these three phenomena in a sequence of causality. Accordingly, our understanding of a situation becomes equal with our 'discovery' of 'what has caused it'.

Jung repeatedly warned us against the limitations and even dangers of a causal-reductive epistemology, especially when applied to our understanding of human nature. The difficulty with this type of epistemology is that, although it is used quite appropriately with reference to many facets of our lives, it has severe limitations when applied to complex and multi-dimensional phenomena, particularly in the realm of human relations, because these do not follow a neat sequential causality; instead, one incident or behaviour influences the other in inexorable circularity that maintains the totality of the *pattern*. This is why systemic thinking uses the term 'circular epistemology' to refer to the approach that appreciates that all elements of a system influence each other in such a way that the distinct *pattern* of the system is maintained.

Applying these considerations to destructiveness, we can see how a causal-reductive approach to understanding is obviously inadequate. However, such an approach is appealing both to our ordinary human 'rationality' and to our psychological need to locate one clear cause, which we can then blame for being responsible for the destructiveness. This process is the basis of scapegoating, according to which blame is placed on someone else, thus making us feel pure and free from the mess of destructiveness. Moreover, a causal-reductive approach to destructiveness has another 'benefit': by forcing us to commit another epistemological error (that of ignoring the role of the observer in the process of observation), it releases us from the burden of having to examine our own position and involvement in it, and even possible indirect complicity. Analytical psychology – as well as modern systemic and ecological approaches – helps us to understand that not only are the phenomena we observe inter-related among themselves, but also, we, as observers, are part of the same larger system. Yet it is comforting to place ourselves in the position of seemingly 'objective observers' and 'detached scientists' whenever we deal with the painful chaos of destructiveness.

Psychologising and Pathologising Destructiveness

Returning to the position that mental health professionals assume in relation to destructiveness, we can now understand more fully why it is appealing to be assigned the privileged position of 'objective commentators'. Society faces a painful paradox because, on the one hand, the phenomenon of destructiveness itself – being widespread and affecting most of us – demands to be comprehended, and on the other hand, what professionals seem to offer (with their 'neat' theories) does not provide any satisfactory explanation. It is within the context of this disturbing dilemma that society recruits mental health professionals to find a way out. Thus, it could be argued that what mental health professionals tend to provide may not be, after all, an understanding of the phenomenon of destructiveness but an anaesthetisation of the pain derived from the incomprehensibility of it. Within the cloud of the inherent epistemological confusion and the anguish emanating from the unintelligibility of this complex phenomenon, the theories advanced by mental health experts in attempting to understand destructiveness may ultimately amount to being not much more than ornate psychologisations and pathologisations which are intended to ease the experienced distress. Thus, unwittingly, I would argue that we are used by society, as experts, to explain away the disturbing complexity of destructiveness and replace it with sanitised theories. Under the pressure of these circumstances, we tend to employ causal-reductive approaches that isolate destructiveness and treat it as if we were investigating something exotic, out there, which has nothing to do with us and our own condition. Once we slip into that position, then inevitably our identity as mental health professionals imposes on our observations the additional constraints that we have already identified; that is, the pathology–health polarity; this results in us combining the causal-reductive approach within the oppositional narrative of pathology and destructiveness which then together produce an inevitable psychologisation and pathologisation of destructiveness. By no means do I wish to suggest or imply in any way that destructiveness is 'normal' or acceptable. But is placing it on the 'normal'–'abnormal' polarity the only way out? Is psychologising and pathologising destructiveness the only possible way to deal with it? Perhaps the first step towards such a deeper understanding would be for us to appreciate the fact that we are indeed trapped and imprisoned by and within these constraints wherein the pathology narrative occupies a key position.

Two examples can illustrate this process. The first is the case of the breakup wars in the former Yugoslavia, when western journalists, stunned by the incomprehensibility and ferocity of the conflict and the lack of any obvious 'causes', went out of their way to explain it away in terms of 'historical' reasons. In brief, it was argued that ancient animosity among the different ethnic groups was covered up by the Tito regime, and once that collapsed, the hostilities broke out. Then, mental health professionals came to add authenticity to this causal-reductive explanation by translating it into psychological language; consequently, the 'covering up' was equated with the intrapsychic process of 'repression', and the whole 'explanation' of the war was thus articulated within a pathological context.

Yet historical evidence shows that there was never ethnic purity in that region (Malcolm 1994) – and therefore, the emphasis on the 'repression' interpretation is suspect. In contrast, no historical explanation was advanced in the case of the Gulf War when the emphasis was on the demonisation of one leader, which was used as the justification for the war; obediently, mental health professionals followed the predominant narrative imposed by society and went on explaining away the pathological traits of Saddam Hussein.

What I am trying to convey here is that a paradox exists that prevents us from developing a deeper understanding and appreciation of the multi-faceted complexities of destructiveness and violence. If we could afford to avoid moralising, without ignoring the moral dimension, and equally avoid psychologising, without ignoring the psychological dimensions, we could perhaps begin to contemplate that destructiveness may, after all, be a tragic facet of the human condition.

The Tragedy of Destructiveness

Returning to the three over-arching factors, we now see that: (1) our position as mental health professionals is coloured by the Societal Discourse of the Expert and we are thus entrapped within the dichotomous narrative of oppositionality (pathology–health); (2) as mental health professionals, we seem to be predisposed to approaching destructiveness as a personal (and indeed an intrapsychic) phenomenon, especially as a result of employing (mostly in an imperceptible way) the comforting but debilitating linear-causal epistemology; and (3) in terms of aetiology, we would be divided according to the party lines of our theoretical perspectives. This division would be based on the debate that can be expressed roughly as follows: Is aggression an innate quality or drive, or it is a reaction to external circumstances? We know that, as analysts, we do not seem to have a good track record when it comes to addressing wider social issues, and this may be the result of our internal over-involvement with this debate. By being preoccupied with our own theoretical debates, we seem to miss glaring external factors such as environmental pressures, socio-political realities, and historical legacies. Spasmodic expressions of concern about social issues appear not only to have little effect but they also exemplify the fact that we do not have, as yet, any approaches that could join together in a seamless way our theory, clinical practice and social concerns as citizens, which may then provide us with coherent responses to the outbursts of destructiveness and violence.

As a consequence of the resulting epistemological confusion, narratives about destructiveness and war seem to occupy a paradoxical position in the western world; usually, they are formulated as referring to either distant and irrelevant phenomena or to noble and heroic events. Stories of pain, terror, and atrocities are told as if they belonged to the past or as if they provide us with abstract moral lessons. This could be the reason why, when new conflicts erupt, we tend to react with perplexity. In this manner, we seem destined not to learn from history but react with horror every time new armed conflict and war break out.

Ultimately, destructiveness may be a *tragic* facet of our human condition: It is difficult to find any other word to describe it more appropriately, especially if we connect it with the original meaning of the word. It is well known that 'tragedy' comes from *'tragos'* (goat), related to the rituals performed by men dressed as goats in ancient Greek ceremonies. However, it may be less known that *'tragos'* is an onomatopoeic word which comes from *'trogo'* (to eat) and *'traganizo'* (to crunch). Thus, goats with their voracious appetite crunch anything in sight, regardless of whether it is good or bad for them. Indeed, there is something tragic in this potential self-destructiveness involved in the goat munching away at everything indiscriminately. Equally, impulsiveness and lack of discrimination are key characteristics of destructiveness.

By emphasising the tragic aspect of destructiveness, we bypass the impossible questions as to whether it is inevitable or not, innate, or reactive, personal, or collective. By acknowledging the tragic element of destructiveness, we accept that as human beings, we are part of the tragedy of destructiveness and our task then becomes the endeavour to locate ourselves as individuals in a meaningful way within it, rather than explain it away.

Ordinariness and the Ecology of Violence

One of the consequences of perceiving destructiveness as a tragic aspect of being human is that our position has the possibility of being shifted and, consequently, the arrogance of the 'detached expert' can be replaced by compassion. From this new perspective, we may admit that destructiveness has always been part of our society. According to Heraclitus, 'war is the father of all things': conflict and strife, oppositionality and destructiveness are part of life, and in addition to their obvious negative effects, they may also have a positive function. Therefore, we do not need to be fixed in a position from which we simply condemn destructiveness and need to distance ourselves from it. Usually, we tend to ascribe all kinds of bizarre and conspiratorial theories behind manifestations of destructiveness in order to distance ourselves from these allegedly inhuman phenomena. Yet, tragically, they may be much more 'ordinary' than we would like to admit; by this, I do not mean that destructiveness is 'normal' (but it certainly should not be characterised as 'abnormal', either).

Lamentably, in the final analysis, violence could be seen as being much more of an 'ordinary' and human response that we would like to accept, and this realisation makes us feel most uncomfortable. Accepting the ordinariness of destructiveness does not support any argument that it is innate or inevitable; it merely acknowledges that given the conditions in society and the vicissitudes of our existence, this phenomenon is, tragically, fairly common and ordinary.

This new perspective allows us to move away from the impossible dilemma of how to avoid pathologising destructiveness without 'normalising' it, which would imply condoning it. The way out is to create a new narrative within which the emphasis is on the 'ordinariness' of destructiveness rather than its evaluation as either 'normal' or 'pathological'.

Hannah Arendt's famous expression 'the banality of evil' aptly and precisely conveys this recognition – that is, violent and destructive acts may often be tragically more ordinary than we care to admit. It is indeed very threatening for us to accept that acts of destructiveness can be human and are committed by ordinary people; instead, we seem to be compelled to attribute inhuman and diabolical motivation to their perpetrators.

In his novel *Captain Corelli's Mandolin*, Louis de Bernières, in a most sensitive way, expresses his thoughts about the ordinariness of destructiveness. In describing the situation when soldiers, after a period of anxious inactivity, experience action; he writes:

> We found that there is a wild excitement when the tension of waiting is done with, and that sometimes it transforms itself into a kind of demented sadism once an action is commenced. You cannot always blame the soldiers for their atrocities, because I can tell you from experience that they are the natural consequences of the inferno of relief that comes from not having to think any more. Atrocities are sometimes nothing less than the vengeance of the tormented.
>
> (de Bernières 1995, p. 61)

The author does not judge or blindly condemn violence in an outright and dismissive fashion but, in a remarkably brave way, attempts to understand it. He characterises violence as a 'natural consequence'; not as good or bad, not as pathological or normal. Although he is not advancing a psychological theory, his expression 'natural consequences' implies very clearly that atrocities, the actions themselves, are part of a bigger whole, and one cannot simply focus exclusively on them ignoring the rest. It also means that the soldiers themselves are brutalised and placed, unmercifully, within a cycle of violence which they did not instigate, and which had begun before them. Violent acts are not committed, necessarily, by perverted individuals but by ordinary people who are caught up in tragic circumstances; most human beings are capable of violence. This narrative emphasises what could be called 'the ecology of violence' rather than individual motivation. Moreover, this perspective locates us not as detached observers but makes us acknowledge that violence is also part of us – part of our lives.

According to Jung, of course, destructiveness is part of our shadow. We cannot deny that our psychological and social worlds contain 'shadow' and violent elements. However, as human beings, we have the capacity and resilience to be creative and constructive, especially when we are fully aware of our own shadow and destructive elements. The way Jung accounts for these 'tragic circumstances' is by means of the archetypal shadow which, when it is activated in a particularly powerful and gripping way, inevitably interconnects the wider collective with the personal (and indeed, intrapsychic) dimension; under these conditions, it is very unlikely that individuals will be able to resist this archetypal 'radiation' and to maintain their own personal integrity. Then, destructiveness reigns supreme and the archetype overwhelms all personal individualities. It is, indeed, frightening to acknowledge that, essentially, this phenomenon is 'ordinary', despite its devastating effects.

Archetypal Fascination

The power of the archetype is manifested both as overwhelming and as fascinating. This phenomenon is equally common and ordinary. Destructiveness and violence are endemic in our society. Violence sells and it strongly influences our art, literature, film, theatre, and even fashion and design – so much so that nowadays, it is appropriate to refer to 'designer violence'. Destructiveness becomes chic, cool, and hip. There is an ever-increasing ingenuity in how to present violence in most artistic forms. Films, theatre, and computer games are full of violence, euphemistically called 'action'.

In a poem entitled *Reyerta* (*The Fight*), Federico García Lorca described a fight with most vivid descriptions of horrible and violent acts, in a beautifully poetic way: referring to the killing of the hero of the poem, he writes that he '. . . rolls down the slope dead, his body laden with lilies and a pomegranate on his brow. Now he rides a cross of fire on the highway to death'. This lyric language is used to describe ghastly acts of excruciating violence when a person was knifed to death by '. . . clasp-knives, rendered beautiful with enemy's blood, glinted like fishes' (Lorca 1960, pp. 37–38). This exemplifies the dazzlingly mesmerising effect of the archetypal fascination (Papadopoulos 2000, 2006, 2010), especially the one that destructiveness exerts.

Accordingly, it is imperative not to ignore the fascination and attraction of destructiveness, and to avoid exclusively concentrating on condemning it. Such an emphasis is inappropriate and futile – and yet, it seems irresistible. The tragic fact is that violence has an exhilarating and even 'liberating' effect on us, for at least four reasons.

1. In the act of violence, as de Bernières (1995) astutely observed, there is a 'relief' for 'not having to think any more'. The 'not thinking' here does not refer exclusively to the cognitive thinking or the Jungian 'thinking' function; instead, it would be more apt to relate it to Bion's idea of the 'capacity to think', which for him is the ability to create a space to reflect when one is overwhelmed by pressing impulses. Thus, not having space to reflect provides a tragic kind of 'relief' insofar as the person eradicates the pain caused by difficult considerations and by the awareness of inner contradictions and conflicts.
2. Gripped by a destructive archetype, individuals achieve a 'false wholeness' insofar as they are driven by 'acting out' behaviour. Although 'acting out' is frowned upon, we fail to appreciate the attraction it has by mobilising the totality of the personality to act in unison. In acting out, there is no division between thoughts, feelings, and actions: all become one, and action flows. This is tragically similar to the coveted aim of many spiritual practices which, after many years of hard and disciplined training, aspire to achieve moments or a state of being whereby action flows and inner divisions are abolished. By no means do I suggest that spiritual meditation and impulsive acts of destructiveness are the same; however, their striking similarities should not be ignored.
3. Destructiveness can have a certain kind of chilling purity which is extremely seductive. This is difficult to convey, but it can be approximated once we connect it with the process of archetypal possession whereby one polarity, in its extreme form and devoid of personal content, dominates the individual. This has a numinous quality because, regardless of the content of the archetype itself, the fact remains that it is a distilled form of an image in its pure state, and it has an unmistakable and most refreshing authenticity. Any purity exerts a powerful attraction and makes people wish to continue connecting with it. Under these conditions, people offer themselves to protect this purity from any threat of contamination and this is the tragic background from which acts of atrocities can be committed in the name of defending – and indeed 'cleansing' – the perceived source of purity. In other words, if a group of people identify with a collective within the context of archetypal possession, they will be willing to sacrifice themselves in order to guard this purity and, in doing so, they will be in a position to do just anything to prevent its pollution. This complex dynamic is close to the 'Luciferian light' of evil, which may attract certain people under these circumstances.
4. Individual identity is subsumed by a wider collective identity. Ordinarily, within the context of personal identity, individuals are burdened by a variety of conflicts and inner struggles. Accordingly, they feel refreshed whenever they (at least temporarily) shed this identity by connecting with wider forms of identity. Society has special sanctioned times when, places

where, and conditions whereby this happens automatically, such as in sports competitions when a person becomes a fan, blissfully abandoning personal history and identity to join with the crowd of supporters of a team. There is an exhilarating and liberating effect when one becomes a tiny speck in the sea of the 'red' or 'green' colour of their team. There are numerous such sanctioned settings where this process of emphasising and connecting with the collective aspects of our identity happens in an innocuous and even growthful and useful way: professional associations, community activities, and family events are just a few of the many examples. However, under certain conditions when acts of destructiveness are played out on the societal plane, people become polarised and, as Jung put it, 'willy-nilly', a collective identity is imposed on them (Jung, 1928).

The tragedy is that there are some paradoxical 'benefits' of this imposition of a collective and a cancellation (however temporary) of the personal identity (Papadopoulos 1997, 1999). Both victim and oppressor, or victim and saviour, glow in the righteousness of their 'cause'. It is self-evident that a victim will feel self-righteous because, having been subjected to brutalisation, they do not need to get in touch with any personal shadow material; after all, these people are not responsible for the calamity that had befallen them, and therefore they deserve our unconditional support, sympathy, love, and eternal care. Although it may be less obvious why the 'oppressor' may feel liberated, it becomes clearer when we remember that perpetrators of violence never consider themselves as such; instead, they see themselves as serving their cause 'selflessly' – they defend the purity of their collective (as previously outlined). There is no need to develop the idea that the 'saviour' feels liberated and infatuated by the role that is assigned to them, as the inflationary and hubristic potential of this position is fairly clear.

Towards an Ecology of Destructiveness

A serious appreciation of all these considerations leads to an approach that emphasises the 'ecology of destructiveness' rather than isolated explanations and simplifications. Not only are the various roles that destructiveness imposes on people (victims, survivors, saviours, oppressors, observers, critics, etc.) interconnected among themselves within archetypal scenarios, but the epistemology we assume with regard to this issue demands (as we have seen) an ecological approach; within such an approach, we are enabled to appreciate how we cannot be detached observers or commentators, how destructiveness is tragically more 'ordinary' than we care to admit, and that it has dark and shadow attractions.

This perspective can also provide the most appropriate framework to examine the impact destructiveness has on us.

The psychological literature has many notable accounts of the nature and meaning of the traumatic experiences engendered by brutalising war situations, but this is not the place to review them and comment on them; instead, I wish to concentrate on some observations which emerge out of an ecological approach to destructiveness, and which are based on my work with survivors of violence and disasters in numerous and varied settings.

To begin with, the very identity of a person is not limited within the strict boundaries of an individual organism or, even less, an intrapsychic location.

This means that survivors of atrocities are challenged in more complex ways than just damage to their identifiable personal material possessions and human relationships. Accepting that the identified aspects of our individuality (name, gender, nationality, profession, various group affiliations, etc.) are based on a complex mosaic of what could be called an 'identity substratum'

(Papadopoulos 1997), under these difficult conditions, individuals are likely to be challenged a great deal more than what is readily evident.

This mosaic substratum of identity consists of many elements which form a coherent whole but which we tend to take for granted, as we are less aware of their presence and their significance. These include such elements as the imperceptible realities of belonging to various collective narratives and set systems of meaning that predispose us to how we perceive events and experiences, and how we shape our epistemological presuppositions. This whole includes a host of elements, e.g. the particular 'geographical landscape and milieu', the fact that we are surrounded by particular types of architectural designs, etc. Ordinarily, we do not think about these elements as parts of our identity because they are quite basic, and we take them for granted. However, in their totality they form a background mosaic upon which the more tangible aspects of our identity are based (Papadopoulos 1997, p. 15).

Therefore, when our exposure to severe forms of destructiveness shakes us to the core of our being, all these elements are also affected in predictable and unpredictable ways, and their implications should also be considered.

Moreover, exposed to destructiveness, there are two other aspects of our identity that become challenged. When a neighbour attacks a neighbour and kills members of their family and destroys their home, the victim does not lose only the family members and their house or even the friendship of the neighbour, but they also lose what is usually referred to as 'faith in humanity'. This is not an abstract faith or irrelevant ideology; without it, a person 'can no longer trust that human values such as friendship, loyalty, respect, decency, love can possibly exist' (Papadopoulos 1997, p. 13), and this kind of loss has far-reaching implications because it affects the very identity of the individual. This is akin to the phenomenon of 'moral injury' (Papadopoulos 2020), which appreciates that the effects of destructiveness are not limited to physical and psychological injuries but also include wounding of our sense of morality and human dignity. The narratives of people I have been working with convincingly convey the devastating range of impacts this type of injury has on people, especially because they are the least discernible.

The second type of challenge that people in these tragic situations invariably experience is connected with the first one and has to do with our capacity to predict life. Ordinarily, we seem to trust ourselves to judge the source of danger, and we take the necessary precautions; however, under these catastrophic and unpredictable conditions, we tend to struggle to trust our ability to assess risks and therefore additional unsettledness is experienced. This interferes with our capacity to 'read life' (Papadopoulos 1997, p. 14) and can have severe implications because it affects not only our personal identity but our very position within our socio-ecology.

Overall, if we understand these reactions as 'normal reactions to abnormal circumstances' (Papadopoulos 2021c), we do not need any pathological polarities to help us understand them. Moreover, we appreciate that this 'disintegration of the identity substratum' (Papadopoulos 1997, p. 14) may be a temporary response – part of a required 'frozenness', variations of which we, as carers, are also likely to experience under the intensity of our close proximity to these kinds of pain.

By adopting the narrative of 'frozenness', instead of pathologising terms such as 'dissociation', 'regression' etc., we appreciate that this is a human, understandable, and 'ordinary' reaction. In addition, we realise that, similarly with the actual process of physical thawing, each object requires its own tempo, intensity, distance from the source of heat, and overall conditions for successful thawing. 'Thawing is a delicate process which may damage the frozen item if not used appropriately' (Papadopoulos 1997, p. 15). Ultimately, the act of freezing also serves the positive function of preserving. In these contexts, what can be preserved are various forms of human values and strengths that may not be able to survive, if they were exposed to the harshness of the reigning destructiveness around at that time.

Severe Forms of Collective Adversities

Whenever we consider forms of destructiveness at a wider societal scale, such as wars, it is appropriate to use more precise terminology, in order not to be misled by inappropriate images and their implications. The term that I consider most fitting is 'Severe Forms of Collective Adversity' (SFCA), which clearly identifies what we are referring to, without requiring any further clarification. The importance of needing to introduce this term comes from my critique of the generally accepted term 'traumatic events' (Papadopoulos 2018, 2021a). The simple reason for questioning this term is its evident tautology: 'traumatic events' are defined as the events that cause 'trauma', and 'trauma' is defined as that which 'traumatic events' cause. Such an elementary error in logic would certainly be avoided by any beginner student of logic, and it is of great concern (as well as of relevance) that it has emerged as the predominant discourse in this field.

The grave disadvantages inherent in the term 'traumatic events' are not limited to its logical error, but also include the confusion between the very nature of the events with the way they are experienced, as well as the way that the characteristics of the events are conflated with the impact they have on people, under various conditions. Such gross confusion severely curtails any judicious examination of the complexities of these crucial phenomena (Papadopoulos 2007).

In effect, SFCA consist of acute – but also chronic – catastrophic disruptions to what is: (1) familiar; (2) predictable; and (3) comprehensible. This combination has drastic effects on all implicated parties; that is, not only on those who are directly damaged but also on all those who reach out to assist them, as well as everyone in society who becomes aware of these SFCA.

Three groups of SFCA effects can be identified. The most expected (and obvious) effects include destruction of life and property; disturbances in forms of social functioning; disruption of geographical, physical, and social ecology; and at the human level, various forms of losses, distress, suffering, trauma, disorientation, frozenness, etc. The less expected (noticeable) effects include the various forms of disruption of the fabric of society and human relationships, e.g. 'moral injury'. Then, the least expected effects include the far-reaching implications of all the previously discussed effects, as well as another category that can best be characterised as *epistemological* effects: that is, the very way our understanding of events and experiences is imperceptibly shaped. Insofar as they are almost unexpected, this last group of effects can have the most devastating consequences.

All three groups of effects have a truly *overwhelming* impact on everybody. This is due to the unique nature of the SFCA, which combines; (1) widespread and concrete destruction that is readily noticeable; (2) the fact that the phenomena are highly emotionally charged; and (3) the enormity of pressing needs – that is, to urgently do something to ameliorate the situation, or at least, to stop further destructiveness. This combination is lethal and has an overpowering and paralysing effect on all implicated parties.

In turn, the impact of such an overwhelming effect is that our *capacity to process* events and experiences becomes diminished. Hence, my rhetorical expression 'the first casualty of trauma is complexity' (Papadopoulos 2002, 2021a, 2021b; Papadopoulos & Gionakis 2018). Losing the ability to hold onto complexity is the most definite and inevitable outcome of our exposure to the destructiveness of the SFCA, regardless of whether one is traumatised or not, or of the degree of the severity of their traumatisation. Such overwhelmingness does not allow us 'space to think' or to reflect, leading us to resorting to what I refer to as 'impulsive conceptualisation' and 'epistemological acting out' (Papadopoulos 2021a). This means that when we are overwhelmed and our capacity to grasp complexity is severely compromised, we are driven to quickly concoct ways of understanding that, inevitably, are going to be oversimplified. The most common form of oversimplification is polarisation, and this is the reason that all SFCA phenomena:

> are riddled with polarised perceptions, opinions, judgements, actions, etc. . . . . [SFCA survivors], for example, are perceived as either 'traumatised' or 'resilient', as 'dangerous' or 'vulnerable', as 'welcome' or 'unwelcome'; those who work with them are equally perceived in sharply polarised ways, as being either 'for' . . . or 'against' them; moreover, they even perceive themselves as either 'omnipotent' (i.e. able to help . . . [SFCA survivors] with everything) or 'impotent' (i.e. unable to help them with anything) . . . . Even the professional discourses tend to be polarised, such as positive psychology emphasising the growthful potentialities of the adversity ('strengths-based' approaches) as opposed to 'trauma-informed' practices focusing on the pathological dimensions of human experiences in the . . . [SFCA] contexts.
>
> (Papadopoulos 2022, pp. 277–278)

In the light of this discussion, the term 'traumatic events' proves the point that all SFCA discourses tend to be polarised and are the product of oversimplification, lacking complexity. It is undisputed that there are endless lists of factors that affect the way people experience events, from personal history, strengths, and weaknesses to supporting systems and the meaning attributed to the events, plus a host of wider societal discourses and socio-political, cultural, legal, and economic factors, among others (Papadopoulos 2021a). Accordingly, the unquestionable and *a priori* fixing of the adjective 'traumatic' to the events betrays the lack of complexity and the diminished capacity to process events and experiences (Papadopoulos 2005). Surely, the same event can be experienced by different individuals under different conditions in different ways. The most obvious example is that the same acts of destructiveness are perceived by some as heroic and by others as acts of terrorism. This truism is denied by the fixed polarisation that the term 'traumatic events' conveys.

In effect, such a lack of complexity distorts reality, and if our initial conceptualisations of the relevant phenomena distort reality, what chance do we have to plan and act effectively?

Archetypal Epistemology

This discussion shows the importance of the epistemological effects of the destructiveness of war and other types of SFCAs. As it was already emphasised, especially because such effects are least expected and are, invariably, unnoticed, their impact can be even more disastrous. Moreover, from a Jungian perspective, the polarisation that emerges in such contexts requires additional attention. In short, in relation to SFCA, combining our analysis of destructiveness with the accompanied archetypal fascination and in the context of the resulting epistemological effects, we realise that, after all, the overwhelmingness of the SFCA pushes us into the archetypal realm.

A Jungian approach helps us develop new insights into what has been discussed in this chapter. To begin with, the extreme forms of polarisation that the destructiveness of SFCA engenders can be understood as expressions of archetypal polarities. We see these clearly conveyed in the archetypal personifications, on the one hand, and the archetypal abstractions and generalisations, on the other. Examples of the first are when we say that Hitler or Saddam Hussein are the archetypes of evil, and examples of the second are when we say that a certain conflict is 'a war between good and evil'.

A proper understanding of a Jungian application to such contexts should help us to seriously be warned that the very nature and complexities of destructiveness are almost inevitably going to push us into archetypal positions which, although they are going to be seductively dazzling, will blind us to the specificities of the realities we are engaged in, and unavoidably will lead us to destructive outcomes. Therefore, the usefulness of a Jungian approach should be the development of the awareness that in such situations, most likely, we are going to become puppets of archetypal scenarios, and our grip on the complexities of the contexts we are in is going to be fairly feeble. We should expect to be swept away by the fascinating attractiveness of the archetypal constellations, whilst deluding ourselves that we are in full control of our perceptions, judgement, and destiny. All these are the results of the snares of the archetypal fascination that lures us into seemingly deeper understanding of the phenomena of SFCA whilst, in reality, it distances us further away from the painful incomprehensibilities and human ordinariness of destructiveness.

Therefore, archetypal epistemology helps us study the effects that archetypal polarisation has on our formation of our very presuppositions that, in turn, give rise to the more specific formulation of our perceptions and thoughts and judgements in a given situation. In effect, archetypal epistemology is a specific form of the epistemological effects that SFCA have on us. Along with the other epistemological effects, archetypal epistemology diminishes our capacity to process events and experiences, weakens our ability to grasp complexity, and is primarily characterised by sharp polarisation.

Finally, a prudent understanding of the archetypal epistemology should help us comprehend the complexities of the reverse relationship between the archetypal impulse and personal agency. The closer we are to the archetypal realm, the more intoxicatingly 'liberated' we feel, precisely because of our distance away from the painful messiness of personal and societal conflicts and historical specificities. However, such a position reduces our personal agency to a dangerously low level. Conversely, the closer we are to our own grounded personal and societal conflicts, painful struggles and suffering, the greater our agency is likely to be for implementing actual changes.

Jung was fully aware of this reverse relationship and accepted that the suffering involved in this struggle should not only not be rejected but should be valued. This is in sharp contrast to most 'trauma approaches' that aim to eliminate all suffering and discomfort, aspiring for some state of abstract 'happiness'. Jung explicitly stated that 'Suffering is not an illness; it is the normal counterpole of happiness' (Jung 1943, §179). Moreover, he reminds us that 'happiness is itself poisoned if the measure of suffering has not been fulfilled. Behind a neurosis there is so often concealed all the natural and necessary suffering the patient has been unwilling to bear' (Jung 1943, §185) because, we may clarify, they were seduced by archetypal flights.

Note

1   This chapter is based on an earlier paper, and it is published here with substantial modifications and with the kind permission of the publishers: (Papadopoulos, R. K. [1998]. Destructiveness, atrocities and healing: Epistemological and clinical reflections. *Journal of Analytical Psychology*, 43, 455–477).

References

de Bernières, L. (1995). *Captain Corelli's Mandolin*. London: Random House, Minerva.

Jung, C. G. (1928). The Swiss line in the European spectrum. In *The Collected Works of C. G. Jung*, Volume 10, edited by Sir Herbert Read, Michael Fordham and Gerhard Adler, translated by R. F. C. Hull. London: Routledge and Kegan Paul.

Jung, C. G. (1943). Psychotherapy and a philosophy of life. In *The Collected Works of C. G. Jung*, Volume 16, edited by Sir Herbert Read, Michael Fordham and Gerhard Adler, translated by R. F. C. Hull. London: Routledge and Kegan Paul.

Lorca, F. G. (1960). *Lorca: Penguin Poets. Introduced*, edited and translated by J. L. Gili. Harmonsdsworth: Penguin Books.

Malcolm, N. (1994). *Bosnia. A Short History*. London: Macmillan.

Papadopoulos, R. K. (1997). Individual identity and collective narratives of conflict. *Harvest: Journal for Jungian Studies*, 43(2), 7–26.

Papadopoulos, R. K. (1998). Destructiveness, atrocities and healing: Epistemological and clinical reflections. *Journal of Analytical Psychology*, 43, 455–477.

Papadopoulos, R. K. (1999). Working with Bosnian medical evacuees and their families: Therapeutic dilemmas. *Clinical Child Psychology and Psychiatry*, 4(1), 107–120.

Papadopoulos, R. K. (2000). Factionalism and interethnic conflict: Narratives in myth and politics. In *The Vision Thing. Myth, Politics and Psyche in the World*, edited by Thomas Singer. London and New York: Routledge.

Papadopoulos, R. K. (2002). Refugees, home and trauma. In *Therapeutic Care for Refugees. No Place Like Home. Tavistock Clinic Series*, edited by R. K. Papadopoulos. London: Karnac.

Papadopoulos, R. K. (2005). Political violence, trauma and mental health interventions. In *Art Therapy and Political Violence. With Art, Without Illusion*, edited by D. Kalmanowitz & B. Lloyd. London: Brunner-Routledge.

Papadopoulos, R. K. (2006). Terrorism and panic. *Psychotherapy and Politics International*, 4(2), 90–100.

Papadopoulos, R. K. (2007). Refugees, trauma and adversity-activated development. *European Journal of Psychotherapy and Counselling*, 9(3), 301–312.

Papadopoulos, R. K. (2010). Extending Jungian psychology. Working with survivors of political upheavals. In *Sacral Revolutions: Cutting Edges in Psychoanalysis and Jungian Analysis*, edited by G. Heuer. London: Routledge.

Papadopoulos, R. K. (2018). Trauma and umwelt. An archetypal framework for humanitarian interventions. In *Cultural Clinical Psychology and PTSD*, edited by Andreas Maercker, Eva Heim, & Laurence J. Kirmayer. Göttingen: Hogrefe.

Papadopoulos, R. K. (Ed.) (2020). *Moral Injury and Beyond. Understanding Human Anguish and Healing Traumatic Wounds*. London and New York: Routledge.

Papadopoulos, R. K. (2021a). *Involuntary Dislocation: Home, Trauma, Resilience and Adversity-Activated Development*. London and New York: Routledge.

Papadopoulos, R. K. (2021b, April). The approach of synergic therapeutic complexity with involuntarily dislocated people. *Systemic Thinking & Psychotherapy*, (18). https://hestafta.org/en/journal/systemic-thinking-psychotherapy/issue/18/the-approach-of-synergic-therapeutic-complexity-with-involuntarily-dislocated-people

Papadopoulos, R. K. (2021c). Families migrating together. In *The Oxford Textbook of Migrant Psychiatry*, edited by D. Bhugra. Oxford: Oxford University Press.

Papadopoulos, R. K. (2022). Therapeutic complexity. In *Seeking Asylum and Mental Health*, edited by C. Maloney, J. Nelki, & A. Summers. Cambridge: Cambridge University Press.

Papadopoulos, R. K. & Gionakis, N. (2018). The neglected complexities of refugee fathers. *Psychotherapy and Politics International*, 16(1).

Papadopoulos, R. K. & Hildebrand, J. (1997). Is home where the heart is? Narratives of oppositional discourses in refugee families. In *Multiple Voices; Narrative in Systemic Family Psychotherapy*, edited by Renos K. Papadopoulos & John Byng-Hall. London: Duckworth.

Chapter 12

Tales of Trauma, Terror, and Awe

Counter-Trauma, Counter–Adversity Activated Development, and Mutual Transformations in the Clinical Setting With Survivors of Collective Violence

Elias Winterton

Abstract

Clinical work with survivors of collective violence, war, torture, and human rights violations can profoundly affect therapists and helpers. Despite various factors within the therapist's psyche and social conditions that can mitigate the risk of secondary traumatization, specific precautions in clinical practices are essential for effective survivor treatment. This chapter explores the transformative impact of witnessing survivors' experiences on analysts using analytical psychology and relational psychoanalysis. Drawing from diverse clinical material, the author creates a conceptual framework, synthesizing Jungian, post-Jungian, and relational psychoanalytical perspectives. The chapter discusses trauma survivors' archetypal defensive mechanisms, adaptive processes, and coping strategies. It examines transference and countertransference patterns, intersubjective responses, and clinical phenomena that pose occupational hazards, leading to secondary traumatization and burnout. The concept of counter–Adversity Activated Development provides a valuable framework for clinicians and mental health professionals working with trauma survivors, aiming to enhance their well-being and reduce the risk of secondary traumatic stress.

During the years of my training to become a Jungian analyst, I moved to live in a different country, facing a period of transition after a recent conflict and earlier many years of political oppression and violence. At the beginning of a group inter-vision session, I was under the impression that my colleagues were waiting for my clinical case presentation with an edgy expression in their eyes. I thought that they guessed I was probably having hard times working as a psychotherapist with severely traumatized patients, and possibly sickened by stories of abuse, cruelty, humiliation, and pain. The story of Hector was a typical portrayal of what his generation had experienced, witnessing the systematic abuse of their fathers, uncles, and grandparents, and kidnappings, disappearances, and political torture. Surviving through the war, he had lost friends and family members in violent manners, then he fled and returned home where he found a mass of ashes and destruction. They seemed enthralled by the sequence of brave actions in the story of Hector, standing out the background of appalling historical events, civil unrest, and their impact on individuals' lives. Awaiting my presentation of the clinical material and discussion of the untold traumatic wounds of Hector, they appeared genuinely surprised and perhaps slightly disappointed when they understood that Hector's mental pain and the reason to request therapeutic support originated when he fell madly in love with the "wrong" woman. It looked quite implausible that this young man did not pay attention to the ordeal of war but felt destroyed by the impossibility to breach the loyalty towards his family, wife and children, and traditional values to pursue his desire and romantic passion, as it would lead to an unrepairable loss. Trying to solve the cognitive dissonance experienced by the

DOI: 10.4324/9781003390039-28

group, the supervisor made a comment that shined the concrete meaning of human resilience, "after death, we thrive for life".

The psychological consequences of collective violence in the aftermath of war can follow multiple trajectories, as Renos K. Papadopoulos has theorized in a unique model, moving beyond the classical notion of resilience (Bonanno et al., 2011) and focusing with special attention on positive changes prompted by the phenomenon called adversity-activated development (AAD) (Papadopoulos, 2007, 2013). Based on his pioneering work with refugees and displaced people, he considers that subjective experience plays a central role to determine the psychological and mental health conditions of individuals. Humans are equipped with a "psychological immune system" that can respond in different ways, interacting with both internal protective and vulnerability factors, and external stressors and "perceived" support. Survivors can bounce back, but their hardwired resilience can be detrimentally reduced when their internal set of resources is damaged (Papadopoulos, 2007, p. 302). Nevertheless, the assumption that every person who has experienced a devastating event has been traumatized by hardship is incorrect. Escaping a dichotomic logic, which assumes that the survivor's status corresponds to being traumatized versus non-traumatized, Papadopoulos exposes a post-Jungian approach and articulates a multi-layered process that may determine negative, and/or neutral, and/or positive responses. Papadopoulos (2013) points out that mental pain, or "ordinary human suffering", and distressful psychological reactions reflect normal forms of experience and subsequent adverse situations, demystifying the medicalization of trauma. Only a minority of those exposed to catastrophic circumstances are likely to develop pathological responses or psychiatric disorders such as post-traumatic stress disorder (Linley & Joseph, 2004; Joseph & Linley, 2008).

Resilience after potentially traumatic events (PTEs) is one of the most compelling concepts in the contemporary traumatic stress research (Nugent et al., 2014). Defined as "a stable trajectory of healthy functioning after a highly adverse event" (Bonanno, 2004, p. 20), resilience, examined through advanced latent trajectory modeling, involves various independent predictors, including cognitive flexibility, adaptability, perceived social support, active coping strategies, moral compass, emotional regulation, balance, perspective, hope, bravery, zest, sense of control, and learning from experience. Resilience is seen as a biopsychosocial and spiritual phenomenon, evolving through interactive exchanges between individuals and their environment. It operates on a continuum opposite to risk and distress, strengthened by connections and relationships with others (Greene, 2002). Theoretical models of resilience are continually refined, either highlighting its dynamic process that can change across and within individuals over time or emphasizing it as a trait-like construct.

AAD (Papadopoulos, 2007, 2013) refers to the positive outcomes that are a direct result of being exposed to life-threatening events. The process of transforming human sorrow into positive development has constantly constellated the history of the human condition. Jung grasped the constructive side of psychopathological symptoms as signs of the unconscious attempt of the individual to correct an imbalance in their inner world (Jung, 1931, 1943, 1951). His attempt has been further advanced in the last two decades by theorists that have extensively investigated the phenomenon of post-traumatic growth (PTG) (Affleck & Tennen, 1996; Tedeschi & Calhoun, 2004; Long et al., 2021; Barnicot et al., 2023), a positive effect of having experienced a traumatic event, from natural disasters to life-threatening situations, with particular emphasis on those involving interpersonal violence. The idea of AAD manifests itself as:

> positive, "growthful" developments which are a direct result of the experiences gained from being exposed to adversity/"trauma", entailing new elements that did not exit prior to

the adversity . . . it can occur even during the traumatic experience, and act silently, out of consciousness leading to the emergence of a new self-organization.

(Papadopoulos, 2007, p. 307)

In the past decade, the academic community has invested resources in interdisciplinary studies to comprehensively explore trauma, resilience, and PTG (Aslam, 2013). Researchers have developed complex designs to include neurobiological, psychosocial, cultural, and societal factors contributing to determining psychopathological and/or growth outcomes (Barnicot et al., 2023) to help understand how to devise appropriate interventions in different settings where the integrity and survival of human life is threatened. Drawing on the latest developments in neurosciences, the neurobiological changes triggered by the impact of PTEs on individuals' wellbeing, shows that behavioral and biological stress responses leading to either pathological or resilient adaptations are "largely influenced by feelings of safety that come about through relationships with others, spiritual or place-based connections" (Matheson et al., 2020, p. 3). This finding can have crucial implications to inform interventions to prevent negative outcomes after the exposure to PTEs in different types of settings, including therapeutic treatment with survivors.

This chapter intends to explore in depth through the clinical lens of Jungian, post-Jungian psychology and relational psychoanalysis different facets of traumatic and positive developments in survivors of collective violence, war, and torture, and how the therapeutic process, transference, and countertransference may create in the therapist, and helpers, similar experiences mirroring both traumatic and growth aspects. Psychoanalysis has extensively focused attention on phenomena of secondary or vicarious traumatization in therapists and other groups of helpers working with survivors, also defined as compassion fatigue or counter-trauma (Davies & Frawley, 1994; Gartner, 2017). Secondary traumatization, or secondary traumatic stress (Figley, 1995), refers to post-traumatic stress symptoms that therapists and other helpers can experience as a collateral effect of using empathy to foster the relationship with the patient (Gartner, 1999). Much less recognition has been devoted to the opposite phenomenon, known as vicarious resilience (Hernandez-Wolfe, 2018), or counter-resilience (Gartner, 2014, 2017).

This chapter attempts to describe an emerging strand of the therapeutic and clinical work with survivors, defined as countertransference responses of AAD experienced by therapists (and helpers). It is explained as a secondary phenomenon implying potential beneficial impact, and transformative growth of therapists mirroring the AAD experience of their patients through the catalyzing effect of joint moments of awe within the dyad.

1. To die, survive, or thrive: the experience of terror and awe in traumatic contexts.

If someone had seen our faces on the journey from Auschwitz to a Bavarian camp as we beheld the mountains of Salzburg with their summits glowing in the sunset, through the little barred windows of the prison carriage, he would never have believed that those were the faces of men who had given up all hope of life and liberty. Despite that factor – or maybe because of it – we were carried away by nature's beauty, which we had missed for so long.

(Frankl, 1959, p. 52)

The concept of AAD offers a complex epistemological tool complementing the literature on resilience, to understand how people can preserve dignity, find meaning, and learn from appalling conditions of vulnerability, degradation, injury, and loss (Papadopoulos, 2007, p. 308). Identifying the strengths and facilitating factors that trigger AAD responses remains under-investigated. Analytical psychology unveils archetypal mechanisms safeguarding the psyche from trauma's harm (Kalsched, 1996, 2013), shedding light on the self-care system preserving the personal spirit. The transcendent function catalyzes a multi-layered process of development and healing, integrating dissociated psyche parts into a unified whole. Jung explained how consciousness and the unconscious are in a relationship of dynamic opposition, which can be conflictual or creative. When creative, the resulting psychic tension sets in motion "the remarkable capacity for change of the human soul" (Jung, 1966b, CW 7, par. 360). This inner psychic tension can apply in front of devasting traumatizing circumstances, to create meaning of inevitable and uncontrolled suffering and death, as the Viennese psychiatrist Victor Frankl (1959, p. viii) called "the will to meaning". Holocaust concentration camp survivor Frankl dedicated his life to developing logotherapy, a method to heal by creating meaning and experiencing positive aesthetic emotions like beauty, practicing tragic optimism, and maintaining a moral attitude. Contemporary research in self-transcendence tends to focus on the emotion of awe (Keltner & Haidt, 2003), an approach that indirectly taps into the Jungian idea of the numinous and the positive and negative experiential polarities associated with it. Coined originally by Rudolf Otto (1917), the term "numinous" was intended to describe a new category of value within the sense of "holy", as well as referred to a state of mind. Otto and Jung used this term to indicate both a quality inherent to an object or an experience that comes over a person, often inadvertently,

> a dynamic agency or effect not caused by an arbitrary act of will. The numinosum – whatever its cause may be – is an experience of the subject independent of his will. The numinosum is either a quality belonging to a visible object or the influence of an invisible presence that causes a peculiar alteration of consciousness.
>
> (Jung, CW 11, para. 6)

The numinous contains both positive and negative qualities of the experience that can occur simultaneously in any encounter with the divine. Otto described the qualities of the *mysterium tremendum et fascinans* as positive (sublimity, awe, excitement, bliss, rapture, exaltation, entrancement), and negative (being overwhelmed, fear, trembling, weirdness, eeriness, humility, an acute sense of unworthiness, urgency, stupor, blank wonder, bewilderment, horror, mental agitation, repulsion, and haunting, daunting, monstrous feelings) (Otto, 1958, p. 37). The concept of numinous served as an empirical basis for Jung, extracting the experience of holiness from theory and translating it into feelings, sensations, and powerful events – a reflection of Jung's personal encounter with the divine.

In the clinical setting, Jung saw the numinous coming as an uncontrollable and "outside conscious volition" powerful force generating from the collective unconscious and demanding his patients to grapple with their conflicts, while challenging the organization of their lives (Jung, CW 8). The emotional and spiritual experience of the numinous has the crucial role, according to Jung, to make archetypal dimensions accessible to consciousness: "It is through numinosity that the archetypes seize the soul and produce faith" (Jung, CW 5, par. 344).

The Jungian idea of the numinous has been scrutinized through the psychoanalytic and the contemporary Jungian perspectives (Casement & Tacey, 2006; Tacey, 2013), who demonstrated how this concept does not strictly attain to the realm of religious matters, but it has been incorporated with a multi-disciplinary approach in the humanities, and in both secular and scientific views of the world. The numinous appears structured across two opposite archetypal polarities, negative and positive, described as Apolline and Dyonisian pathways (Giaccardi, 2006), or outlined in the dynamic relation between masochism as the shadow and complementary aspect of religious veneration and worship (Gordon, 2002).

The path connecting mental (and physical) pain and transcendence towards extraordinary experiences of amazement, awe, and ecstasy is described psychoanalytically in religious rites and rituals, and in ascetic and mystical practices (Konečni, 2005), and in the phenomenology of political events such as terrorism (Papadopoulos, 2006) and some aspects of the psychic space organized around states of twoness observed in the phenomenon of torture (Luci, 2017).

The numinous aspect of the psychoanalytic process and the therapist's experiences activated by trauma and adversity in countertransference will be explored in the next section.

The Emotion of Awe and the Numinous

According to Emily Dickinson, beauty and truth are both different sides of the same whole, as they are both brethren in the pursuit of higher ideals, and bound to the inevitability of oblivion and the ultimate futility of human pursuits.

The domain of the numinous remained an appanage of philosophers and religious scholars until the early 2000s (Gallagher et al., 2015). The neuroscience of awe started with the landmark paper by Keltner and Haidt (2003), presenting a "conceptual approach to awe" (ibidem, p. 303). They suggested that awe experiences are characterized by two aspects: "perceived vastness", and a "need for accommodation" (ibidem).

Awe is a complex emotion playing a pivotal role in religious transformations, aesthetics, and political change, raising a fleeting interest in psychologists (Shiota et al., 2014; Schneider, 2008, 2018). Awe is defined in English as "fear or dread, mingled with veneration, and the attitude of a mind subdued to profound reverence in the presence of supreme authority, moral greatness or sublimity, or mysterious sacredness" (Keltner & Haidt, 2003, p. 298). Echoing Otto's (1958) essay, and Jung's theoretical speculations on the numinous, these researchers distinguished different "zests" of awe: threat (Stellar et al., 2017), beauty, ability, virtue, and supernatural causality. In recent years, neuroscientific approaches have focused on the empirical study of awe (Chirico & Yaden, 2018; Gottlieb et al., 2018). The more recent findings considered that awe corresponds to an altered state of consciousness or "self-transcendent experience" (STE). Emotion researchers consider awe as a compound positive state of mind (Frijda, 1986), overlapping with wonder (Gallagher et al., 2015), and aiming to induce a process of elaboration of new information about the environment. The evolutionary origin and functions of awe have profound implications for understanding its role in human development. Initially, the focus was on the primordial origin of awe in promoting social cohesion and obedience to cultural norms "a feeling (that) is likely to involve reverence, devotion, and the inclination to subordinate one's own interests and goals in deference to those of the powerful leader" (Keltner & Haidt, 2003, p. 307). More recently, researchers advanced the "nature-first" hypothesis, pointing at the link between awe and natural scenes (Chirico & Yaden, 2018).

Evolutionary theories show that awe has an important fitness-enhancing function, holding the ability to induce people to use cognitive accommodation in information-rich environments, demanding to update their mental schemas and to use their analytical abilities rather than to rely on mental shortcuts when they are confronted by stimuli that violate their current understanding of the world (Allen, 2018; Anderson et al., 2018).

In conclusion, the neuroscientific empirical findings on the experience of awe provide new ground to comprehend the original idea of the numinous theorized by Jung (1956, 1960). Awe is described a complex state of mind which entails bodily sensations as part of the emotional experience, and the autonomous nervous system seems to respond according to a variety of psychophysiological patterns depending on the type of awe experienced (Gordon et al., 2017). On a psychological level, awe diminishes a person's sense of self, or "small self-effect" (Bai et al., 2017), whereby people shift away from their concerns, focus on a more holistic comprehension, and become more humble (Stellar et al., 2017). Moreover, awe experiences appear associated also to relational responses, generating forms of pro-social behaviors, empathy, perspective taking, and generosity (Zhang et al., 2014; Prade & Saroglou, 2016).

The awe-led effect to self-diminishment is often reported by both religious and non-religious individuals describing their spiritual experiences, which in turn give rise to humility, expanding their vision (Saroglou et al., 2008) and perception of time, and provoking profound cognitive accommodation that can create life-lasting changes in how people view themselves and the world. These awe-filled transformational moments, are well represented by an autobiographical account by Viktor Frankl, contained in the essay "Reflections on Mystery and Awe" (Elkins, 2001). Surviving the Holocaust and losing his family and wife, Frankl recounts a memory shortly after his liberation.

One day, I walked through the country past flowering meadows, for miles and miles, toward the market town near the camp. Larks rose to the sky, and I could hear their joyous song. There was no one to be seen for miles around; there was nothing but the wide earth and sky and the larks' jubilation and the freedom of space. I stopped, looked around, and up to the sky – and then I went down on my knees. At that moment there was very little I knew of myself or of the world – I had but one sentence in mind – always the same: "I called to the Lord from my narrow prison, and He answered me in the freedom of space". How long I knelt there and repeated this sentence memory can no longer recall. But I know that on that day, in that hour, my new life started. Step for step I progressed, until again I became a human being.

(ibid., p. 141–42).

This intense moment of awe may be the most important, is described as provoking an irreversible transformative experience of life, reasonably comparable to an AAD response (Papadopoulos, 2007).

Theory of emotions describe awe as a complex state of mind, embracing terror and fear on the one hand, and amazement on the other side (Plutchik & Conte, 1997). This bipolarity resembles the archetypal structure of the numinous experience. Trauma-related psychic mechanisms may involve terror, despair, numbness, disorganization, and splitting, and may fall into the "dark side of awe". These phenomena are predominant in primitive group dynamics whereby the individual self tends to identify with the group (Freud, 1921), and according to contemporary research, they can also predict harmful behaviors directed at outgroups, including discrimination, aggression, and genocidal tendencies (Bai et al., 2017; Gordon et al., 2017). In post-Jungian perspective, Papadopoulos (2006) provided a compelling interpretation of the driving motivations for terroristic actions, under the effect of fascination generated by Pan's archetypal unipolarity.

In psychotherapy, trauma can transform into awe during moments of truth, as both members of a dyad deeply connect and create a new understanding of the painful, terrifying, and appalling experiences shared by survivors of collective violence. The significance of awe in psychotherapy settings with survivors of collective violence can be understood as a critical element facilitating pathways to personal transformation and AAD. Awe also acts as a protective agent for the therapist against burnout and secondary traumatization.

Transpersonal psychologists highlighted the centrality of the experience of awe in therapy as an agent of change and theorized the conditions conducive to promote this emotional experience, aiming to sustain deep and lifelong changes in patients (Bonner & Friedman, 2011, 2016). Their recommendations resonate harmoniously with Jungian clinical approaches, emphasizing the importance of incorporating spirituality in the therapeutic process. Awe serves as a primary gateway to the spiritual dimension, and its emergence in therapy depends on the unique dyadic combination formed by the individuality of the patient and therapist, including timing, manner, and forms.

Awe can strengthen the bond between therapist and patient, serving as a powerful emotion that enhances the sense of connection and relatedness to others. In the analytical process, trauma-related states of mind can be framed within the spectrum of the numinous, especially when the truth of the traumatic experiences can emerge beyond the defensive walls of dissociative mechanisms, emotional dysregulation, and projective identifications (Gold, 2017; Steele, 2017; Itzkowitz, 2017). In these moments, when the therapist is sufficiently sensitive and cognizant, the numinous emotions of trauma can be transposed onto their awesome side, unfolding the transformational potential of human-inflicted pain and anguish.

2. Counter-trauma and counter-AAD.

The major mode of therapeutic action in the Jungian clinical method is based on the unique relational equation created by the encounter of the individuality of the analyst/therapist and of the patient (Jung, 1966a). This cornerstone confers to analytical psychology an optimal premise offering fundamental principles, different tools, and useful techniques for treating individuals who suffer the effect of traumatic events related to collective violence, war, and torture (Van Eenwyk, 2001; Luci, 2022).

In his groundbreaking work, anticipating contemporary views on intersubjectivity in relational psychoanalysis, Jung (1929) observed the ways through which the personality of the analyst contributes to analytic process and the mutual transformative process activated by the unconscious identity established by the dyad, using the anthropological term participation mystique. He matured his original views on the dynamics of transference and countertransference later (Jung, 1946), using alchemy as a model showing different levels of projective identification and transference, identifying stages in its evolution and resolution.

Jung enhanced the understanding of transference by incorporating the archetypal dimension into the analytic process. This process, guided by the unconscious drive of individuation, seeks to expand and flourish in psychological depth and complexity. The Self, serving as an overarching archetype, encompasses all other archetypes. The analytic setting becomes the stage where the final aim of the human psyche unfolds, creating meaning and progressing towards a teleological, prospective dimension. This process opposes regressive projections of unconscious material from personal past experiences. The dynamic tension between intrapsychic directions of autonomy, differentiation, and relatedness creates the ground for analytic growth in the analysand. Transference and countertransference constitute multileveled web of transecting relationships, interpersonal and intrapsychic, conscious, and unconscious, occurring simultaneously within and between analyst and patient (Kirsch, 1995; Samuels, 2006), leaning towards augmented self-awareness and capacity to integrate unconscious, dissociated, and traumatic experiences. Anticipating the emphasis on mutuality and intersubjectivity of the modern psychology of self, Jung gained insight from his own emotional vulnerability and mental suffering, into the reciprocal, concurrent process of emotional transmission between therapist and analysand. He defined the archetype of "the wounded healer" as a phenomenon whereby "the analyzed wounds affect the wounds of the analyst. The analyst either consciously or unconsciously passes this awareness back to his analyzed, causing an unconscious relationship to take place between analyst and analyzed" (Jung, 1966b, par. 422), and as a condition for the effective therapeutic action is the self-awareness of the analyst of their own wound: "It is his own hurt that gives the measure of his power to heal" (Jung, 1951, par. 239)

The relevance of the wounded healer analytical model is particularly useful to explore and address: (1) the possibility of the occurrence of countertransference and counter-trauma responses, in relation to different conditions including personal, social, and cultural factors related to the therapist; (2) the increased risk of developing secondary/vicarious traumatization and burnout working with survivors; and (3) the potential emergence of counter-resilience and counter-AAD responses.

Countertransference Responses in Trauma Therapists

Empathic attunement represents the major enabler of the therapeutic process, and with trauma survivors, the emotional contagion triggered by dissociated can lead to pathogenic countertransference reactions, and/or activate defense mechanisms (Lansen & Haans, 2004; Wilson, 2004). The experience of the therapist is strongly influenced by contextual and environmental factors, such as the political context, a volatile or hostile working environment, adequate support, and resources to help the survivors. As an example of the augmented risk of secondary traumatization, Chilean psychotherapists working in a threatening environment with victims of political repression described in depth their strong countertransference reactions. This indicates an additional risk for those operating in similar conditions and facing a higher probability of identifying with the victims (Comas-Díaz & Padilla, 1990).

The literature on the countertransference of trauma therapists includes working with patients bearing the wounds of different types of traumatic experiences, including physical, psychological, and sexual abuse, incest, pedo-pornography, natural disasters, critical incidents, torture, war, migration, political and collective violence (Gartner, 2017; Wilson & Drozdek, 2004).

Under the pressure of powerful unconscious traumatic contents and dissociated emotions, the countertransference reactions can massively influence the body experience of both patient and analyst (Luci, 2022; Zoppi, 2017), and often elicit defensive strategies aiming to avoid traumatic and disturbing contents, taking emotional distance, numbing, difficult boundaries management, overinvolvement, dysregulated affect, a sense of role reversal, disconnection from others, anger, a sense of betraying victims, and the disruption of long-held belief (Wilson, 2004).

Therapists with secure attachment can possibly cope better with their countertransference reactions and maintain an optimal attunement (Talia et al., 2020) towards the traumatized patients throughout the process of exploring overwhelmingly disturbing material. Research in psychotherapy has explored the therapist's state of mind in relation to adult attachment representations. Findings indicate that secure therapists are more effective in creating a secure space, aiding patients in symbolizing their internal states and making sense of them. These therapists demonstrate better self-monitoring abilities and empathetically link their countertransference reactions to the patient's internal states. This encourages sensitive exploration without coercion or detachment from their inner experience. Therapists working with survivors of war, torture, and collective violence face the challenge of developing counter-traumatic responses to their patients, such as unconsciously distancing from or denying traumatic material mirroring the defensive mechanisms of the traumatized patient (Howell, 2017). Therapists with dismissive attachment might be more prone to minimize or shift attention from the patient's painful experiences, while preoccupied therapist might over engage in the relationship, blurring psychic boundaries, and be unable to maintain the right distance from the traumatic experience and ultimately display symptoms of traumatic stress (Wilson, 2004). Gartner (2017) reflects on the importance of the empathic attunement with the emotions surrounding the traumatic material, being able to recognize them and to present them to the patient without feeling overwhelmed by horror.

The archetype of the wounded healer provides a useful framework to help therapists, especially those in training and supervision, to be aware of the risk of developing counter-trauma, or vicarious traumatization (Figley, 1995). Remaining aware of parallel responses of vicarious AAD or counter-resilience (Hernandez-Wolfe, 2018) exhibited by trauma victims on the face of frightening, heart-rending experiences, can contribute to the effectiveness of the therapeutic action. Pearlman and Saakvitne (1995) note that both counter-traumatic and counter-resilient responses can appear simultaneously, sometimes intensifying one another. Transcending the numinous polarity of terror and moving into moments of awe, letting new meaningful and life-changing dimensions surface, results from the therapist's capacity to fully participate in the real substance of the patient's intrapsychic horrifying experiences and accompanying external events. It involves joining the patient in the process of transformation. The unconscious relationship, the participation mystique in the therapeutic dyad, can lead to mutual personal growth and AAD. The psychoanalytical work of Gartner with sexually abused boys (1999, 2017) offers a powerful example of this process. The author sustains that the curative essence of the therapist is being able "to take the other person's experience into one's own, to survive what was done to the patient, and what the patient does and has done as a result" (Howell, 2017, p. 101).

Truth and Beauty: A Gateway From Trauma to Awe

Counter-traumatic responses occur in parallel and are complementing counter-AAD responses. The therapeutic relationship offers the space for a mutual process of transformation, when both therapist and patients can encounter in what Daniel Stern calls "now moments" (Stern, 2004) when the optimal empathic attunement of the dyad can kindle healing, for instance when the patient observes the therapist being temporarily dazed in response to the patient's disclosure of trauma. When trauma survivors can see that the therapist allows horror into them, they feel that they know these horrors, and that they are true. Truth in trauma treatment implies the capacity of the therapist to bear painful visceral and body experiences (Luci, 2022), and patiently find the way to link these unspeakable atrocities to ideas, words, and meanings. The truth of the therapist's vicarious experience allows the patient to feel an authentic empathic link, attempting to understand what they often perceive as unknowable. These moments of truth can elicit a sense of awe, and deep connection, in the words of a trauma therapist: "counter resilience would not be possible if it were not for what could be called 'the aesthetics of truth'" (Howell, 2017, p. 100). The correspondence between truth and beauty is recognized by poets: according to Emily Dickinson (1890, RW 449), "beauty and truth, themselves are one"; and paraphrasing Keats, "beauty is truth and truth is beauty". The awesome side of trauma can shine in the beauty of the therapeutic couple "sharing something that is truly felt, even if horribly painful or, paradoxically, ugly, is uplifting for both members of the dyad" (Howell, 2017, p. 100). The upcoming clinical material will exemplify the described countertransference processes.

A Tale of Trauma, Terror, and Awe

Anastasia found haven for her mother and three kids, fleeing just three days before the unspeakable horrors during the siege of her hometown. She approached the therapy setting with a mix of curiosity and fear, as part of the assistance package from the national organization supporting asylum seekers. Anastasia, appearing older than her 37 years, smiles to project a positive image while concealing distress from chronic pain since her husband's death. She speaks about her daily routine, including housekeeping, baking, walking, and caring for her mother and young children. Listening to this extensive list of tasks, I feel drained and fatigued as they put strain on her weak body. The only moments when Anastasia seems unable to impress a constructive sense in our meetings is when she touches upon the difficult relationship with her children, who appear anxious, reckless, and disobedient, and make her very angry, triggering memories of the war, which make her fearful and helplessness. With a shred of shame, she confesses that she freezes due to feeling hatred towards her kids and the impulse to physically abuse them. Despite the disturbing revelation, I felt engaged and could understand the intense anger her children raised in her, along with her unexpected need to evacuate it violently. Anastasia seemed surprised when she saw that in my eyes there was no trace of critical judgment for being a "bad mother", but rather a profound comprehension and sadness. She then took courage and disclosed that on one occasion, she realized that she had harmed her oldest son, likely hitting him with a wooden spoon, but she could not remember exactly when and for how long. Without comments, I joined her in a long silence, feeling a burning sharp pain in my chest. Conscious of being close to tears, I watched her gaze go far from the room, with a frightened expression on her face. I stayed in these long moments, until Anastasia returned to stare at me. I said that she never told me what happened to her before leaving her country. Anastasia said, "now that I know that you can feel what I feel, and you are not scared of it . . ." and nodded repeatedly at me.

It was hard to remember for how many days, when the strikes started, Anastasia and her family had spent in the basement of their apartment; the perception of time changes when secluded in a dark room, as they cannot discriminate clearly between day and night. She had tried to make the space comfortable, but it was cold, extremely humid, and immersed in twilight, making it challenging to entertain the kids. All sounds outside provoked a sense of alarm, danger, and fear, conditioning all her reactions, thoughts, and feelings. As the therapeutic relationship unfolded, Anastasia reflected on the first instinct for a human being to seek protection and find refuge when terrified to die. For many days, she was convinced that remaining calm and patient, waiting within the walls of her apartment, was the best option to survive. After the first round of bombardments ended, she ran out with two neighbors to collect packs of food from the destroyed local supermarket. She witnessed dead bodies, destruction, and unthinkable things, including two cruel killings of innocent people by the occupying enemy forces. She persuaded herself that she could befriend some of them, and in this way they would have helped her, seeing not a hostile opponent but a vulnerable single mother. When four of them came to visit her place one evening, they were visibly drunk. They threaten to cut her children's throats. Petrified in terror, she promptly asked her mother to stay in the basement with the children and to sleep until her return. She tried to talk with them as long as possible, praying inwardly that they would fall asleep. Then the ordeal started. She remembers that for long hours she endured being beaten and raped with hard objects, and that she was penetrated by the four of them repeatedly, until the day after. When she woke up, they had left her place. Covered in blood and wounded, she became aware of the pain only an hour later as the shock in her body faded away. Initially, she washed herself, knowing that the cold water could not cleanse the dirtiness that infiltrated her soul during the night. Then, she went to open the door of the basement and saw her mother and children, alive, anxious but safe, and she was overwhelmed by wonder and awe. She thought that it was incredibly beautiful that none of her perpetrators had touched the kids, and her sacrifice and the tortures suffered were repaid. At this point of her story, I was enchanted by the emotional shift that had occurred. Similarly to the oblivion of the pain of labor after giving birth, Anastasia appeared able to forget the night of breathtaking violence and to find solace in the sense of beauty of finding herself alive with them. I was resonating with mind and body with her emotions. To conclude her story, smiling in tears, she defined that night as a blessing, because it made her free from fear, able to react and leave her town before the massacre of many inhabitants. In a strange way, she could express gratitude towards her brutal perpetrators, indirectly responsible of the gift of a second life. It is difficult to convey the massive emotional and somatic responses I experienced listening to Anastasia's harrowing story, especially my internal participation in the transformation of horror into awe when she recalled her enlightening moment of insight recognizing that she survived thanks to the violence endured, and her perpetrators.

The deepest moments of connection and engagement with Anastasia were allowed by the mutual recognition when both of us were aware that we were joining together a moment of truth in front of disturbing and dissociated traumatic material. As a therapist, I felt that despite the traumatic separation of parts of the self and the acting out of her anger and terror, Anastasia was able to process and mentalize thoughts, feelings, and intentions, integrating them to make sense of a catastrophic event as an activator of a positive new life change. Working with survivors of war, torture, and collective violence can often resonate with miserable, shame-ridden, or livid parts of the self, perceived as not collectively acceptable. The therapeutic experience can validate the feelings and experiences of both the therapist and the patient in a regenerating, transformative, and fulfilling way when they contribute to constructing moments of truth, matching affects and meanings.

Conclusions

In this chapter, the author aimed to discuss important aspects of the therapeutic action in the clinical work with traumatized survivors of war, torture, and collective violence. Based on the clinical Jungian model of the wounded healer, the mutual engagement in transference and countertransference is explored through the perspective of trauma and counter-traumatic responses, and in parallel of adversity-activated development (AAD) and counter-AAD. The role of numinous emotional experiences of awe is discussed as a crucial gateway towards the integration and transformation of the traumatic experience, and its translation into a life-changing new mindset. This model is essential for understanding one aspect of working with severely traumatized individuals, particularly for training and supervision purposes. Its application can mitigate the risk to develop secondary traumatic stress and burnout in therapists and identify new ways to enhance intervention with specific groups of vulnerable survivors enhancing the experience of awe and AAD responses.

References

Affleck, G., & Tennen, H. (1996). 'Construing benefits from adversity: Adaptational significance and dispositional underpinnings'. *Journal of Personality*, 64(4), 899–922.

Allen, S. (2018). *The Science of Awe*. Berkeley, UC: Greater Good Science Center.

Anderson, C. L., Monroy, M., & Keltner, D. (2018). 'Awe in nature heals: Evidence from military veterans, at-risk youth, and college students'. *Emotion*. https://doi.org/10.1037/emo0000442

Aslam, N. (2013). 'What doesn't kill me makes me stronger. Are the adverse life events the prerequisite for maturation and growth?'. *JAMC*, 24(2), 1–2.

Bai, Y., Maruskin, L. A., Chen, S., Gordon, A. M., Stellar, J. E., McNeil, G. D., & Keltner, D. J. (2017). 'Awe, the diminished self, and collective engagement: Universals and cultural variations in the small self'. *Journal of Personality and Social Psychology*, 113(2), 185–209. https://doi.org/10.1037/pspa0000087

Barnicot, K., McCabe, R., Bogosian, A., Papadopoulos, R., Crawford, M., Aitken, P., Christensen, T., Wilson, J., Teague, B., Rana, R., Willis, D., Barclay, R., Chung, A., & Rohricht, F. (2023, February 17). 'Predictors of post-traumatic growth in a sample of United Kingdom mental and community healthcare workers during the COVID-19 pandemic'. *International Journal of Environmental Research and Public Health*, 20(4), 3539. https://doi.org/10.3390/ijerph20043539. PMID: 36834236; PMCID: PMC9965513

Bonanno, G. A. (2004). 'Loss, trauma, and human resilience: Have we underestimated the human capacity to thrive after extremely aversive events?'. *American Psychologist*, 59(1), 20–28. https://doi.org/10.1037/0003-066X.59.1.20

Bonanno, G. A., Westphal, M., & Mancini, A. D. (2011). 'Resilience to loss and potential trauma'. *Annual Review of Clinical Psychology*, 7, 511–535. https://doi.org/10.1146/annurev-clinpsy-032210-104526. PMID: 21091190

Bonner, E. T., & Friedman, H. (2011). 'A conceptual clarification of the experience of awe: An interpretative phenomenological analysis'. *The Humanistic Psychologist*, 39(3), 222–235. https://doi.org/10.1080/088732

Bonner, E. T., & Friedman, H. (2016). 'The role of awe in psychotherapy: Perspectives from transpersonal psychology'. *Voices: The Art and Science of Psychotherapy*, 52, 62.

Casement, A., & Tacey, D. J. (eds.) (2006). *The Idea of the Numinous: Contemporary Jungian and Psychoanalytic Perspectives*. London, New York: Routledge.

Chirico, A., & Yaden, D. B. (2018). 'Awe: A self-transcendent and sometimes transformative emotion'. In H. C. Lench (ed.), *The Function of Emotions* (pp. 221–233). Cham: Springer International Publishing. https://doi.org/10.1007/978-3-319-77619-4

Comas-Díaz, L., & Padilla, A. M. (1990). 'Countertransference in working with victims of political repression'. *American Journal of Orthopsychiatry*, 60(1), 125–134. https://doi.org/10.1037/h0079189

Davies, J. M., & Frawley, M. G. (1994). *Treating the Adult Survivor of Childhood Sexual Abuse: A Psychoanalytic Perspective*. New York, NY: Basic Books.

Dickinson, E. (1890). *Poems*. Boston, MA: Roberts Brothers.

Elkins, D. N. (2001). 'Reflections on mystery and awe'. *The Psychotherapy Patient*, 11(3–4), 163–168. https://doi.org/10.1300/J358v11n03_12

Figley, C. (1995). *Compassion Fatigue: Coping with Secondary Traumatic Stress Disorder in Those Who Treat the Traumatized*. NY: Brunner/Routledge.

Frankl, V. (1959). *Man's Search for Meaning*. London: Hodder & Stoughton.

Freud, S. (1921). 'Group psychology and the analysis of the ego'. In S. Stratchey (eds.), *The Standard Edition of the Complete Psychological Works of Sigmund Freud (1920–1922)* (Vol. XVIII). London: Hogarth Press.

Frijda, N. (1986). *The Emotions*. Cambridge, UK: Cambridge University Press.

Gallagher, S., Reinerman-Jones, L., Janz, B., Bockelman, P., & Trempler, J. (2015). *A Neurophenomenology of Awe and Wonder*. https://doi.org/10.1057/97811374960588

Gartner, R. B. (1999). *Betrayed as Boys: Psychodynamic Treatment of Sexually Abused Men*. New York, NY: Guildford.

Gartner, R. B. (2014). 'Trauma and countertrauma, Resilience and counterresilience'. *Contemporary Psychoanalysis*, 50(4), 609–626. https://doi.org/10.1080/00107530.2014.945069

Gartner, R. B. (2017). *Trauma and Countertrauma, Resilience and Counterresilience. Insights from Psychoanalysts and Trauma Experts*. London, New York: Routledge.

Giaccardi, G. (2006). 'Accessing the numinous: Apolline and Dionysian pathways'. In A. Casement & D. J. Tacey (eds.), *The Idea of the Numinous: Contemporary Jungian and Psychoanalytic Perspectives*. London, New York: Routledge.

Gold, S. N. (2017). 'Growing together. A contextual perspective on countertrauma, counterresilience, and countergrowth'. In R. B. Gartner (ed.), *Trauma and Countertrauma, Resilience and Counterresilience. Insights from Psychoanalysts and Trauma Experts* (pp. 112–125). London, New York: Routledge.

Gordon, A. M., Stellar, J. E., Anderson, C. L., McNeil, G. D., Loew, D., & Keltner, D. J. (2017). 'The dark side of the sublime: Distinguishing a threat-based variant of awe'. *Journal of Personality and Social Psychology*, 113(2), 310–328. https://doi.org/10.10bonn37/pspp0000120

Gordon, R. (2002). 'Masochism: The shadow side of the archetypal need to venerate and worship'. In A. Samuels (ed.), *Psychopathology: Contemporary Jungian Perspectives* (pp. 237–254). London: Karnac Books.

Gottlieb, S., Keltner, D. J., & Lombrozo, T. (2018). 'Awe as a scientific emotion'. *Cognitive Science*. http://doi.wiley.com/10.1111/cogs.12648

Greene, R. R. (2002). *Resilience: Theory and Research for Social Work Practice*. Washington, DC: NASW Press.

Hernandez-Wolfe, P. (2018). 'Vicarious resilience: A comprehensive review'. *Revista de Estudios Sociales*, 66, 9–17. https://doi.org/10.7440/res66.2018.02

Howell, E. (2017). 'Speaking to and validating emotional truth in the jury-build Self'. In R. B. Gartner (ed.), *Trauma and Countertrauma, Resilience and Counterresilience. Insights from Psychoanalysts and Trauma Experts* (pp. 97–111). London, New York: Routledge.

Itzkowitz, S. (2017). 'The interpersonal-relational field, counter trauma, and counter resilience'. In R. B. Gartner (ed.), *Trauma and Counter Trauma, Resilience, and Counter Resilience. Insights from Psychoanalysts and Trauma Experts* (pp. 45–65). London, New York: Routledge.

Joseph, S., & Linley, P. A. (2008). *Trauma Recovery and Growth: Positive Psychological Perspective on Posttraumatic Stress*. John Wiley & Sons, Inc.

Jung, C. G. (1929). 'Psychotherapy today'. In *Collected Works* (Vol. 16). London: Routledge.

Jung, C. G. (1931). 'The aims of psychotherapy'. In *Collected Works* (Vol. 16). London: Routledge, Kegan Paul.

Jung, C. G. (1943). 'Psychotherapy and a philosophy of life'. In *Collected Works* (Vol. 16). London: Routledge.

Jung, C. G. (1946). 'The psychology of the transference'. In *Collected Works* (Vol. 16). London: Routledge.

Jung, C. G. (1951). 'Fundamental questions of psychotherapy'. In *Collected Works* (Vol. 16). London: Routledge.

Jung, C. G. (1956). 'Symbols of transformation'. In *Collected Works* (2nd ed., Vol. 5). Princeton, NJ: Princeton University Press.

Jung, C. G. (1960). 'The structure and dynamics of the psyche'. In *Collected Works* (Vol. 8). Princeton, NJ: Princeton University Press.

Jung, C. G. (1966a). 'Psychology of the transference'. In *Collected Works* (Vol. 16). London: Routledge.

Jung, C. G. (1966b). 'Two essays on ANalytical psychology'. In *Collected Works* (Vol. 7). Princeton, NJ: Princeton University Press.

Kalsched, D. (1996). *The Inner World of Trauma: Archetypal Defences of the Personal Spirit* (1st ed.). Routledge. https://doi.org/10.4324/9781315788081

Kalsched, D. (2013). *Trauma and the Soul: A Psycho-Spiritual Approach to Human Development and Its Interruption*. Routledge, Taylor & Francis Group.

Keats J., in Abrams M. H. (1968) *Ode on a Grecian Urn in Twentieth Century Interpretations of Keats's Odes* Editor Jack Stillinger. Englewood Cliffs: Prentice-Hall

Keltner, D., & Haidt, J. (2003). 'Approaching awe, a moral, spiritual, and aesthetic emotion'. *Emotion and Cognition*, 17, 297–317. https://doi.org/10.1080/02699930302297

Kirsch, J. (1995). 'Transference'. In M. Stein (ed.), *Jungian Analysis* (pp. 170–209). Chicago-La Salle: Open Court.

Konečni, V. J. (2005). 'The aesthetic trinity: Awe, being moved, thrills'. *Bulletin of Psychology and the Arts*, 5(2), 27–44.

Lansen, J., & Haans, T. (2004). 'Clinical supervision for trauma therapists'. In J. P. Wilson & B. Drožđek (eds.), *Broken Spirits: The Treatment of Traumatized Asylum Seekers, Refugees, War and Torture Victims* (pp. 317–353). Brunner, Routledge.

Linley, P. A., & Joseph, S. (2004). 'Positive change following trauma and adversity: A review'. *Journal of Traumatic Stress*, 17, 11–21.

Long, L. J., Phillips, C. A., Glover, N. et al. (2021). 'A meta-analytic review of the relationship between posttraumatic growth, anxiety, and depression'. *Journal Happiness Studies*, 22, 3703–3728. https://doi.org/10.1007/s10902-021-00370-9

Luci, M. (2017). *Torture, Psychoanalysis and Human Rights*. London: Routledge.

Luci, M. (2022). *Torture Survivors in Analytic Therapy: Jung, Politics, Culture*. London: Routledge.

Matheson, K., Asokumar, A., & Anisman, H. (2020). 'Resilience: Safety in the aftermath of traumatic stressor experiences'. *Frontiers in Behavioral Neuroscience*, 21(14), 1–21. https://doi.org/10.3389/fnbeh.2020.596919. PMID: 33408619; PMCID: PMC7779406

Nugent, N. R., Sumner, J. A., & Amstadter, A. B. (2014). 'Resilience after trauma: From surviving to thriving'. *European Journal Psychotraumatology*, 1, 5. https://doi.org/10.3402/ejpt.v5.25339. PMID: 25317260; PMCID: PMC4185140

Otto, R. (1917, 1958). *The Idea of the Holy*. New York: Oxford University Press.

Papadopoulos, R. K. (2006). 'Terrorism and panic'. *Psychotherapy & Politics International*, 4(2). https://ojs.aut.ac.nz/psychotherapy-politics-international/article/view/201

Papadopoulos, R. K. (2007). 'Refugees, trauma, and adversity-activated development'. *European Journal of Psychotherapy and Counselling*, 9, 301–312. https://doi.org/10.1080/13642530701496930

Papadopoulos, R. K. (2013). 'Extending Jungian psychology. Working with survivors of political upheavals'. In G. Heuer (ed.), *Sacral Revolutions Reflecting on the Work of Andrew Samuels – Cutting Edges in Psychoanalysis and Jungian Analysis*. London: Routledge. https://doi.org/10.4324/9781315787305

Pearlman, L. A., & Saakvitne, K. W. (1995). *Trauma and the Therapist: Countertransference and Vicarious Traumatization in Psychotherapy with Incest Survivors*. W. W. Norton & Company.

Plutchik, R., & Conte, H. (1997). *Circumplex Models of Personality and Emotions*. Washington, DC: American Psychological Association.

Prade, C., & Saroglou, V. (2016). 'Awe's effects on generosity and helping'. *Journal of Positive Psychology*, 11(5), 522–530. https://doi.org/10.1080/17439760.2015.1127992

Samuels, A. (2006). 'Transference/countertransference'. In R. K. Papadopoulos (ed.), *The Handbook of Jungian Psychology. Theory, Practice, and Applications* (pp. 177–195). London: Routledge.

Saroglou, V., Buxant, C., & Tilquin, J. (2008). 'Positive emotions as leading to religion and spirituality'. *Journal of Positive Psychology*, 3(3), 165–173. https://doi.org/10.1080/17439760801998737

Schneider, K. J. (2008). 'Rediscovering awe: A new front in humanistic psychology, psychotherapy, and society'. *Canadian Journal of Counselling*, 42(1), 67–74.

Schneider, K. J. (2018). *The Phenomenology of Awe*. https://www.psychologytoday.com/us/blog/awakening-awe/201806/the-phenomenology-awe

Shiota, M. N., Thrash, T. M., Danvers, A. F., & Dombrowski, J. T. (2014). 'Transcending the self: Awe, elevation, and inspiration'. In *Handbook of Positive Emotions* (pp. 362–377).

Steele, K. (2017). 'Lessons I never wanted to learn'. In R. B. Gartner (ed.), *Trauma and Countertrauma, Resilience and Counterresilience. Insights from Psychoanalysts and Trauma Experts* (pp. 126–136). London, New York: Routledge.

Stellar, J. E., Gordon, A. M., Piff, P. K., Cordaro, D., Anderson, C. L., Bai, Y., Maruskin, L. A., & Keltner, D. (2017). 'Self-transcendent emotions and their social functions: Compassion, gratitude, and awe bind us to others through prosociality'. *Emotion Review*, 9(3), 200–207. https://doi.org/10.1177/1754073916684557

Stern, D. N. (2004). *The Present Moment: In Psychotherapy and Everyday Life*. W. W. Norton & Co.

Tacey, T. (2013). *The Darkening Spirit: Jung, Spirituality, Religion*. London: Routledge.

Talia, A., Muzi, L., Lingiardi, V., & Taubner, S. (2020). 'How to be a secure base: therapists' attachment representations and their link to attunement in psychotherapy'. *Attachment and Human Development*, 22(2), 189–206. https://doi.org/10.1080/14616734.2018.1534247

Tedeschi, R., & Calhoun, L. (2004). 'Posttraumatic growth: Conceptual foundations and empirical evidence'. *Psychological Inquiry*, 15, 1–18. https://doi.org/10.1207/s15327965pli1501_01

Van Eenwyk, J. R. (2001). 'Jungian perspectives on the etiology and treatment of torture'. *Journal of Contemporary Psychotherapy*, 31, 181–197. https://doi.org/10.1023/A:1013916121307

Wilson, J. P. (2004). 'Empathy, trauma transmission, and countertransference in posttraumatic psychotherapy'. In J. P. Wilson & B. Drožđek (eds.), *Broken Spirits: The Treatment of Traumatized Asylum Seekers, Refugees, War and Torture Victims*. Brunner, Routledge.

Wilson, J. P., & Drozdek, B. (2004). *Broken Spirits: The Treatment of Traumatized Asylum Seekers, Refugees and War and Torture Victims* (1st ed.). Routledge. https://doi.org/10.4324/9780203310540

Zhang, J. W., Piff, P. K., Iyer, R., Koleva, S., & Keltner, D. J. (2014). 'An occasion for unselfing: Beautiful nature leads to prosociality'. *Journal of Environmental Psychology*, 37, 61–72. https://doi.org/10.1016/j.jenvp.2013.11.008

Zoppi, L. (2017). 'Chilled to the bone: Embodied countertransference and unspoken traumatic memories'. *Journal of Analytical Psychology*, 62(5), 701–709. https://doi.org/10.1111/1468-5922.12357. PMID: 28994466

Outro

The Age of Hypocrisy – From the Suspension to the End of Certainties: War as Reset

Stefano Carpani

Abstract

Is peace a mere pause between wars, a suspended time? This chapter delves into the intricate relationship among war, societal shifts, and psychological perspectives, drawing insights from Homer, C.G. Jung, and early 20th-century literary figures. Examining the narratives of writers such as Joseph Roth, Paul Valéry, and William B. Yeats, the chapter argues that war symbolizes the end of an era – a loss of dignity and integrity, resulting in a world characterized by disorder and decay. Addressing psychosocial facets, it explores the decline of spirituality and the emergence of technological advancements as contributors to societal shifts. Central to the analysis are Jung's insights into the pre-World War I era, framing war as a manifestation of deteriorating spiritual quality and the ascent of a "godless movement." The chapter investigates how the loss of spirituality parallels the decline of civilizations, fostering a chaotic and war-prone world. A central theme is the exploration of war as an expression of hopelessness and helplessness, marked by hedonism, decadence, and a loss of societal values. Drawing parallels between the interwar years in Berlin and contemporary society, the chapter scrutinizes the superficial embrace of hedonistic trends, emphasizing the need for meaningful reflection and meditation in an age dominated by hyper-individualization. In conclusion, this chapter offers a multidimensional exploration of war, portraying it not solely as a historical and political phenomenon but also as a psychological and spiritual reset mechanism. It underscores the intricate interplay between human nature, societal dynamics, and the evolving contours of modernity.

War and Literature

To explore the topic of war it is inconceivable without looking at Homer. According to Rosenthal's (2012) study on Homer: (1) war it is not a human but a divine one, orchestrated by the gods; (2) war transcends human control, being predetermined, predestined, and inescapable; (3) war possesses an intoxicating allure, coursing through us with adrenaline, and even in its ferocity, it can be perceived as noble and remarkably beautiful; (4) war is the arena where men carve their legacies in glory, binding them together in a unique and irreplicable manner; and (5) in Homer's work, humans are depicted as primitive beings.

The wars of the 20th century – as if Homer's description had not been sufficient, with its insistent, almost obsessive focus on the "many graphic descriptions of spearing, disemboweling and the like. Veins, arteries, tendons, and vital organs are described in detail as countless lives are lost"[1] – made us understand (also with the development of cinema), as if it was necessary, that wars are *hell*. Of course, it is hell – what else could it be (although not in Dante Alighieri's way)? It would be naïve not to acknowledge this stark reality, just as it is naïve to entertain the notion that "there should be no wars". After all, wars have been a recurring feature

DOI: 10.4324/9781003390039-29

of human history since time immemorial, and their persistent existence underscores a certain grim realism.

Can peace be considered a pause between wars: a suspended time? Yet, the question persists: What is war? Why is it challenging to prevent, stop, and avoid wars? Why have humans been embroiled in conflict for millennia? When viewed through the lens of the 21st century, what other dimensions can we ascribe to war? Is it conceivable to regard war as a reset mechanism? And if so, a reset of what, precisely?

James Hillman (2005, p. 259), claimed:

War belongs to our soul as an archetypal truth of the cosmos. It is a human work and inhuman horror, and a love that no other love has been able to overcome. We can open our eyes to this terrible truth and becoming aware of it, devote all passionate intensity to undermining the enactment of war, strengthened by the courage that culture processes even in the dark ages to continue to sing as it resists war. We can understand it better, postpone it longer work to gradually remove it from the support of a hypocritical religion. But the war as such will remain until the gods themselves leave.

Simone Weil (Rosenthal, 2012), upon examining Homer's portrayal of war, highlights that the genuine worth of the *Iliad* lies in its author's capacity for realism and his unsentimental portrayal of a world consumed by chaos. In this realm, brutality is acknowledged matter-of-factly by leaders, soldiers, and even the gods themselves.

Both Homer and Hillman share the belief that war is a divine decree, destined to persist until the gods withdraw. In contrast, Simone Weil's perspective suggests that war is synonymous with chaos. I would lean more towards the idea that chaos ensues when the gods depart. Contrary to Homer and Hillman, war may have little to do with gods themselves but rather with their absence signifying: a time when the world loses its spiritual moorings and drifts away from the guidance of spirituality, enters war. Therefore chaos. Nevertheless, how can Homer's and Hillman's proposals be plausible in a society where God is deceased (Nietzsche), or in a post-World War II world in which, if Auschwitz exists, God does not (Levy, 2014)?

In *The Crisis of the Mind* (1919), poet Paul Valéry expresses profound disillusionment during World War I, underlining that contemporary societies have come to understand that we – like those before us – are mortal, and that now, we sense that the fragility of our civilization mirrors that of life itself. In a single sentence, he hints at the potential shift of intellectual power to America or other continents, signifying a transformation away from Europe's historical centrality. Also Yeats in his own *The Crisis of the Mind* (1919) reflects on the post-World War I era, exploring the state of the world and the human condition during a tumultuous time.

Joseph Roth, hailing from the town of Brody in East Galicia, currently Ukraine, and formerly an Austrian-Hungarian province, dedicated much of his work to understanding what Zweig would call *The World of Yesterday*. In *Radetzkymarsch*, his character Skowronnek says (Roth, 2010, p. 250):

Today, not even the emperor takes responsibility for his monarchy. In fact, it seems that even God no longer intends to take responsibility for the world. It used to be simpler before! Everything was secure. Every stone stayed in its place. Life's roads were well-paved. Solid roofs rested on the walls of houses. But today, Mr. District Captain, today the stones lie haphazardly in the streets, disorderly and dangerously piled up, and the roofs are full of holes letting rain into the houses.

Then Skowronnek proclaims, "Everyone must now decide for themselves which path to take and what kind of home to build".[2] This concept has proven true in Western societies since the 1950s, and sociologists Beck and Beck-Gernsheim (2002) termed it "homo optionis".

Roth adds, again employing his character Skowronnek, (Roth, 2010, p. 250):

> When my children don't listen to me, I simply try not to lose my dignity. There's nothing else to be done. Sometimes I observe them while they sleep. Their faces then seem completely foreign to me, almost unrecognizable, and I realize that they are strangers, belonging to an era that is yet to come and that I will not know. My children are so small! One is eight years old, the other ten, and their faces are round and rosy when they sleep. Yet, in their sleep, those faces have something very cruel about them, as if the cruelty of the looming time, of the future, is already appearing in their sleep. I do not wish to know them, to understand them.

War as the End of *The World of Yesterday*

To gasp the rupture between the modern world[3] and subsequent eras – which scholars have labeled post-modern, late, second modern, and second-late-modern, a crucial point might be when a man feels, "My children don't listen to me", signaling a loss of dignity, that the modern world comes to an end. This, of course, is a reality today also for women. It is losing dignity, or as Beebe (1992) calls it "integrity" that can, as Zweig suggests, prompt the end of *The World of Yesterday* (2014) (as this person knew it) and leading to the continued degradation of it.

It is in the preface to Beebe's *Integrity in Depth* that Rosen (1992, p. XI) underlines: "both integrity and the Self are spiritual concepts that unify and facilitate transcendence and transformation". Beebe reinforces this idea, asserting that "Integrity must be pursued as a desideratum in itself" (1992, p. 15), defining it as "the stage of being untouched" (1992, p. 6). In agreement with Grudin (1999, pp. 73–75), Beebe believes that "integrity may be defined as psychological and ethical wholeness, sustained" (1992, p. 17). Hence, I propose that it is the loss of dignity/integrity, that composed yet illusory demeanor, that marks the entry into a new era – one so well described by Roth and Zweig, characterized by the dismantling of the father as authoritative and its sliding into an authoritarian entity, the questioning of tradition as the sole means to sustain society (and its nostalgia), and the recognition of the illusory nature of "order" as a defense against anarchy. I argue that the loss of dignity/integrity leaves a vacuum into which anarchy – which inherently disregards and dismantles the integrity of individuals – arises.

In Roth's work, a predominant theme is the presence of broken men, with women, when they appear, depicted as prostitutes, servants, or sick or deceased mothers[4]. This represents a fundamental and underexplored dimension in understanding the decline and fall of the Austro-Hungarian, Russian and German empires and Europe at large, leading to the World Wars.[5] From a psychological standpoint, the first half of the 20th century appears to be a period where the feminine aspect, and generative potential, remained marginalized. Perhaps research that begins investigating the illness and death of Sissi, Empress of Austria and Queen of Hungary, could shed light on this aspect. When diversity of perspectives is lacking, the balance between the masculine and feminine, the well-being of both (declines or is negatively affected and) the world inevitably loses its equilibrium and falls into decay and war.

Zweig, in his renowned work *The World of Yesterday* (2014, p. 210), observes the dawn of the 20th century, noting the remarkable acceleration of the world's pace. He underlines "discoveries and inventions followed one another, quickly becoming common knowledge because, finally, when the collective interest was at stake, nations showed greater solidarity".[6] Consequently, he remarks that people of that era began to sense the emergence of European solidarity and a shared European national consciousness, prompted by the pride in the rapid succession of technological and scientific achievements (2014, p. 210). He underscores the growing perception of the senselessness of borders in a time when aircraft could easily transcend them: "How artificial these customs and border guards seem, how much they contradict the spirit of the times, which unmistakably aspires to unity and universal brotherhood!"[7] Zweig concludes that during those years, "each of us drew energy from the general enthusiasm of the era, witnessing our individual self-confidence growing and intensifying along with the collective one" (2014, p. 210).[8]

These snapshots offer a precise view of the shift from the 19th to the 20th century and, equivalently, a portrayal of the shift from the 20th to the 21st century, characterized by the far-reaching influence of globalization. As we entered the 21st century, we moved beyond what Beck (1992) termed the "risk society" into what Italian sociologists Giaccardi and Magatti (2023) call the "age of shocks", whereby reality is increasingly shaped by our relationship with technology (algorithms) and we face a barrage of climatic, political, migration, and war-related shocks, leaving little room to enter depth. This era, I propose, grapples with the concept of God, transcending Nietzsche's assertion of His death while affirming His non-existence (Levy, 2014). Technology – bypassing Jung's saying that "the gods have become our diseases" (CW13, Para. 54) – rises to the level of a deity: creator, sustainer, and terminator of life.

The decrease of spirituality serves as a common thread during the transitions from the 19th to the 20th century and from the 20th to the 21st century. Therefore, we must reflect on what has transpired and how Europe – once characterized by a sense of brotherhood, sisterhood, and unity – twice descended into warfare. Why did Europe transition from what Zweig describes as the "era of universal confidence" to the one that followed, characterized by decline and obscurity? Of course, the historical reasons are clear, and this is not the place to delve into them at length. The causes of this transition are complex and multifaceted. Among the psychosocial causes that led to World War I and World War II, we can identify several factors, including the following.

1. An insatiable hunger for power, influence, ego supremacy, and financial gains from those in charge (and only later territory).
2. The dynamic of once-fraternal rulers becoming enemies.
3. The aging of prominent and authoritative figures like Emperor Franz Joseph.
4. Wilhelm II and his ambition over the future of Europe.[9]
5. The armies of that era which had languished in a state of lack of conflicts for more than 70 years, leading to the deterioration of their soldiers' readiness. As Roth (2010) elucidated, the Austro-Hungarian army, beginning in the mid-19th century, was in a perpetual state of anticipation for war. During this extended period of waiting, soldiers immersed themselves in decreasing training and increasing partaking in gambling, prostitution, corruption, and drunkenness, and experienced a loss of honor.
6. While the concept of "Austria Felix" in our collective consciousness is associated with the cultural contributions of Franz Joseph's empire, the darker aspects are often overlooked. As society began to resemble the declining Roman Empire, fundamental values gradually eroded.
7. Many influent individuals at the courts in Vienna began to believe, even though the emperor was hesitant, that war might be preferable to being invaded by those who were enamored with Marx, Engels, and Lenin (as Roth suggests); therefore, by the "movement of the Godless" (Jung, CW10, para. 372) in a post-tsarist Russia.

War as a Psychosocial Stance

What interests me are the psychosocial factors at the core of this book, particularly C.G. Jung's perspective on the pre-World War I era. In *Wotan* (C.W.10, para. 371), Jung emphasizes that before 1914, people saw the idea of war between civilized nations as a far-fetched notion in a rational, internationally organized world. This notion, in my view, symbolizes the biblical "fat cows" period – an era of abundance destined to end, giving way to the "lean cows" times marked by economic difficulties and social crises, ultimately leading to chaos.

This aligns with Zoja's intuition and queries that Russia could serve as "therapy" for the West (2023). It revolves around the seismic shift that occurred at the turn of the 20th and 21st centuries between those enamored with the past, the "world of yesterday",[10] and those who embrace the future, the "the world of tomorrow". This metaphorical earthquake is about *reset* versus *movement*. Those yearning for a *reset* do not embrace modernity well, embodying the irony of Karl Valentin's quip: "In the past, the future used to be better"[11], conveying a nostalgic and yet melancholic sentiment. Perhaps Putin's agenda is rooted in this sentiment, aiming to reset the current world and revive a cherished era lost with the fall of the Soviet Union. An illusionary ideologized world in which only "the future used to be better", contrasting the harsh reality faced by the majority. While those wishing for *movement* represent a new conception of modernity, characterized by openness to transforming the world through human invention (Giddens, 1998, p. 94).

Observing the years before World War I in Austria, Germany, and Russia is a clear example of the unsustainability of an outdated economic, social, and cultural model based on the abundance and hedonism of a privileged few. Both Jung and Zweig concur that during this period, the world was becoming more open and interconnected but was also losing its spiritual aspect.

Analyzing the causes of World War I, Jung (CW10, para. 372) points out that "significant things have been happening. We need feel no surprise that in Russia the colourful splendours of the Eastern Orthodox Church have been superseded by the Movement of the Godless",[12] While acknowledging the low spiritual quality of the scientific reaction, Jung suggests that the spread of 19th-century scientific enlightenment was inevitable in Russia. Importantly, Jung does not frame this turn as merely a political development but rather as a spiritual one. He asserts that the loss of spirituality and the emergence of the godless movement can only result in decay.

Jung's and Zweig's ideas support the consideration that early 20th-century Europe experienced a decline into obscurity due to the erosion of spirituality and the emergence of a world driven by enlightenment and technology.

War as Hopelessness and Helplessness

This era gave rise to a decadence marked by a pronounced hedonistic and Dionysian undertone (Nietzsche's Dionysian concept[13] doesn't apply here). Zweig (2014, p. 332) offers valuable insights into this early 20th-century period. His analysis of the interwar years is not only illuminating but perhaps even prophetic for our time, despite being written retrospectively in his own era:

> Along the Kurfürstendamm,[14] adolescents with heavily made-up faces strolled by, wearing tightly cinched waistlines, and they weren't all just sex workers; high school students were always on the lookout for opportunities to earn some money. In the dimly lit corners of certain establishments, you could spot State secretaries and wealthy financiers fondly and shamelessly caressing young, drunken sailors. Not even Rome in Suetonius's time witnessed

orgies quite like the dances where hundreds of men in women's attire and women in men's clothing gyrated under the indulgent gaze of the police. With the precipitous collapse of values, a sort of madness seized the previously orderly bourgeois classes. Young girls boasted about their perversions, and being suspected of virginity at the age of sixteen was considered a shame in a Berlin school of that era. Everyone wanted to share their adventure, and the more bizarre it was, the more interesting. But the most disgusting aspect of this eroticism was its chilling insincerity.

The German orgiastic frenzy, fueled by hyperinflation, was essentially a feverish mimicry. You could see it in the faces of those bourgeois young girls from good families who would have preferred to wear braids instead of cropped, boyish hair and eat apple pie with whipped cream instead of drinking liquor. It was quite evident that the masses found this state of constant hyper-excitability unbearable, these precarious balancing acts on the tightrope of inflation, while everyone longed for a bit of peace, order, and bourgeois security. The republic was secretly hated, not because it suppressed that unrestrained freedom, but on the contrary, because it let the reins too loose. Anyone who lived through those apocalyptic months and years, experiencing the bitterness and disgust, foresaw that a backlash – a tremendous reaction – would eventually come.

Zweig's portrayal echoes Putin's attempt to imbue war with a higher purpose: fighting the decline of Western society to protect the values of Great Russia. Both in today's Western world and 1900s Berlin, freedom is often associated with aesthetic choices, extending beyond Beck and Beck-Gernsheim's *homo optionis* concept (2002). Today, there is a return to that decadent aesthetic proposed of Huysmans, where exposing one's body without restraint is taken for granted. Some may label the streets as a burlesque spectacle, but this is not the core issue. Fashion and body coverage have evolved through the ages. The concern lies in a superficial embrace of the hedonistic and Dionysian without depth. This trend becomes a temporary affiliation for those seeking meaning and a sense of belonging, only to shift to another group or trend in a few years. In this neo-decadent society, there is little room meaningful reflection and *meditation*.

In my book *Absolute Freedom* (Carpani, 2024), I discussed Jung's perspective on self-knowledge. Jung (CW14, para. 497) claimed that "what I call coming to terms with the unconscious the alchemists called 'meditation'". He cited Ruland, who defined meditation as "an Internal talk of one person with another who is invisible, as in the invocation of the Deity, or communion with one's self, or with one's good angel". In today's hyper-individualized, hyper-liquid super-society, I propose (Carpani, 2023b, 2024) a renewed need for internal talk, with Jungian psychoanalysis being one avenue for this. Jung (CW14, para. 498) claimed that modern meditation methods primarily enhance are "concentration and consolidating consciousness, but have no significance as regards effecting a synthesis of the personality". Instead, they tend to shield consciousness from the unconscious and suppress it, rendering them therapeutically ineffective. Jung (CW14, para. 498) proposed analysis as a form of meditation, even though few individuals have experienced the transformative effects of this, and fewer consider using the objective clues provided by dreams as a meditation focus.

More than 70 years after Jung's essay, little has changed. While meditation had a bad reputation in the West during Jung's time, today, contemporary meditation and contemplation practices such as lunchtime yoga, daily morning meditation, and holistic retreats have become fashionable. However, these practices do not encourage meaningful internal dialogue. Instead, they facilitate a momentary calmness, like recharging a smartphone battery, before returning (recharged) to the hustle of affluent society. They are indispensable in our society but do not provide a complete disconnect.

Therefore, I agree with Jung's view that:

No one has time for self-knowledge or believes that it could serve any sensible purpose. Also, one knows in advance that it is not worth the trouble to know oneself, for any fool can know what he is. We believe exclusively in doing and do not ask about the doer, who is judged only by achievements that have collective value . . . Western man confronts himself as a stranger and that self-knowledge is one of the most difficult and exacting of the arts.

(CW14, para. 498)

The issue with the Berlin and Vienna of the early 20th century, as described by Zweig and Roth, is their inability to foster a synthesis of personality. The same can be said for contemporary neo-decadent society. What Zweig and Roth articulated about the early 1900s regarding the erosion of values, even during the seemingly open and experimental Roaring Twenties, offers insight into the current Western situation. It's worth pondering whether such openness and experimentation, as described by Bauman, are common features in societies on the brink of decline or implosion, indicative of a sense of decadence. These behaviors may serve as a form of compensation for the fear of death and the end of an era. This openness and experimentation may be the initial steps toward decay and decadence, a somewhat superficial way of living, lacking depth and soul.

War as Reset

I strongly believe that the description of the interwar years (1918–1939) and the contemporary era aligns with Freud's perspective in *Civilization and Its Discontents* (2002). In this seminal work, he delved into humanity's desire to escape the constraints of the bourgeois world and its rules to unleash primal instincts. This phenomenon initially manifests in the *hinterland* of society and – especially – the human psyche, as depicted by Arthur Schnitzler in his significant work *Traumnovelle* (2011/1925). It subsequently infiltrates everyday life, as documented by Zweig's accounts of Berlin (2014) and Roth's descriptions of Vienna (2010).

It is important to remember that the two World Wars had different motivations. World War I was characterized by a sense of detachment from reality, still fueled by an "illusion" of a "dream of a better, fairer, and more peaceful world" (Zweig, 2014, p. 242). Therefore, the nation-states, after the initial transformation of the early 1900s, realized their divergent worldviews, leading to the belief that only armed conflict could reset the cherished status quo and its universal dissemination. In contrast, World War II – even larger than World War I – had a spiritual rationale: it aimed to protect freedom and moral goodness (Zweig, 2014, p. 242). In essence, I propose, both World Wars where reset attempts aimed at restoring a deteriorating order and set of values, striving to revive the "world of yesterday" against the fear of the "world of tomorrow" (whatever that may signify).[15]

I see the invasion of Ukraine by Putin's Russia as having elements of both World Wars. On one hand, it is "detached from reality" and driven by the "illusion" of creating a "better, fairer, and more peaceful world". On the other hand, this war has a spiritual rationale, akin to World War II, a battle to safeguard freedom and moral goodness of Russia against the depravation of the West.

German historian and journalist Joachim Fest, in *Der zerstörte Traum* (*The Shattered Dream: On the End of the Utopian Age*, 1992), contended that Nazism and communism both emerged in response to the challenges of bourgeois industrialization and the erosion of traditional values, as explored by Zweig. This contrast with Zoja's proposal (2023), since both Russia and Western Europe – although taking different paths and timelines – embraced industrialization that began in the late 19th century. The distinction arises from the fact that in Russia, the ideas of Marx and Engels gained significant traction as an antidote to the misdeeds of the royal family, while in many other countries, notably Germany and Austria, the transition from monarchy to republic (after World War I) gained momentum in a democratic context.

But if both Nazism and communism emerged as responses to the erosion of traditional values, to what does Putin's anachronistic and 20th-century–like war with Ukraine respond? In light of what we have witnessed – consider, for instance, the invasions of Iraq by U.S. presidents George H.W. Bush and George W. Bush – it's become clear that modern wars conducted by contemporary empires are not necessarily driven by the highest moral values, the preservation of freedom, the pursuit of peace, or a wish for a better world.

One can interpret Putin's actions as an attempt to rectify (and reset) Ukraine's adoption of Western values, as I underlined in each interview of Part I of this book. On this issue, Western media frequently emphasize that Putin invaded Ukraine not only to "demilitarize and de-Nazify" the pro-Western Ukraine but also to suppress the Q+ movement and rights,[16] and end the queer movement, aligning with Moscow patriarchs like Kirill. In fact, an article titled "Russian Church Leader Appears to Blame Gay Pride Parades for Ukraine War" was published on *The Moscow Times* on March 7, 2022, highlighting that "Patriarch Kirill said the war is about 'which side of God humanity will be on' in the divide between supporters of gay pride events – or the Western governments that allow them – and their opponents in Russian-backed eastern Ukraine".

Is Putin, with Kyirill's support, as antagonistic toward Q+ individuals as Nazism and communism were toward queer (of whatever sort, therefore those non-aligned) in the 20th century? History demonstrates regimes' aversion to queerness in the public sphere, emphasizing a preference for strict law and order. In the interviews in Part I of this book, it is underscored that Putin's primary objective is obstructing Ukraine from fully adopting specific European values, as championed by figures like Habermas and Derrida (Borradori, 2004), rather than condemning Western decadence. Like any dictator, Putin is intolerant of queerness and seeks its elimination, deeming it a threat to a nation and its culture if allowed to proliferate. Historical instances, such as the end of the Roman, French, Russian, and Austrian empires – where the ruling classes engaged in nocturnal debauchery which have nothing to do with to Nietzsche's Dionysiac or Jung's synthesis of personality – illustrate this viewpoint.

In the shift from modernity to the contemporary era, the Q+ movement and individuals have gained a more pervasive and mainstream status to the point that today, in Western society, many consider themselves to have some aspects of Q+. However, concurrently, the true significance of queerness often becomes obscured or diluted, especially when it is exploited by corporations for profit. Likewise, in my view, the Q+ movement and individuals have consistently played a fundamental role in the advancement and transformation of every society. Nevertheless, challenges arise in the 21st century when authorities struggle to harness and regulate this movement. What was once led by a few individuals has now evolved into a collective of millions of individuals. Being staunchly opposed to repression by the authorities within the state, I want to shift my focus to broken queerness, which occurs when queerness does not lead to the synthesis of personality.

War as a Remedy for Unhappiness (When the Real Enemy is Happiness)

To emphasize the significance of queerness in today's society, I'll recount a case involving a male patient from my paper "I tacchi a Spillo della Patty" (Carpani, 2023a, 2024). In discussing Patty's experience as a queer individual, he initially conveyed feelings of emptiness and stagnation during our first session. It is essential to avoid conflating queerness with a sense of emptiness or disorientation.

From our initial meeting, Patty – donned in work coveralls, pearl-colored nail polish, and high-heeled women's shoes – signaled a self-awareness of a predicament and an active pursuit of resolution. Patty articulated a discomfort in conventional environments and a preference for queer spaces, a choice deserving acknowledgment and respect. Yet, the significance behind wearing high heels and nail polish warranted exploration. When queried about the meaning of queerness, Patty characterized it as experimentation, rejecting the binary cultural impositions, asserting, "If I like a pair of women's shoes, I'll wear them".

Does experimentation, lacking profound insight, resemble a mere superficial aesthetic pursuit akin to the decadentists, symbolists, and Parnassians, as illustrated by Joris-Karl Huysmans in his 1884 work *À Rebours*? The protagonist, Des Esseintes, embodies eccentricity, seclusion, and physical ailments, vehemently rejecting 19th-century bourgeois society. Huysmans portrays Des Esseintes' quest to withdraw from society, seeking solace in a self-constructed artistic realm. The narrative extensively catalogues his complex aesthetic preferences, reflections on literature, painting, matters of faith, and heightened sensory experiences.

Could the scenario which I have outlined (experimentation devoid of insight) be the potential outcome or conclusion of today's queer movement? A movement drained of its original meaning and inhabited by individuals seeking purpose and belonging? Is the emergence of the broken-queer movement a reflection of a historically contextualized and neurotic lifestyle, reminiscent of Huysmans' protagonist and the Berlin and Vienna depicted by Zweig and Roth? Does this narrative parallel the fate of many in Western society – eccentric, isolated, deeply averse to contemporary norms, yearning to withdraw from societal obligations while seeking solace in an idealized artistic realm? Is Berlin today, like in the 1920s, the focal point of such a phenomenon? While these questions may not hold significance for Patty, being primarily theoretical, what Patty emphasized in our initial meeting is crucial: a sense of emptiness. This emptiness, referred to by Patty, is necessary, if not pleasant.

Patty's emptiness suggests a connection between their current state and a sense of not yet being fully born, existing in an embryonic state akin to Osiris, a phallus-less mummy in a dream fragment. Neither fully alive nor dead, Patty finds solace in the queer world. The imperative for Patty is to differentiate and prioritize developing an individual personality. Currently undifferentiated, dreamwork and insights from the soul (Carpani, 2024) unveil psychological undifferentiation and a maternal pseudo-integration, symbolized by pearl-colored nail polish and high heels. Our goal is not to alter gender identity or queerness for societal conformity, but to unveil a hidden connection consciously expressed through nail polish and heels, and unconsciously in dreams.

From the Suspension to the End of Certainties: The Age of Hypocrisy

The COVID-19 pandemic offers a lesson akin to war. In early 2020, I noted (Carpani, 2022, p. XXV) that the pandemic swiftly instigated profound and enduring shifts in how we live and work, reshaping both our internal and external states and challenging established certainties

(Carpani, 2022, p. XXV). This upheaval marked a departure from the accustomed certainties of modern life, defined by a relentless pursuit of success and identity at the expense of self-exploration (Carpani, 2022, p. XVIII). The suspension of certainties, initially daunting, evolved into a space for reflection and the cultivation of patience for those courageous enough to explore their own Pandora's box (Carpani, 2022, p. XXIX). Hence, I emphasized the contemporary necessity for renewed internal dialogue, fostering individuation in a second-late modern society (Carpani, 2022, p. XXIX).

Regrettably, my earlier optimism proved naïve, and Michel Houellebecq's foresight – "the world will be the same, only a little worse" – rings true. The invasion of Ukraine marks a shift from the previously termed "suspension of certainties" to what I now label as the "end of certainties". These fading certainties echo Jung's insight on war between civilized nations becoming a fable in our rational, internationally organized world (C.W.10, para. 371).

I asserted (Carpani, 2022, p. XXIX) and maintain that the pandemic, exposing us to profound uncertainty, presents an opportunity for change and new beginnings. It prompts a shift from outer to inner focus, from extroversion to introversion, and encourages finding balance between the inner and outer realms. Both the pandemic and the war serve as a reckoning and an opportunity for reset. This reset, as proposed in my previous work (Carpani, 2022, p. XXIX), envisions a departure from a linear world based on hopes and expectations, welcoming the contemplation of interiority and spirituality, not necessarily religion. This shift aims to foster creativity from the soul and oppose a worldview based solely on competition.

My call, unheard then and now, stresses the importance of acknowledging that creativity and creative fantasy foster fluidity and pluralism. If we prioritize fluidity over individualization and liquidity, there's a chance to counter anxiety, depression, suicidality, and anomie. Yet, I believe, the roots of this anachronistic war, reminiscent of the 20th century, echo the reasons Roth's character Chojnicki shouted in outdoor rallies (Roth, 2010, p. 150):

This Empire is destined to decline. As our emperor closes his eyes, we fall into a hundred pieces. The Balkans will be stronger than us. All peoples will establish their own sordid little states, and even the Jews will appoint a king in Palestine. In Vienna, the stench of sweat from the democrats is so strong that you can't even stay on the Ringstrasse anymore. Workers wave red flags and no longer want to work. The mayor of Vienna is a drunkard. The priests are with the people, and sermons in Czech are delivered in churches. At the Burgtheater, Jewish filth is staged, and to put it mildly, every week a Hungarian latrine manufacturer becomes a baron. I warn you, gentlemen, that if we don't act now, it's over. Who lives will see!

Roth underscores that those who heard Chojnicki laughed and dismissed him, failing to comprehend his message (Roth, 2010, p. 151). Regardless, Chojnicki, a parliament deputy critical of his own institution, consistently won re-election with financial, political, and unexpected advantages (Roth, 2010, pp. 150–151). In contemporary times, akin to Roth's era, these circumstances provide scant optimism for the future. This, in my view, constitutes true societal decay, distinct from the perceived degeneration often attributed to the queer community by conservatives.

Conclusions

As Europeans straddling the 20th and 21st centuries, marked by the Berlin Wall's fall in 1989, our identities are shaped by Erasmus' Europe. Having studied in various countries, forming cross-cultural relationships, and building families, we once firmly believed war to be inconceivable.

A rhetorical question lingered: "If war broke out between now-friendly nations, where would I stand?" This question became crucial. Before Russia's invasion of Ukraine, many, having dismissed war as implausible, also rejected the idea of military service. However, post-February 24th 2022, a profound transformation occurred. Individuals reverted to a 20th-century notion of the "idea di nazione" (idea of nation and nationalism), and identity, as Chabod (2021) elucidated.

While I am aware of the risk to be perceived again as naïve and that Michel Houellebecq might certainly be revealed to be right again, we as Europeans have to work for individual and collective change, even individuation. We have to work for those ideals described by Roth and Zweig at the beginning of the 20th century. For a true and united Europe of sisters and brothers of whatever gender, sex, origin as well as ethnicity. In fact, Europe it is de facto queer!

Ultimately, I strongly believe that the pandemic and war beg for a reset, which consists of the ability to ask each other: Who am I? How do I want to live? Who are we? How do we want to live? Therefore, paraphrasing Giegerich (2010, p. 234),[17] it could be said when looking at the world and history that *this is it* and that *it is the world that contains everything it needs within itself* and that our continuous seeking is our running away from our own fulfillment. *Following* Sholem Aleichem (d. i. Schalom Rabinowitsch, 1921, p. 27), it could be said that "everything happens as it should happen; for if things should be different from what they are, they would be otherwise!"

I firmly believe that the pandemic and war call for a reset – a reflection on who we are individually and collectively, and how we aspire to live.

Postscript

In early October 2023, while bedridden with COVID-19 after completing this Outro, the timing coincided with media attention shifting from the ongoing Ukrainian/Russian conflict to the Middle East. This shift followed Hamas' attack on October 7 and Israel's counter-attack.

Reflecting on this events, John Beebe's following words come to mind.

> the attempt to suppress the diverse, desires to make the most happiness out of what we have is the same energy that wants to spoil what we have with war so that no one can enjoy it, as if we must start again or turn back from the self that knows what's good for itself to a dictated regulation of pleasure so that even more war can seem a remedy for unhappiness when the real enemy all along has been happiness.[18]

The conflict could certainly be seen as a desire for a reset, with Palestinians yearning to return to their world pre-1948, and Israelis seeking a secure place in the Promised Land. Until these contrasting hopes reconcile, change seems elusive. However, a reset should not only be considered from a political perspective but also from a psychological standpoint. Jung aids in elucidating this, as I will demonstrate shortly.

Although for many, the 20th century was what W.H. Auden called "The Age of Anxiety" (1948), for many it was the century of fear (Camus, 1946, p. 257). Today, people have begun to describe themselves in "The Age of Liquidity" (see Bauman, 2000, 2007; Beck, 1992) or "The Age of Uncertainty", but how do we describe the century in which that sense of dislocation from even anxiety is unfolding? Since many do not even stir themselves to comment, as if nothing particular were happening, would it not make sense to describe the 21st century as the century of hypocrisy? This would at least nominate the kind of denial of reality and projection of responsibility that has making consensual orientation to the stresses around us all but impossible.

To be sure, recognition is generally given to the extraordinary condensation and rapid occurrence of significant global events within the preceding two decades – terrorist attacks, wars, financial crises, and a pandemic, to name a few. These occurrences have undeniably exerted a profound influence on our way of life. I propose that, particularly within the initial quarter of this century, it is discerned as the epoch characterized by *hypocrisy*, whose highest pinnacle is attained by analyzing the ethos of the international community in relation to recent historical events such as (just to name a few examples): 2003's U.S. invasion of Iraq, 2021's U.S. abandonment of Afghanistan, 2022's Russian invasion of Ukraine and Israeli's ongoing apartheid of Palestine, as well as Gaza´s disproportionate reaction and invasion, after the events of October 7, 2023. I see the roots of these afore mentioned occurrences in what Camus called fear when referring to the previous century. Therefore, I believe that hypocrisy is a consequence of fear, because fear is always linked to evil. And evil, as Jung underlines in *The Red Book* (Liber Secundus 67/68, p. 310) is about God´s and men's relation to how failure to engage with certain realities, and even to imagine they will not exist if we do not dwell on them, is contributing mightily to our own unaccounted for suffering:

> The God suffers when man does not accept his darkness. Consequently men must have a suffering God, so long as they suffer from evil. To suffer from evil means: you still love evil and yet love it no longer. You still hope to gain something, but you do not want to look closely for fear that you might discover that you still love evil. The God suffers because you continue to suffer from loving evil. You do not suffer from evil because you recognize it, but because it affords you secret pleasure, and because you believe it promises the pleasure of an unknown opportunity. So long as your God suffers, you have sympathy with him and with yourself. You thus spare your Hell and prolong his suffering. If you want to make him well without engaging in secret sympathy with yourself, evil puts a spoke in your wheel – the evil whose form you generally recognize, but whose hellish strength in yourself you do not know. Your unknowing stems from the previous harmlessness of your life, from the peaceful passage of time, and from the absence of the God. But if the God draws near, your essence starts to see the and the black mud of the depths whirls up. Man stands between emptiness and fullness. If his strength combines with fullness, it becomes fully formative. There is always something good about such formation. If his strength combines with emptiness, it has a dissolving and destructive effect, since emptiness can never be formed, but only strives to satisfy itself at the cost of fullness. Combined thus human force turns emptiness into evil.

Jung's portrayal (useful to look at the early 21st century) is a striking tableau where "God suffers" due to humanity's reluctance to acknowledge its darkness. Men grapple with a "suffering God" and, in a nuanced perspective, either "suffer from evil" or, perhaps more aptly put, succumb to its influence through hypocrisy, which means to undermine any sense of evil as crisis by simply denying that it means that much to us. By contrast, those who acknowledge that they are suffering from evil have accepted the complexity of their relationship to it. Instead of hypocrisy, individuals who own their relationship to evil can admit that in a way they both "love evil" and yet, paradoxically, "love it no longer". This is the hypocrisy in the 21st century.

Jung's exploration of what today is referred to as the political double standard reveals a reluctance to scrutinize closely, driven by the fear that such examination might unveil a lingering love for evil amid the hope for gain. Hence, men are drawn to love evil, enticed by its clandestine pleasure and the belief that it holds the promise of an unknown opportunity – whether in resisting Russia, seeking Holocaust reparations, or other endeavors. The pivotal notion here is that men "prolong their suffering", intricately linked to the disruptive force of evil, likened to a spoke in the wheel. This obstruction arises from an unconscious, formidable strength within oneself, particularly when considering the impact of "unknowing", rooted in the prior harmlessness of life, the tranquil flow of time, and the absence of a guiding deity.

As underlined in this chapter, the early 20th and 21st centuries both are marked by peace and transformation, followed by despair and destruction. So, is there a solution to the problem of evil? Jung suggests (*The Red Book* – Liber Secundus 67/68, p. 310) that when "God draws near", humanity's strength becomes potent when combined with fullness, imbuing it with a formative quality. He underscores that evil prevails and becomes pervasive when "strength combines with emptiness", leading to a dissolving and destructive impact. The latter, coming up empty in the face of being undermined by the sheer strength of evil, its awful obduracy, has been our experience since the commencement of the 21st century. Can we replace emptiness with fullness? This stands as the sole solution to a potential "another catastrophe".

Notes

1 Rosenthal (2012, p. 15), translated into English by Stefano Carpani.
2 Roth, (2010, p. 250), translated into English by Stefano Carpani.
3 As I have written elsewhere (Carpani 2004),

> modernity is the term used to refer to the ways of living, or social organizations, which appeared in Europe around the 17th century and extended their influence to most of the world. . . . an essential element of modernity is the notion of change and progress.

> As I have written elsewhere (Carpani 2024),

> As underlined by historians Maiken Umbach and Bernd Huppauf (2005, p. 8), modernity is "a matter of movement, of flux, of change, of unpredictability" rather than something static. Modernity is the period that corresponds to the beginning of modern society.

> According to sociologist Anthony Giddens (1998, p. 94), modernity "is associated with a certain set of attitudes toward the world, the idea of the world as open to transformation by human invention". Modernity evolves into what Beck and Beck-Gernsheim (2002) call "reflexive modernization" or "second modernity", what Giddens (1990) calls "high" or "late" modernity and what sociologist Zygmunt Bauman (2000) calls "liquid" modernity. This is characterized by the intensification and speeding up of aspects such as reflexivity (Beck et al. 1994) and the reduction of space and time separation (Giddens 1990). Elsewhere (Carpani 2024), I proposed that se live in a second-late-modernity. Giaccardi and Magatti (2022, pp. 80–81) term this: supersociety. This is (translated from Italian by the author):

> (1) constituted, first, by the intensity, density and extension of technical mediation in the relationship with reality; (2) with the supersociety, the framework of interdependencies is such that it is unrealistic to think separately about social organisation and the planetary ecosystem; (3) the super-society is qualified by the level reached in the human's capacity for self-production. . . . What is new is that super-society tends to incorporate the entire human organism, in all its biological and cognitive dimensions, within its own dynamics.

4 While, on the contrary, the representation of the broken men, the absent or dry or violent father, has been described at length. See, for example, Zoja and Samuels.
5 This could be an entire book! There is very interesting research that suggest that where women are able to study, earn an income, become leaders, etc., society improves. See: https://www.unwomen.org/en/what-we-do/economic-empowerment/facts-and-figures
6 Translated by Stefano Carpani.
7 Translated by Stefano Carpani.
8 Translated by Stefano Carpani.
9 While he may not have been consciously aware of his own inferiority complex, for which he compensated with his grandiose aspirations, this drove him to surpass his grandfather, Wilhelm I. This ambition can also be interpreted as an attempt to compensate for physical disabilities, as viewed from an Adlerian perspective.

10 And those today we can include ISIS, Al-Qaeda, the Taliban, contemporary populists, and – of course – Putin.

11 Quote attributed to Karl Valentin: "Die Zukunft war früher auch besser". No certain biographical information about it. Retrieved from the internet on 18 September 2023.

12 Is this not something similar to what some of the reasons addressed by Putin – although he also addressed the depravation of Western society – to justify the invasion of Ukraine?

13 According to the German philosopher, as reminded by the Italian philosopher and activist Sofia Righetti (2018):

> Dionysus is the god of music, ecstasy, and wine, of bacchanals, voluptuous pleasure, sensuality that leads to ecstasy, and passion that elevates to sublime rapture. It's an intense and enthusiastic 'yes' to life, an ardent and eager acceptance that ultimately tears apart the veil of Maya. It encompasses sensuality but also the pleasure derived from the senses: touch, taste, and wild dances driven by ethereal music that permeates the flesh and becomes vibrant, insatiable, and mad with a vivid and penetrating love for life and all that it encompasses. It exalts the energetic impulses that manifest in everything that makes us feel alive, present, and powerful. It rejects sensory chastity, modesty, and moderation. Nietzsche is here to tell us to embrace life in its entirety, to savor it to the core (to quote Henry David Thoreau) and have no regrets for experiencing it in all its sensory potency. This is achieved through study, certainly, but above all through everything that makes us feel alive, energetic, pulsating, and intoxicated by life. This way of relating to reality is possible through deep and philosophical awareness, which melds and merges with the wisdom contained in books.

14 The Kurfürstendamm, often referred to as the Ku'damm, stands as the central artery of Berlin's western city center and ranks among the most renowned streets globally (both in the 1920s and today). It is considered Berlin's foremost shopping destination, similar to the Champs-Élysées of Paris and New York's 5th Avenue.

15 An additional unsolved issue arose after the end of World War II, whereby the United States asserted an inflated right to bring or restore freedom and moral goodness – thus leading to another reset, one that invariably aligned with its own economic and global dominance interests.

16 Previously known as LGBTQIA+. See: https://www.lgbtq.sociology.cam.ac.uk/

17 Original quote: "There is nothing to be sought, nothing that would be somewhere else, be in the future or in the transcendence"; "It is this real life of hers that contains everything it needs within itself"; and "her seeking is he running away from her fulfilment".

18 Private conversation.

Bibliography

Aleichem, S. (1921). *Die Geschichten des Milchhändlers*. Berlin-Wien: Eliasberg.

Auden, W. H. (1948). *The Age of Anxiety*. London: Random House.

Bauman, Z. (2000). *Liquid Modernity*. Cambridge: Polity.

Bauman, Z. (2007). *Liquid Times*. Cambridge: Polity Press.

Beck, U. (1992). *Risk Society: Towards a New Modernity*. London: Sage.

Beck, U. (1994). 'The reinvention of politics', in Beck, U., Giddens, A. and Lash, S. (Eds.), *Reflexive Modernization: Politics, Tradition and Aesthetics in the Modern Social Order*. Cambridge: Polity Press.

Beck, U., and Beck-Gernsheim, E. (2002). *Individualization: Institutionalized Individualism and its Social and Political Consequences*. London: Sage.

Beebe, J. (1992). *Integrity in Depth*. College Station: Texas A&M University Press.

Beebe, J., and Carpani, S. (2023). An epistolary conversation. *Journal of Analytical Psychology*, 68(4), 638–664. Co-Editors-in-Chief: Carolyn Bates and Arthur Niesser. Oxford: Wiley Publishing.

Borradori, G. (2004). *Philosophy in a Time of Terror: Dialogues with Jurgen Habermas and Jacques Derrida*. Chicago: University of Chicago Press.

Camus, A. (1946). *The Stranger*. New York: Knopf.

Carpani, S. (2004). *The Formation of Narratives of Self-Identity. A Study of the Turkish Community in Berlin*. Unpublished M.Phil. thesis. Univeristy of Cambridge.

Carpani, S. (2022). The fall of the Berlin wall: Complex theory and the numinous in the development of history (a neo-Jungian approach). In *Individuation and Liberty in a Globalized World: Psychosocial Perspectives on Freedom after Freedom*. London: Routledge.

Carpani, S. (2023a). I tacchi a Spillo della Patty. In *Sensi migranti. Le identità del contemporaneo. Atti del Convegno "Ibridazioni e contaminazioni" (Circolo della Stampa, Torino)*. Edited by S. Candellieri and D. Favero. Bergamo: Moretti & Vitali.

Carpani, S. (2023b). *Absolute Freedom: The 'I+I' (Individuation + Individualization) as a Metanarrative of Self-Development in a Second-Late-Modern Society*. Unpublished PhD thesis. University of Essex.

Carpani, S. (2024). *Absolute Freedom*. London: Routledge.

Carpani, S., and Luci, M. (2023). *Lockdown Therapy: Jungian Perspectives on How the Pandemic Changed Psychoanalysis*. London: Routledge.

Chabod, F. (2021). *L'idea di nazione*. Bari: Laterza.

Fest, J. (1992). *Il Sogno Distrutto*. Milano: Garzanti.

Freud, S. (2002). *Civilization and Its Discontents*. London: Penguin Classics.

Giaccardi, C., and Magatti, M. (2022). *Supersocietà*. Milano: Il Mulino.

Giddens, A. (1990). *The Consequences of Modernity*. Cambridge: Polity Press.

Giddens, A. (1998). *The Third Way: The Renewal of Social Democracy*. Cambridge: Polity.

Giegerich, W. (2010). *The Soul Always Thinks*. New Orleans: Spring Journal.

Grudin, R. (1999). *The Grace of Great Things*. New York: Ticknor & Fields.

Hillman, J. (2005). *A Terrible Love of War*. London: Penguin.

Huysmans, J.-K. (1884). *À Rebours*. Internet.

Jung, C. G. (1979 [1955]). *Collected Works, Volume 14: Mysterium Coniunctionis*, 2nd Edition. Princeton, NJ: Princeton Univerisity Press.

Levy, P. (2014). *La Tregua*. Torino: Einaudi.

Lyotard, J.-F. (1984 [1957]). *The Postmodern Condition: A Report on Knowledge*. Minneapolis: University of Minnesota Press.

Righetti, S. (2018, November 26). Quote from the link. www.filosofemme.it/2018/11/26/limportanza-del-dionisiaco/

Rosen, D. H. (1992). Preface. In *Integrity in Depth*. Edited by J. Beebe. College Station: Texas A&M University Press.

Rosenthal, R. (2012). *Ethics and War in Homer's Iliad*. Talk given in March 2012 by Carnegie Council President, at the annual Maine Humanities Council Winter Weekend Seminar, at Bowdoin College, Brunswick Maine. Retrieved on the Internet on 18.9.23. https://www.carnegiecouncil.org/media/article/ethics-and-war-in-homers-iliad#:~:text=By%20making%20war%20a%20choice,come%20from%20above%20and%20below

Roth, J. (2010). *La Marcia di Radezky*. Roma: Newton Compton Editori.

Samuels, A. (1985). *The Father: Contemporary Jungian Perspectives*. London: Free Association Books.

Samuels, A. (1989). *The Plural Psyche*. London: Routledge.

Samuels, A. (1993). *The Political Psyche*. Oxford: Taylor & Francis.

Samuels, A. (2001). *Politics on the Couch: Citizenship and Internal Life*. London: Profile Books.

Samuels, A. (2016). *The Plural Psyche: Personality, Morality and the Father*. London: Routledge.

Schnitzler, A. (2011 [1925]). *Traumnovelle*. Kindle.

Umbach, M. and Huppauf, B. (2005). *Vernacular Modernism: Heimat, Globalisation, and the Built Environment*. Stanford: Stanford University Press.

Valéry, P. (1919). *The Crisis of The Mind*. Internet.

Zoja, L. (2016). *Il gesto di Ettore: Preistoria, storia, attualità e scomparsa del padre*. Torino: Bollati Boringhieri.

Zoja, L. (2018). *The Father: Historical, Psychological and Cultural Perspectives*. London: Routledge.

Zoja, L. (2023). *Russia as a 'Therapy' for the West?* In Doppiozero. Retrieved from the Internet on 18.8.2023. https://www.doppiozero.com/la-russia-come-terapia-per-loccidente

Zweig, S. (2014). *Il Mondo di Ieri*. Milano: Garzanti.

Contributors

Editors

Stefano Carpani, Ph.D., is an Italian psychoanalyst and sociologist (member and lecturer of the C.G. Jung Institute Zürich, and postgraduate of the University of Cambridge) working in private practice in Berlin (DE) and online. He serves as scientific consultant to Pacifica Graduate Institute Extension and International Studies (USA). He initiated the YouTube series *Breakfast at Küsnacht, Lockdown Therapy* and *War as Reset*, and co-created *Psychosocial Wednesdays* (currently serving as its chairperson). He initiated and curates *Jungianeum: Initiatives for Contemporary Analytical Psychology and neo-Jungian Studies*, the book series titled *Re-covered Classics in Analytical psychology, Neo-Jungian Studies*, and *JUNGIANEUM/Yearbook*. For the Italian magazine *Doppiozero*, he hosts a column titled "Cultivating the Soul in the SuperSociety." Among his edited books: *Breakfast at Küsnacht* (Chiron, 2020 – IAJS Best Edited Book nominee); *Anthology of Contemporary Classics in Analytical psychology: The New Ancestors* (Routledge, 2022 – GRADIVA Best Edited Book nominee). His most recent books are The New Myth of Analysis (Jungianeum, 2024) and *Absolute Freedom* (Routledge, 2024).

Ludmilla Ostermann, M.A., is a Berlin- and Northrine-Westfalia-based journalist and editor focusing on political, social and scientific topics. She currently writes for different German media outlets, including German Press Agency (dpa) and the University of Bielefeld. Here she contributed to a series of interviews about the war in Ukraine. As an online editor, she is creating content for Futurium, a museum in Berlin dedicated to future topics since 2019. She is in charge of digital contents. As a member of the founding team, her focus there is on scientific journalistic contributions on research and socio-political topics. She completed a traineeship at the daily newspaper *Westfalen-Blatt* (Bielefeld) and devoted herself to political and social issues in the East Westphalian province from 2009–2019. Since 2022, she is back at the Westfalen-Blatt (Bielefeld) as freelance journalist, news manager, and digital consultant. Ludmilla Ostermann holds bachelor's degrees in German and English studies, and a Master's degree in history from Bielefeld University.

Part I

Father Giuseppe Bettoni was born in Tavernola Bergamasca (Italy) and belongs to the order of the Sacramentine Fathers. In 1984, he was in Milan as Vicar of the Parish of Saint Angela Merici, which he served as parish priest from 1996–2010, and in 1989 he began to deal with the problem of drug addiction among the young people of the neighborhood. In that year, a group of volunteers gathered around him, motivated by the desire to do something concrete to help young people escape from the slavery of drugs, accompanying them on the difficult path to therapeutic communities. Over time, the attention of the volunteers and Father Giuseppe shifted to human immunodeficiency (HIV)-positive children; many of them have contracted the disease from their parents and the families in which they live are often unable to look after them. Father Giuseppe and the volunteers work daily with them and their HIV-negative brothers and sisters. In those years the group, grew stronger until, in 1991, it officially gave birth to the Arché association, with the mission of "inventing hope every day" for HIV-positive children and their families. In 1997, the Home for Mothers and Children with HIV opened in Milan, but after the early 2000s, when the HIV emergency returned thanks to antiretroviral treatment, Arché's focus shifted to the mother and child unit in conditions of psychological and social distress. In 2000, Father Giuseppe Bettoni became provincial vicar and president of the European Conference of the Sacramentine Fathers, a position he held until 2012. Arché became a foundation in 2013 and continues to this day in its daily work aimed at vulnerable children and families in the construction of social, housing, and working autonomy by offering support and care services. Arché house opened in 2019.

Joseph Cambray, Ph.D., is a North American Jungian analyst, CEO/President, and Provost at Pacifica Graduate Institute. He is also past president of the International Association for Analytical Psychology (IAAP); he served as the U.S. editor for the *Journal of Analytical Psychology* and is currently on the editorial boards of the *Journal of Analytical Psychology*, *The Jung Journal: Culture & Psyche*, and *Israel Annual of Psychoanalytic Theory, Research and Practice*. He has been a faculty member at Harvard Medical School in the Department of Psychiatry at Massachusetts General Hospital, Centre for Psychoanalytic Studies; adjunct faculty at Pacifica Graduate Institute. His numerous publications include the book based on his Fay Lectures, *Synchronicity: Nature and Psyche in an Interconnected Universe*, and a volume edited with Linda Carter, *Analytical Psychology: Contemporary Perspectives in Jungian Psychology*.

George B. Hogenson, Ph.D., is a North American Jungian analyst. He received his Ph.D. in philosophy from Yale University and his M.A. in clinical social work from the University of Chicago. He is a diplomate Jungian analyst in private practice in Chicago, where he works primarily with adults dealing with life transitions, dream work, and trauma. He serves on the editorial board of the *Journal of Analytical Psychology*, was the vice president of the IAAP and is the author of *Jung's Struggle with Freud* (1983), as well as numerous chapters on archetypal theory, synchronicity and the nature of symbols.

Professor Verena Kast, Ph.D., studied psychology at the universities of Basel and Zurich, and trained as a psychoanalyst at the C.G. Jung Institute Zurich. She received her doctorate in 1973 with a thesis on "Creativity in the Psychotherapy of C.G. Jung." In 1982, she habilitated at the University of Zurich with a study on the "Significance of Grief in the Therapeutic Process." From 1973–2008, she taught at the University of Zurich. She has lectured and lectures at various universities in Switzerland and abroad, and at the C.G. Jung Institute Zürich. She is past president of the Swiss Society for Analytical Psychology, past president of the International Society for Analytical Psychology, and past president of the C.G. Jung Institute Zürich. She was chairwoman of the International Society for Depth Psychology, co-director of the Lindau Psychotherapy Weeks. She has written and continues to write books and essays in the field of emotions, attachment, and separation, but also symbolism. Her love belongs to the imagination and its development.

Dmitry Kotenko is a Russian analytical psychologist at the Moscow Association of Analytical Psychology. His primary research interests are the dynamic life of Russian cultural complexes, the archetypes of life and death, and the variety of shapes they have taken in recent history. His research interests also include the healing of collective trauma and the effects new media have on privacy and human behavior. Kotenko is the founder of the APRG (Analytical Psychology Research Group) project, which organizes and conducts group studies of collective processes in Russia. He lives and works in Moscow.

Elana Lakh, Ph.D., is an art therapist-supervisor and Jungian analyst, and a member of the Israeli institute of Jungian psychology in honor of Erich Neumann. She holds a doctorate in cultural studies. She is the author of *The Origins of Evil in the Human Psyche: A Jungian Reading of Creation Mythologies* (Carmel, 2017, in Hebrew). She teaches and practices art therapy in Jerusalem and internationally, specializing in treatment of sexually abused individuals. She researches archetypal aspects of creation mythologies and of paintings. She is a human rights activist, trying to raise awareness to the facts and effects of the Israeli occupation of Palestinian lives for the past 30 years, and writes about the Israeli-Palestinian conflict from a Jungian perspective.

Tine Papič is a Slovenian Jungian training analyst and supervisor. He studied computer and information science and philosophy at the University of Ljubljana and psychotherapy science at Sigmund Freud University Vienna, and trained as a Jungian analyst in the router program. He is currently president of the Slovenian Society for Analytical Psychology and a member of the steering group of the Analysis and Activism movement. He is also a co-founder, lecturer, and organizer of the Jungian program at Sigmund Freud University Vienna Ljubljana branch and one of the founding members of Open Institute for Psychotherapy Ljubljana.

Iryna Semkiv is a Ukrainian psychologist and IAAP router, trained at the Polish Association of Jungian Psychoanalysis (developmental group of IAAP), as well as a lecturer at the psychotherapy department (Ukrainian Catholic University, L´viv, Ukraine). She is currently a *war refugee* in Poland after she escaped Ukraine in March 2022 with her son.

Pastor Vickie Sims is a retired Church of England priest who served as Archdeacon of Italy and Malta between 2016 and 2019. Sims was educated at Iowa State University (USA) and ordained in 2003. After a curacy in Grantham, she was at St Andrew, Coulsdon from 2005–2014. She was Chaplain of Milan with Lake Como and Genoa from then until her retirement in February 2019.

Murray Stein, Ph.D., is a North American Training and Supervising Analyst at the International School of Analytical Psychology Zurich (ISAP-ZURICH). He was president of the IAAP from 2001–2004 and President of ISAP-ZURICH from 2008–2012. He has lectured internationally and is the editor of Jungian Psychoanalysis and the author of *Jung's Treatment of Christianity, In MidLife, Jung's Map of the Soul, Minding the Self, Outside Inside and All Around*, and *The Bible as Dream*. The first volume of his collected writings, titled *Individuation*, has recently been published. He lives in Switzerland and has a private practice in Zurich.

Caterina Vezzoli is an Italian Jungian analyst in private practice in Milan (Italy). She trained as a psychologist at the University of Padua and at the C.G. Jung Institute in Zurich, and is a training analyst there and at the Centro Italiano di Psicologia Analitica (CIPA). She has served as director of studies and as honorary secretary and treasurer of CIPA's Milan Institute, as well as its national treasurer and vice president. She taught for many years in the department of psychiatry at the University of Milan. She has done extensive research on children's dreams, child analysis, the psychology and physiology of sleeping and dreaming, and the associations experiment, and has written extensively on these topics. She has also been a member of the IAAP international committee for coordination and development of Jungian child analysis and is the IAAP liaison for the developing group in Malta. She is president and director of the Philemon Foundation.

Dmytro Zaleskyi is a Ukrainian MD, psychiatrist, and Jungian analyst. He is an Individual member of the IAAP since 2010. Zaleskyi was born in 1964. In 1988, he graduated from the Kyiv Medical Institute with a degree in medicine from the Department of Psychiatry. He worked at the Kyiv City Psycho-neurological Hospital №1, and then at the All-Union Research Center for Radiation Medicine at the USSR Academy of Medical Sciences. From 1986–1991, he worked as a liquidator of the consequences of the Chernobyl disaster as a medical assistant, neurologist, and psychotherapist. He is first president of the Official Development Group of IAAP in Ukraine. He served in the Armed Forces of Ukraine for mobilization, as Chief of his battalion's medical service in 2015–2016. He is currently serving the armed forces of Ukraine as chief of his battalion's medical service.

Luigi Zoja, Ph.D., is past president of CIPA and past President of IAAP. His former teaching activity took place at the School of Psychiatry of the Faculty of Medicine, State University of Palermo, at the University of Insubria (Italy), and at the University of Macao (China). He published several books in English: *Drugs, Addiction and Initiation* (1989/2000), *Growth and Guilt* (1995), *The Father* (Gradiva Award 2001), *The Global Nightmare. Jungian Perspectives on September 11*, (2002), *Cultivating the Soul* (2005), *Ethics and Analysis* (2007), *Violence in History, Culture and the Psyche* (2009), and *Paranoia. The Madness that Makes History* (2017).

Part II

Henry Abramovitch, Ph.D., is founding president and senior training analyst at the Israel Institute of Jungian Psychology in Honor of Erich Neumann, Professor Emeritus at Tel Aviv University Medical School and Past President of Israel Anthropological Association. He has served on ethics and program committees of the IAAP and is active in Israel Interfaith Encounter Association. He teaches and supervises Routers in the IAAP Developing Groups in Eastern Europe and Kazakhstan. He is author of *The First Father* (2010), *Brothers and Sisters: Myth and Reality* (2014), *Why Odysseus Came Home as a Stranger and Other Puzzling Moments in the Life of Buddha, Socrates, Jesus, Abraham and other Great Individuals* (2020), and with Murray Stein, the play *The Analyst and the Rabbi* (2019). His special joys are poetry, dream groups, and the holy city of Jerusalem. Email: henry.abramovitch@gmail.com

Natalia Bolycheva, Russia, is a scholar in culture studies, analytical psychologist, psychoanalyst, and certified body-oriented therapist. She is a member of the Professional Community of Analytical Psychologists, Moscow. Her interests include Russian cultural unconscious and cultural complexes in group processes and individual psyche, Russian cultural identity, traditional

worldview in modern life, conflicts between new cultural norms and traditional cultural attitudes, modern aspects of child/parent separation, search for initiation in secular societies, etc.
Some of her recent conference papers and publications are: "Split Parent. Frustration as a Separation Phenomenon," "Identity Contradictions: between New Cultural Attitudes and Old Cultural Complexes. The Phenomenon of Borders in the Traditional Culture and the Principle of
'Share,'" "Red and White: Two Sides of the Russian Soul," "Initiation as a Cultural Pattern of
Adolescent Separation," "Puer – a Hero of Modern Time. Liminality as a New Norm," "Vanished Father and His Children," and "Mother Plague and Coronapocalypse. On the Archetypal Sources of Pandemic Images."

Chiara Capri, M.D., is a psychiatrist and Jungian analyst. From December 2019 to March 2020,
she served as the psychiatrist for the European Project ICARE (Integration and Community
Care for Asylum and Refugees in Emergency) at SAMIFO, a public health center for refugees
and asylum seekers. Since 2019, she has worked as a psychiatrist for the European project
"Freedom of Choice," awarded to the Psychoanalytical Institute for Social Research (IPRS).
This project explored personality traits, family structures, psychopathy traits, and resilience
among the children of Mafia members in Calabria, Campania, and Sicily. From 2016–2019,
she was the head researcher for the Sicilian team involved in the Italian validation of the
Historical, Clinical, and Risk Management-20 Version 3 (HCR-20 V3). Currently, she heads
the Early Interventions Center in the West Area of the Mental Health Department – Rome 1,
where she works as a psychiatrist and psychoanalyst for adolescents. Her passion for sociology, history, and criminology has led her to write several books about the Sicilian Mafia and
its victims, including: *Our Lantern. China is Close and Cosa Nostra Knows It* (Navarra Editore, 2010), *Libero. The Entrepreneur Who Didn't Say Yes to Protection Money* (Castelvecchi
Editore, 2011), and *Twenty: '92 and 2012* (Coppola Editore, 2012). Her areas of interest and
research include the psychological aspects of organized crime, the therapeutic use of fairy
tales, myths, adolescence, and sociology.

Roula-Maria Dib, Ph.D., is a holder of the UK Global Talent Visa as an award-winning literary scholar, poet, and editor. Her research interests are at the interstices of literature, modern
poetry and poetics, creative writing, and Jungian psychology. She is the winner of the British
Council's Alumni Awards 2021–2022 for the Culture and Creativity category in the United
Arab Emirates and is the recipient of the American University in Dubai Provost's Award for
Outstanding Literary Achievement 2020. Her book *Jungian Metaphor in Modernist Literature* (Routledge, 2020) was shortlisted as a finalist for the international IAJS book awards,
and some poems from her collection *Simply Being* (Chiron Press, 2021) received Pushcart
Prize nominations. She is the founding editor of the literary and arts journal *Indelible*, and
creative producer of literary event series, Indelible Evenings, as well as Psychreative, a vir-
tual salon for researchers, artists, and writers with a background in Jungian psychology.

Karin Fleischer, Ph.D., is a Jungian analyst, training analyst, supervisor, and founding member of the Uruguayan-Argentinian Society for Analytical Psychology (SUAPA) and is on
the executive board (2018–2021) of CLAPA (Latin American Committee for Analytical
Psychology). She is a licensed psychologist (Universidad de Buenos Aires) and Master of
Science in Dance Movement Therapy (California State University, East Bay, USA), and a
university professor of graduate and postgraduate courses in analytical psychology. She has
introduced Authentic Movement/Embodied Active Imagination in Argentina and in several
Latin American countries, teaching seminars on the Body and Active Imagination, nationally

and internationally, during the past 25 years. She has published academic articles in the *Journal of Analytical Psychology* (2020, 2022) on topics related to collective trauma, dissociation, psychosomatic disorders, the body, and Active Imagination. www.movimientoautentico.com.ar

Kathleen Kirgin, Ph.D., is a depth psychologist and a globally experienced leadership consultant, working with leaders and professionals in personal development and the cultivation of conscious leadership skills. Dr. Kirgin received her Bachelor of Arts from Loyola University (USA), and obtained her master's degree and her doctorate in depth psychology from Pacifica Graduate Institute. The focus of her doctoral research concerns collective trauma, transgenerational wounding within the Irish culture, and the possibility of transforming these cultural wounds. Besides her work with private clients, she is also an adjunct professor at Pacifica Graduate Institute.

Elana Lakh, Ph.D., is an art therapist/supervisor and Jungian analyst, and a member of the Israeli institute of Jungian psychology in honor of Erich Neumann. She holds a doctorate in cultural studies, and is the author of *The Origins of Evil in the Human Psyche: A Jungian Reading of Creation Mythologies* (Carmel, 2017, in Hebrew), based on her doctoral dissertation. She teaches and practices in Jerusalem and internationally, specializing in treatment of sexually abused individuals. She researches archetypal aspects of creation mythologies and of paintings and other artworks. She writes about the Israeli-Palestinian conflict from a Jungian perspective, drawing on 30 years of human rights activism in this context.

Marianne Meister-Notter, Dr. Phil., is a Swiss psychotherapist working with adults, children, and adolescents in her own private practice in Zurich (CH). She is a former vice president of the Curatorium, as well as the head of the "Further Training & Vocational Policy" committee, and is a lecturer, training analyst, and supervisor at the C.G. Jung Institute Zürich in Küsnacht, Switzerland. She is the president of the Ethics Commission of the Association of Swiss Psychotherapists (ASP). Her most recent publication is titled *The Key to Self: Recognition of Self Through Depth-Psychology-Oriented Astrology* (Patmos, 2015 and Chiron, 2022).

Professor Renos K. Papadopoulos is Professor of Analytical Psychology and Director of the Centre for Trauma, Asylum and Refugees, and also a member of the Human Rights Centre of the Transitional Justice Network and of the Armed Conflict and Crisis Hub all at the University of Essex (UK), as well as honorary clinical psychologist and systemic family psychotherapist at the Tavistock Clinic. He is a practicing clinical psychologist, family therapist, and Jungian psychoanalyst. He served on the executive committee of the IAAP, was responsible for setting up the Academic Section and the Developing Groups within IAAP, and he organized the first IAAP Academic Conference (in 2002). He was editor of *Harvest: International Journal for Jungian Studies* for 14 years, founding editor of the *International Journal of Jungian Studies* and co-founder of the International Association for Jungian Studies. His four-volume work *C.G. Jung: Critical Assessments* (1992) remains the lengthiest Jungian book (1750 pages).

Professor Heyong Shen is a professor of psychology at the City University of Macao and South China Normal University, Guangzhou. He is a Jungian analyst (IAAP member), sandplay therapist (ISST), president of China Society for Analytical Psychology (CSAP), and China Society for Sandplay Therapy (CSST). He was a speaker at the Eranos East and West Round

Table Conferences (1997, 2007, and 2019), and the main organizer of the International Conference of Analytical Psychology and Chinese Culture (1998–2018), and founding member of the Garden of the Heart & Soul project. He is the author of *Psychology of the Heart* (Fay Lecture), *C.G. Jung and Chinese Culture, Chinese Culture Psychology*, and numerous articles on the interface between analytical psychology and Chinese culture.

Elias Winterton, Ph.D., is an analytical psychologist and psychotherapist, dedicated to expanding Jungian and post-Jungian ideas beyond the clinical setting. His interdisciplinary approach aims to apply depth psychology to emerging fields of knowledge and experience.

Heba Zaphiriou-Zarifi (GAP, UKCP, IAAP) is a senior Jungian analytical psychologist, training analyst, and supervisor, with a private practice in London. She is regularly invited to give seminars at the C.G. Jung Club London, as well as teaching seminars and Reading Jung groups for members and trainees. She is leadership trained in BodySoul Rhythms® at the Marion Woodman Foundation and founder of The Central London Authentic Movement Practice, where she runs monthly group sessions focusing on embodied active imagination. She integrates philosophy, body/psychotherapy, and the creative arts into her clinical work. Heba has extensive experience in energy-based therapies for trauma-work based on sound vibrations and bodywork. A proponent of Peace with Justice, she consults on psychosocial projects in the Middle East, also working with victims of war. Heba is a speaker at international conferences and a published contributor to academic journals. Her alma mater is the Sorbonne, where she gained two master's degrees and wrote her doctoral thesis in philosophy.

Index

Note: Page numbers in *italics* indicate a figure and page numbers in **bold** indicate a table on the corresponding page.

7 October 2023 204–205

absent father 100–107
active imagination, embodied 178–179;
    experience with 180–181, *180–181*
adjacent possibility 61
adversity-activated development (AAD) 6,
    245, 247–249, 251, 254, 256; *see also*
    counter-Adversity Activated Development
    (counter-AAD)
Afghanistan 29–30, 95–96, 112–113, 122
Agamemnon 45–46
American Civil War 32–33
analysis 131–135, 173–174, 180–183, 226–227;
    confrontation with evil 129–131; confrontation
    with "my death" 127–129; psychic functions
    in early development 175–177; somatic
    dimension in Jungian clinical work 178–180;
    state terrorism and its consequences 174–175;
    the unconscious 227–231
animal behavior 108–115
animus/anima 23–25, 132, 227–231
anti-Semitism 26–27, 36, 120
archetypal epistemology 232–234, 243–244;
    archetypal fascination 238–240; ecology
    of destructiveness 240–242; limitations of
    linear-causal epistemology of destructiveness
    234–235; ordinariness and ecology of violence
    237–238; psychologising and pathologising
archetypes 11, 14–18, 23–27, 46–48, 100–104,
    109–110, 207; archetypal patterns 61, 69; and
    analysis in the shadow of terror 128–131;
    archetypal epistemology 232–245; and cultural
    identity 137–138, 143–144, 149–150; and
    dreams 197–200; frozen in fear because of war
    228–231; and inherited war traumas 165–167,
    170–171; and the Northern Ireland conflict
    214–217; and Sicily 154, 159–160; and

survivors of collective violence 249–254
Argentina 3, 5, 75, 173–174, 179
art 164–171
assistance to refugees 23–31
attachment to land 32–43
attitudes 187
Austria/Austria-Hungary 26–29, 114, 120–121,
    261–264, 267
authentic movement 178–179
authoritarianism 28, 35, 49–50, 100–102
awe 246–250, 255–256; countertransference
    responses in trauma therapists 253–254; and
    counter-trauma 252–253; and the numinous
    250–252; truth and beauty 254

Balkan wars 11, 15
beast/bestial 2, 90, 92–93, 116–118, 137–139,
    142, 205–206, 209, 214
beauty 254
Beck, Ulrich 19, 31, 50–51, 59, 67, 262–263, 265
Berlin Wall 12, 19, 50, 62, 67, 83, 269
Bion, Wilfred 16, 239
Bosnia 11–13, 92–93
Bosnian War 92
Britain *see* Great Britain
Bucha 66, 71–73, 90
Bush, George H.W. 21, 44, 51, 122, 267
Bush, George W. 30, 36, 44, 267

cave paintings 100–107
Center/Edge dynamics 138–140
certainties, suspension of 268–269
China 3–4, 68–69, 111–112, 185, 189–199
Chinese characters 185–187
Christian Era 32, 39
Christianity 29–30, 39–40, 53–57, 129–130, 133,
    148–150, 197, 207–208, 215–217
Christian Russian cosmos 148–150

climate change 32, 40, 69, 197
clinical work 179–180; confrontation with evil
 129–131; confrontation with "my death"
 127–129; counter-Adversity Activated
 Development 246–256; countertransference
 responses in trauma therapists 253–254;
 counter-trauma 252–253; the emotion of awe
 and the numinous 250–252; in the shadow of
 war and terror 127–135; somatic dimension in
 Jungian clinical work 178–181, *180–181*; truth
 and beauty 254
co-construction of meaning 2, 69, 92
cohesion among nations 53–60
Cold War 3–4, 28, 50–51, 61, 65, 147
collective forgetfulness 61
collective psyche 7, 14–16, 131, 212–217
collective trauma 4–6, 15–17, 173–176, 222
collective violence, survivors of 246–250,
 255–256; countertransference responses in
 trauma therapists 253–254; counter-trauma
 252–253; the emotion of awe and the
 numinous 250–252; truth and beauty 254
communication 53, 82, 109, 138, 167, 171, 177
communism 13, 120–121, 266–267
compensation 2, 25, 35, 65–66, 94, 104–105,
 110–111
complexity 232–234, 243–244; archetypal
 fascination 238–240; ecology of
 destructiveness 240–242; limitations of
 linear-causal epistemology of destructiveness
 234–235; ordinariness and ecology of violence
 237–238; psychologising and pathologising
 destructiveness 235–236; Severe Forms of
 Collective Adversity (SFCA) 242–243; tragedy
 of destructiveness 236–237
cooperation 102–104
Corriere della Sera 13, 29, 42, 54, 58, 68–69, 84,
 96, 102, 112, 123
corruption 13, 32, 38, 50, 114–115, 121, 263
counter-Adversity Activated Development
 (counter-AAD) 246–250, 255–256; and
 counter-trauma 252–253; countertransference
 responses in trauma therapists 253–254; the
 emotion of awe and the numinous 250–252;
 truth and beauty 254
countertransference 71–73, 184, 246–254
counter-trauma 248, 252–254, 256
COVID-19 pandemic 19–21, 30, 32–33, 51,
 95–96, 108–109, 116–117, 197–198,
 268–270
creativity 100–107, 164–166, 269
Crimea 1, 3, 65, 90, 141, 145
Croatia 11–12, 53
cultural complexes 79, 136–138, 197–198,
 212–213, 222–223; and depth psychology
 213–214; Good Friday Agreement 218–219;

martyr complex 215–217; negotiating with
 the shadow element 217–218; shared space
 and unified Ireland 219–220; victim complex
 214–215
cultural identity 136–138; Center/Edge dynamics
 138–140; holy war 147; liminality symbols
 140–144; miners 144–145; rite of passage
 145–146; SVO poetry 148–150; Victory Day
 147–148
cultural imagery 185–187
Cyprus 19, 30, 43, 84, 112, 123
Czech Republic 13, 82, 120, 269

death 127–129; the dead and the living 148–150
decadence, Western 32, 35, 267
deception 26, 53, 107
dehumanization 15–16, 116–118
democracy 4–5, 57, 102; and attachment to land
 32–35; and masculine and feminine aspects
 within war 28–30; and medics 74–76, 77;
 and motives of Russia for invading Ukraine
 83–84; psychological perspective 111–114;
 and suspension of certainties 267–269; and
 transformation of values 18–21; and war as
 failure 45–48, 51; and war as part of the human
 condition 94–98
depth psychology 213–214
Derrida, Jacques 13, 44, 52, 102, 113, 122
destructiveness 232–234, 243–244; archetypal
 fascination 238–240; ecology of 240–242;
 limitations of linear-causal epistemology of
 234–235; ordinariness and ecology of violence
 237–238; psychologising and pathologising
 235–236; Severe Forms of Collective Adversity
 (SFCA) 242–243; tragedy of 236–237
dialogue 26, 45–47, 53, 56–58, 83–85, 108–110
diplomacy 4, 20, 23, 56, 69, 105, 276
displacement 169
Donbas 5–6, 136–138; Center/Edge dynamics
 138–140; holy war 147; liminality symbols
 140–144; miners 144–145; rite of passage
 145–146; SVO poetry 148–150; Victory Day
 147–148
Dostoevsky, Fyodor 45, 48
dreams 200; attitudes of war 187–191, **188–189**,
 190–191; cultural imagery of Chinese
 characters 185–187; *I Ching* 191–199, *192*
Dugin, Aleksandr 20, 35

early psychological functions, recovery of
 173–174, 180–183; psychic functions in early
 development 175–177; somatic dimension in
 Jungian clinical work 178–180; state terrorism
 and its consequences 174–175
ecology: of destructiveness 240–242; of violence
 237–238

education, peace 53, 56–57
Eisenhower, Dwight D. 41–42
embodied analysis 173–174, 180–183; psychic
    functions in early development 175–177;
    somatic dimension in Jungian clinical work
    178–180; state terrorism and its consequences
    174–175
emotion 6–7, 23–25, 53–54, 62–63, 71–73, 91,
    92–93, 189, *190*; of awe 250–253
empires 2
energy crisis 53, 58
energy policy 53, 58
Engels, Friedrich 263, 267
epistemology 232–234, 243–244; archetypal
    fascination 238–240; ecology of
    destructiveness 240–242; limitations of
    linear-causal epistemology of destructiveness
    234–235; ordinariness and ecology of violence
    237–238; psychologising and pathologising
    destructiveness 235–236; Severe Forms of
    Collective Adversity (SFCA) 242–243; tragedy
    of destructiveness 236–237
Europe 2; and collective unconscious 64–70;
    dreams 190–191, *191*; frozen in fear because
    of war 228–229; and Israel 116, 120–123;
    masculine and feminine aspects within war 23,
    26–29, 31; and medics 71, 74–77; motives of
    Russia for invading Ukraine 80–85; Northern
    Ireland conflict 219–220; psychological
    perspective108–114; and religion 53–54,
    57–58; Russia's attachment to land 34–35,
    38–39, 42–44; suspension of certainties
    261–264, 267–270; transformation of values
    11–15, 17–22; war as archetype 102–106; war
    as failure 47, 50–52; war as part of the human
    condition 92, 95–98; women in Ukraine 90–91
European Union 12, 53, 64, 67, 69, 82, 220
European values 44, 53, 121, 267
evil 129–131; *see also* good and evil
existential threat 116–117, 146
experience of early trauma 173–174, 180–183;
    psychic functions in early development
    175–177; somatic dimension in Jungian
    clinical work 178–180; state terrorism and its
    consequences 174–175
expression 187

fascination, archetypal 232, 238–240, 243–244
fascism 15, 26, 45, 48, 114, 147–148, 156
father, absent 100–107, 159–160
father complex 11, 16, 230
fatherland 2
fear 226–227; fear of not having enough 92; the
    unconscious 227–231
feminine, the 23–24, 63, 92–93, 103, 118, 128,
    160, 262

Ferrara, Maurizio 19–20, 29–30, 42–43, 51–52,
    96, 106, 112–113, 122–123
France 18, 58, 74, 77, 82, 111
Freud, Sigmund 24–25, 34–36, 46–48, 64–67,
    100–101

Galicia 13, 26, 82, 120, 261
gender 4
gender dynamics of war 45
Germany 1–4, 82, 102; attachment to land 34–39,
    43–44; collective unconscious 65–67; and
    Israel 120–121; masculine and feminine
    aspects within war 26–28; psychological
    perspective 109–110, 114; religion 57–58;
    suspension of certainties 262–267, 275;
    transformation of values 16–17, 22; war as
    failure 45–46
God/gods 18, 24, 42, 46, 64; analysis in the
    shadow of terror 129–130; cultural identity
    148–150; dreams 186–187; frozen in fear
    because of war 228–230; Israel 118; Palestine
    202–210; psychological perspective 111–112;
    suspension of certainties 260–264, 271; war as
    archetype 100–102
good and evil 11, 116, 209, 224, 233, 244
Good Friday Agreement 218–219
Gorbachev, Mikhail 3, 11, 16, 27, 50, 61, 65, 86,
    98, 110
Great Britain 19–20, 30, 42–43, 51, 84, 96–97,
    111–113, 213–219
Great Mother 134, 154, 230; absence of separation
    from 159–160
Greece 25, 45
Greek mythology 45–52, 129

Habermas, Jürgen 13, 44, 102, 113, 122
happiness 268
healing 6–8
heart, psychology of the 200; attitudes of war
    187–191, **188–189**, *190–191*; cultural imagery
    of Chinese characters 185–187; *I Ching*
    191–199, *192*
Heidegger, Martin 145
helplessness 264–266
hero/heroism 35–37, 48–49, 103–105, 136–139,
    144–149
Hillman, James 40–44, 45, 48, 56, 64, 100–101,
    199, 261
history: mafia 154–157; role of 32
Hitler, Adolf 16, 26–27, 35–36, 65–67,
    109–112
holy war 147–148
Homer 160, 260–261
hopelessness 264–266
Houellebecq, Michel 31, 59, 68, 95, 269–270
humanity, loss of 207

Hungary/Austria-Hungary 26–29, 82, 91, 261–263
hypocrisy, age of 268–271

I Ching 191–199
identity 6, 53–54, 82–84, 136–138; Center/Edge dynamics 138–140; holy war 147; liminality symbols 140–144; miners 144–145; rite of passage 145–146; SVO poetry 148–150; Victory Day 147–148
illiberal values 45, 48
imagery, cultural 185–187
imagination: embodied active imagination 178–179, 180–181, *180–181*; leaps of 61
inevitability 100–101
inherited war traumas 164–171
insecurity 92, 156, 170
institutional charters 53
interruption of psychic functions 176–177
Iraq 2, 19–23, 29–30, 36–38, 42–44, 51–52, 68–69, 83–84, 96–97, 106, 112–113, 122–123
Ireland 213; cultural complexes of 214–215; unified 219–220; *see also* Northern Ireland
Irish Republican Army (IRA) 213, 215–218
Islam 123, 130, 133, 207
Islamic State 47, 118
Israel 1–5, 114–124, 129–134, 201–210, 222–225, 270–271
Italy 48–58, 111–113, 154–156

Japan 24, 28, 40–41, 43, 49, 83, 186
Jerusalem 116–118, 127–130, 222–225
Jesus 53–56, 59, 71, 93, 98, 138, 278
Jews/Judaism 26–27, 120–121, 130–134, 201–209, 224, 269
Judaism *see* Jews/Judaism
Job 208–210

Kenya 19, 30, 43, 84, 112, 123
Kirill (Patriarch) 53–55, 101–102, 112, 267
kleptocratic society 32, 37
Korea 2, 23, 29, 41, 83–84, 96
Kozyrev, Andrei 50
Kurdish people 24, 37, 47, 118
Kyiv 53–57, 71–72, 82–83

land, attachment to 32–43
leaps of imagination 61
legends related to mafias 154–157
Lenin, Vladimir 25–27, 263
Leningrad 32, 36, 145
LGBTQ 26, 35, 61
libido 86, 89
liminality symbols 140–144
linear-causal epistemology of destructiveness 234–235

literature 260–262
living and the dead, the 136, 148–150
long-term thinking 108
loss of humanity 207
love 41–42, 45–46, 54–56, 64, 92–99, 100–101, 208–209
love triangle 86
Lviv 53–54, 86–88

mafia 154–161
Magatti, Mauro 50–51, 54–55, 67
mana personality 23, 25
martyr complex 215–217
Marx/Marxism 29, 263, 267
masculine aspects within war 17, 23–24, 36–37, 45–47, 63, 92–93, 103, 108–110, 118
master and apprentice metaphor 71, 75
meaning, co-construction of 2, 69, 92
military capabilities 3, 32
military investments 23, 28, 64
military thinking 23–31
miners 144–145
mother *see* Great Mother
movement, authentic 178–179
multiple crises, convergence of 32–44
murder, sacrificial 201–202, 207–210; 7 October 2023 204–205; Israel vs. Palestine 203; Jungian theory vs. reality 202–203; loss of humanity 207; pressing questions 204; the victimised 205–207
Muslims *see* Islam
Mussolini 20, 48, 110
mutual transformations 246–250, 255–256; countertransference responses in trauma therapists 253–254; counter-trauma 252–253; the emotion of awe and the numinous 250–252; truth and beauty 254
mythology, Greek 45–52, 129

NATO 20, 27–28, 38, 226–229
Nazism 22, 26–27, 32, 112, 116, 120–121, 141–143, 146–147, 266–267
negotiations, peace 94, 217–218
neo-Jungian perspective 154–161
Nietzsche 24, 64, 101, 261, 263–264, 267
Northern Ireland conflict 212–213; and depth psychology 213–214; Good Friday Agreement 218–219; martyr complex 215–217; negotiating with the shadow element 217–218; shared space and unified Ireland 219–220; victim complex 214–215
numinous, the 61–64, 249–254

objectives 32, 44
old empires 61
Old Testament 92, 98

orcs 116, 119, 143, 151n12
ordinariness 237–238

Palestine 201–203, 207–210; 7 October 2023
204–205; Jungian theory vs. reality 202–203;
loss of humanity 207; Palestinian clients
131–135; pressing questions 204; the
victimised 205–207
paranoia 64–66, 108–111, 114, 123
pathologising destructiveness 235–236
patience 47, 108, 114, 208, 230, 269
peace: Good Friday Agreement 218; negotiations
94, 217–218
peace education 53, 56–57
Peace Ireland 212–213; and depth psychology
213–214; Good Friday Agreement 218–219;
martyr complex 215–217; negotiating with
the shadow element 217–218; shared space
and unified Ireland 219–220; victim complex
214–215
perpetrator 130, 206, 224, 256
personality of a mafioso 157–159
personification 108–110, 216–217, 243
pneumothorax 71, 73–74
poetry 148–150
Poland 82, 86–91
polarization 11, 15, 61–63
political power 53, 55, 63, 218
politics 32, 60, 197–198, 203–204, 213
post-traumatic stress 71–73, 104, 247–248
prayer 149, 206–209
pseudologia phantastica 100–107
psychological functions, recovery of 173–174,
180–183; psychic functions in early
development 175–177; somatic dimension in
Jungian clinical work 178–180; state terrorism
and its consequences 174–175
psychologising destructiveness 235–236
psychosocial stance, war as 264
Putin, Vladimir 2–3, 264–267; attachment to
land 32–33, 35–40, 42–43; and collective
unconscious 61–69; and Israel 120–123;
masculine and feminine aspects within war
23–31; motives for invading Ukraine 79–84;
psychological perspective 108–115; and religion
53–55; transformation of values 11, 14–20; war
as archetype 100–107; war as failure 45–52;
war as part of the human condition 94–96

Q+ movement 267
queerness 267–268

reality, Jungian theory vs. 202–203
rebuilding 100–107
recovery of early psychological functions
173–174, 180–183; psychic functions in early

development 175–177; somatic dimension in
Jungian clinical work 178–180; state terrorism
and its consequences 174–175
Red Revolution 111, 121
refugees 23–31
religion 18, 24, 53–56, 61–64, 92–95,
100–101, 121
remedy for unhappiness 268
research, arts-based 164–171
resilience 4–7, 56–57, 67–69, 247–248, 253–254
respect for the other 92, 98
restoration 3–4
rite of passage 145–146
Romania 91, 155
Roth, Joseph 260–263, 266–270
Russia 1–5; attachment to land 32–44; collective
unconscious 61, 65–69; cultural identity
136–150; Donbas liminality symbols in the
space of 144; dreams 196–198; frozen in fear
because of war 226–229; and Israel 116–123;
masculine and feminine aspects within war
23–30; medics 71–76; motives for invading
Ukraine 79–85; psychological perspective
108–115; religion 53–58; suspension of
certainties 262–267, 270–271; transformation
of values 11–22; war as archetype 101–107;
war as failure 45–52; war as part of the human
condition 93–98; and women in Ukraine 86,
90–91
Russian Orthodox Church 26, 53, 55, 61, 101, 137

sacrificial murder 201–202, 207–210; 7
October 2023 204–205; Israel vs. Palestine
203; Jungian theory vs. reality 202–203; loss
of humanity 207; pressing questions 204; the
victimised 205–207
self-regulation 167
separation from the Great Mother 159–160
Serbia 11, 15, 22
Severe Forms of Collective Adversities 242–243
shadow 222–225; negotiating with the shadow
element 217–218; of terror 127–135
shame 11, 67, 94, 209, 265
Sherman, William T. 32–33, 37, 41
Shi hexagram of *I Ching* 191–199
Sicily 154–161
Sinn Fein 212–213; and depth psychology
213–214; Good Friday Agreement 218–219;
martyr complex 215–217; negotiating with
the shadow element 217–218; shared space
and unified Ireland 219–220; victim complex
214–215
Slovenia 1, 11–15, 20, 277
sobornost 148–150
social breakdown 92, 96
societal interdependence 61

somatic dimension in clinical work 178–181, *180–181*

soul, the 59, 128, 193–194, 268–269

Soviet cosmos 148–150

Soviet Union, collapse of 79–85

Srebrenica 92

Stalin 105–106, 123

state terrorism 173–175, 180–183; psychic functions in early development 175–177; somatic dimension in Jungian clinical work 178–180

Stevens, Anthony 11, 14–15, 18

suffering 222–225

survivors of collective violence 246–250, 255–256; countertransference responses in trauma therapists 253–254; counter-trauma 252–253; the emotion of awe and the numinous 250–252; truth and beauty 254

suspension of certainties 268–269

SVO (Special Military Operation) poetry 148–150

Switzerland 1, 28, 30–31, 90, 104, 106, 194

symbols *see* liminality symbols

terror/terrorism 131–135, 173–174, 180–183, 246–250, 255–256; confrontation with evil 129–131; confrontation with "my death" 127–129; countertransference responses in trauma therapists 253–254; counter-trauma 252–253; the emotion of awe and the numinous 250–252; psychic functions in early development 175–177; somatic dimension in Jungian clinical work 178–180; state terrorism and its consequences 174–175; truth and beauty 254

therapists, trauma 253–254

Tolstoy, Leo 45, 48

total war 2

traditional culture 138–139

tragedy of destructiveness 236–237

transference 71–73, 177, 202, 246–248, 252

transformation 3–4, 246–250, 255–256; countertransference responses in trauma therapists 253–254; counter-trauma 252–253; cultural complexes 212–220; the emotion of awe and the numinous 250–252; of individuals 32–44; truth and beauty 254; of values 11–22

trauma 4–5, 173–174, 180–183, 246–250, 255–256; countertransference responses in trauma therapists 253–254; counter-trauma 252–253; the emotion of awe and the numinous 250–252; Jungian arts-based research 164–171; psychic functions in early development 175–177; somatic dimension in Jungian clinical work 178–180; state terrorism and its consequences 174–175; truth and beauty 254

trauma therapists 253–254

Tribushnyi, Dmitriy 143, 145, 148

Troy 45–47

Trump, Donald 27–28, 34–38, 65–66

truth 254

Ukraine 1–5; attachment to land 32–38, 42–44; collective unconscious 61, 64, 68; cultural identity 136, 139–147; Donbas liminality symbols in the space of 140–143; dreams 187, 191, 193–194, 196, 198; embodied analysis 173; frozen in fear because of the war 226, 228; and Israel 116–123; masculine and feminine aspects within war 23–30; medics 71–72, 74–75, 77; motives of Russia for invading 79–84; psychological perspective 108, 110, 113–115; religion 53–54, 57–58; suspension of certainties 261, 266–267, 269–271; transformation of values 11–15, 18–22; war as archetype 101–106; war as failure 47, 49–52; war as part of the human condition 92, 94–98; women in 86–87, 90–91

unconscious 227–231

unhappiness 268

United States 74; attachment to land 32–35, 38–44; collective unconscious 62–69; and dreams 189; frozen in fear because of war 226–228; and Israel 122–123; masculine and feminine aspects within war 23, 28–31; and motives of Russia for invading Ukraine 82–84; and the Northern Ireland conflict 213, 217–218; and Palestine 204–205; psychological perspective 112–114; religion 55–58; transformation of values 12–13, 19–22; war as archetype 102–106; war as failure 47–52; war as part of the human condition 94–98

unity of "the living and the dead" 148–150

Valéry, Paul 260–261

values: European 44, 53, 121, 267; illiberal 45, 48; transformation of 11–22

vassalization of power 53

victim complex 214–216

victims 205–207

Victory Day in Russia 147–148

Vietnam 68–69, 83–84, 109, 112–113

violence 246–250, 255–256; countertransference responses in trauma therapists 253–254; counter-trauma 252–253; ecology of 237–238; the emotion of awe and the numinous 250–252; truth and beauty 254

Weil, Simone 261

West, the 3–4, 34; and dreams 197–198; frozen in fear because of war 226–229; and Israel

116–117, 121–123; masculine and feminine aspects within war 23–28; and motives of Russia for invading Ukraine 80–85; suspension of certainties 264–267; transformation of values 11–14; war as archetype 102–105; and war as failure 45, 50–52; war as part of the human condition 92–97; Western decadence 32, 35, 267; women in Ukraine 90–91

women 23–25, 36–37, 45–47, 53–54, 57, 63, 76, 86–90, 93, 103–104, 109–110, 118

World of Yesterday, The (Zweig) 114, 261–263

worldview 138–139

World War I 94, 110–111, 185, 261–266

World War II 1–3, 147; attachment to land 34–36, 39–41; collective unconscious 64–65; masculine and feminine aspects within war 25–29; psychological perspective 110–114; suspension of certainties 261–263, 266; transformation of values 14–19; war as archetype 101–106; war as part of the human condition 93–96

Wotan 28, 45, 101, 111, 121, 264

Yeats, William Butler 260–261

Yeltsin, Boris 11, 16, 25–27, 36, 50, 105, 110

Yemen 71, 95, 108

Yugoslavia 12–13, 20, 22, 37, 122, 236

Zelenskyy, Volodymyr 14, 29–30, 57, 100, 104, 122

Zionism 202–205, 209

Žižek 18, 20